A CONCISE ECONOMIC
HISTORY OF THE WORLD

A CONCISE ECONOMIC HISTORY OF THE WORLD

*From Paleolithic Times
to the Present*

SECOND EDITION

RONDO CAMERON

New York   Oxford
OXFORD UNIVERSITY PRESS
1993

Oxford University Press

Oxford   New York   Toronto
Delhi   Bombay   Calcutta   Madras   Karachi
Kuala Lumpur   Singapore   Hong Kong   Tokyo
Nairobi   Dar es Salaam   Cape Town
Melbourne   Auckland   Madrid

and associated companies in
Berlin   Ibadan

Library of Congress Cataloging-in-Publication Data

Cameron, Rondo E.
A concise economic history of the world : from Paleolithic times to the present / Rondo Cameron. — 2nd ed.
p. cm.   Includes bibliographical references and index.
ISBN 0–19–507445–9 (c). — ISBN 0–19–507446–7 (pbk.)
1. Economic history.   I. Title.
HC21.C33   1993
330.9—dc20   92–14539

9 8 7 6 5 4 3 2 1

Printed in the United States of America
on acid-free paper

For my grandchildren

Lukas, Margaux Lisle,
Kyla, Graham Zane

Preface

The first edition of this book appeared in May 1989, shortly before the dramatic events that completely realigned the economies and polities of Eastern Europe and the former Soviet Union. The main purpose of this new edition, therefore, is to take account of those events and place them in their historic setting. This is done in a new Chapter 16, which also brings developments in the European Community more nearly up to date, places the war in the Persian Gulf area in historical perspective, and recounts some other recent events. I cannot, of course, guarantee that still newer developments will not occur before publication that will once again render the account of recent events out of date.

I have taken advantage of the opportunity to correct some errors that appeared in the first edition, to update statistical data and the bibliography, and to make minor alterations here and there. In particular, I have fleshed out the treatment of non-Western economies in Chapter 4, which some critics rightly believed was overly concise.

Readers should appreciate, as I do, the assistance of Susan Hannan of Oxford University Press and Pat Richardson of Emory University.

Atlanta   R. C.
January 1992

Preface to the First Edition

This book began as a textbook for an upper division undergraduate course in European economic history. I have, in fact, used drafts of the chapters in my own courses in that subject, and that is the purpose for which I expect the book will be mainly deployed. I hope, however, that it will also serve as a text for courses (increasingly common) in world, international, and comparative economic history, and as an adjunct or supplementary text for courses in American, North American, and Australasian economic history. I would also like to recommend it as a supplementary text for courses in economic development, most of which lack historical dimensions.

After considering a number of titles, I opted for *A Concise Economic History of the World* at the urging of many friends and with the hope that the book will also interest general, nonacademic readers. Partly with that in mind, I have deliberately kept the treatment brief, as the title indicates. That in itself caused many problems of both selection and interpretation: what to include, what to exclude, how to present complex and contentious issues? I have endeavored to present the factual record as clearly and succinctly as possible, but I have not shrunk from offering my own interpretation of controversial topics.

Professional colleagues will note that I have, for the most part, eschewed such long-established terms as *capitalism, mercantilism,* and *the industrial revolution.* While, in professional discourse, these terms can sometimes serve as convenient abbreviations or verbal shorthand for complex series of events or ideas, they lend themselves too readily to stereotyped notions. Too often, in fact, they tempt even seasoned scholars to avoid an analysis of the underlying events or ideas rather than to use these terms as tools for such analysis.

Most of my own research in primary sources has been confined to the nineteenth century and, for some topics, to the eighteenth and twentieth centuries. It should be clear, therefore, that for all the centuries before (and for many topics in the centuries mentioned) I have depended on the research of others. The extent of my dependence is imperfectly reflected in my Annotated Bibliography, which, for the most part, lists only books in English. Besides the items listed there, however, I have also utilized books in languages other than English and, in particular, journal articles and other monographic research in numerous languages.

In addition to library and archival research, personal experience has shaped my views on a number of subjects. I like to think that my boyhood on a Texas farm in the 1930s has improved my understanding of certain topics in agrarian history. In the 1960s I spent two years in South America as a special field representative of the Rockefeller Foundation. In 1976 I traveled extensively in Africa under the auspices of the U.S. Information Agency. I have also traveled widely in Eastern Europe and the USSR, including Soviet Central Asia and Soviet Armenia. These experiences have influenced my perception of the process of economic development in both market and planned economies. A short-term visiting professorship at Keio University in Tokyo in the spring of 1987, kindly arranged by Professor Akira Hayami, enabled me to better my knowledge of Japanese economic history.

Some critics will undoubtedly regard my title as grandiose rather than modest ("the world" versus "concise"). On the other hand, many will charge me with being Eurocentric. My defense is that events in Europe, especially in the last thousand years, have been of determining importance in the making of the contemporary world, the understanding of which this book is intended to facilitate. Chapter 4 presents the bare details of non-Western economies on the eve of European expansion; subsequent chapters discuss Europe's impact on the rest of the world, and vice versa, in the course of European expansion (and contraction).

This book has been underway for a very long time. Passages in Chapter 3 first appeared in the introductory lecture to a course in modern economic history that I taught at Yale University in 1951. Other chapters have been shaped, and reshaped, over the years as a result of my own research, my reading of an ever-burgeoning literature, my students' questions, and my colleagues' suggestions. The final product is very different from the one originally envisaged.

In the many years in which this book has been maturing I have accumulated obligations to persons too numerous to mention, even if I could remember them all. The late Professor Robert S. Lopez of Yale University gave me detailed criticisms on the first three chapters before his untimely death. Other friends and colleagues who have read portions of the manuscript include Professor Jerome Blum, Princeton; Professor Jakob Finkelstein, Yale; the late Professor Michael Flinn, Edinburgh; Professor Rainer Fremdling, Groningen; Professor Kristof Glamman, Copenhagen; Professor John Komlos, Pittsburgh; Ms. Millicent Lambert; Professor David Mitch, University of Maryland-Baltimore County; Professor Clark Nardinelli, Clemson; Professor Jonathan Prude, Emory; Professor Richard Sylla, North Carolina State; Dr. Maria Todorova, Sofia, Bulgaria; and Professor Mira Wilkins, Florida International University. Conversations with colleagues on five continents have helped me to clarify my ideas and to free me from some errors. Without implicating them in any way for what follows, I would like to acknowledge in particular T. C. Barker, London; Ivan Berend, Budapest; J.-F. Bergier, Zurich; Carlo M. Cipolla, Pisa and Berkeley; A. W. Coats, Nottingham; D. C. Coleman, Cambridge; Roberto Cortés Conde, Buenos Aires; François Crouzet, Paris; Wolfram Fischer, Berlin; R. M. Hartwell, Oxford and Chicago; Lennart Jörberg, Lund; Eric E. Lampard, Stony Brook; David Landes, Harvard; Angus Maddison, Groningen; Peter Mathias, Cambridge; William N. Parker, Yale; Carlos M. Pelaez, Rio de Janeiro; T. C. Smout, St. Andrews; Richard Tilly, Münster; Gabriel Tortella, Madrid; and Herman Van der Wee, Leuven. My greatest debt for friendly criticism is to Professor Charles P. Kindleberger, M.I.T., who read the entire manuscript in record time and made copious suggestions, not all of which I have accepted for reasons known to both of us. The personnel of Oxford University Press in New York, especially Marion Osmun, has been patient, understanding, and helpful.

My institutional obligations are also great. The John Simon Guggenheim Memorial Foundation awarded me a fellowship for 1969–70 (postponed to 1970–71) for other purposes, but during the year I completed drafts of the first three chapters. The Rockefeller Foundation invited me to be a guest scholar at its Villa Serbelloni in Bellagio, Italy, for a month in the summer of 1971, during which I touched up these drafts. I spent the academic year 1974–75 as a Fellow of the Woodrow Wilson International Center for Scholars in the Smithsonian Institution, Washington, D.C., where I managed to draft a few chapters while working on another project. In 1980–81 a Humanities Grant from the Rockefeller Foundation, again for other purposes, allowed me time to add bits and pieces to chapters already in draft. Finally, a leave of absence from Emory University in 1986–87 permitted me to complete the bulk of the manuscript.

Ms. Arlene DeBevoise saw the entire manuscript through a word processor, made many valuable suggestions, and helped prepare the index.

Atlanta, Georgia   R. C.
August 1988

Contents

List of Tables

List of Figures

A CONCISE ECONOMIC
HISTORY OF THE WORLD

1

Introduction: Economic History and Economic Development

Why are some nations rich and others poor? This seemingly simple question is directed at the heart of one of the world's most pressing contemporary problems, that of uneven economic development. Only war and peace, population pressure, and environmental salubrity—and thus the survival of the human race—are issues of similar magnitude. Because of unequal economic development, revolutions and coups d'état have occurred; totalitarian governments and military dictatorships have deprived whole nations of political liberty, and many individuals of their personal freedom and even their lives. Millions have died miserably and unnecessarily of starvation, malnutrition, and disease—not because food and other resources were unavailable, but because they could not be delivered to those in need. The United States and several other wealthy nations have expended billions of dollars in well-intentioned attempts to assist their less fortunate neighbors. In spite of these varied efforts, the income gap between the relatively small number of affluent nations and the vast majority of impoverished ones not only persists but grows wider year by year.

The situation appears to be paradoxical. If some nations are rich and others poor, why do not the poor ones adopt the methods and policies that have made the others rich? In fact, such attempts have been made but, in most instances, without striking success. The problem is far more complicated than it appears on the surface. In the first place, there is no general agreement on *which* methods were responsible for the higher incomes of the wealthy nations. Second, even if such agreement existed, it is by no means certain that similar methods and policies would produce the same results in the different geographical, cultural, and historical circumstances of today's low-income nations. Finally, although much research has been devoted to the problem, scholars and scientists have not yet produced a theory of economic development that is operationally useful and generally applicable.

There are various approaches to the study of economic development—fortunately not mutually exclusive. The historical approach employed in this book does not aim at producing a general, universally applicable theory of economic development. Instead, historical analysis can focus, as other approaches cannot, on the *origins* of the presently existing unequal levels of development. A correct diagnosis of the origins of the problem does not, in itself, guarantee an effective prescription, but without such a diagnosis one can scarcely hope to remedy the problem. Second, by focusing on instances of growth and decline in the past, the historical approach isolates the *fundamentals* of economic development, undistracted by arguments over the efficacy or desirability of particular policies for specific current problems. In other words, it is an aid to objectivity and clarity of thought.

Policymakers and their staffs of experts, faced with the responsibility of proposing and implementing policies for development, frequently shrug off the potential contributions of historical analysis to the solution of their problems with the observation that the contemporary situation is unique and therefore history is irrelevant to their concerns. Such an attitude contains a double fallacy. In the first place, those who are ignorant of the past are not qualified to generalize about it. Second, it implicitly denies the uniformity of nature, including human behavior and the behavior of social institutions—an assumption on which all scientific inquiry is founded. Such attitudes reveal how easy it is, without historical perspective, to mistake the symptoms of a problem for its causes.

This book is offered as an introduction to the study of both economic history and economic development. It is not, however, intended to be comprehensive in either area. There are many valid reasons for studying history apart from its potential contribution to solving contemporary practical problems; for a complete understanding of the problem of economic development other methods of study and observation must be employed as well. In this general survey of the economic development of humankind from prehistoric times to the present, certain "lessons of history" are highlighted. Although some historians believe their function is to "let the facts speak for themselves," "facts" respond only to specific questions posed by the analyst who deals with them; posing such questions inevitably involves a process of selection, conscious or unconscious, especially in so brief and synoptic a volume as this.

Before we undertake the historical narrative it is necessary to define certain terms and to formulate some basic concepts to guide the subsequent analysis.[1]

Development and Underdevelopment

In 1990 the average (or per capita) annual income of residents of the United States was approximately $21,000. In Switzerland, the most prosperous country in Europe, it amounted to more than $29,000. For western Europe as a whole, the average was above $18,000. These nations, along with Canada, Australia, and New Zealand, contained less than 14 percent of the world's population but accounted for approximately 58 percent of measurable income and more than 86 percent of the world's manufacturing output. If Japan is included, the figures rise to 16 percent for population, 70 percent for income, and 90 percent for manufacturing output; most of the remainder of manufacturing output was produced in the former Soviet Union and the socialist countries of eastern Europe, which together contained about 8 percent of the world's population.

[1] The remainder of this chapter was written with the needs of noneconomists in mind. Readers with a substantial background in economics may wish to skim it or skip it altogether.

TABLE 1-1. GNP per Capita, Selected Countries, ca. 1989 (in 1989 dollars)

High-Income Economies (avg.)	18,330	Lower-Middle Income (avg.)	1,360
Switzerland	29,880	Mexico	2,010
Japan	23,810	Turkey	1,370
United States	20,910	Thailand	1,220
Germany, Federal Republic	20,440	El Salvador	1,070
United Kingdom	14,610	Egypt	640
Israel	9,790	Bolivia	620
Spain	9,330		
Upper-Middle Income (avg.)	3,150	*Low-Income Countries (avg.)*	330
Greece	5,350	Indonesia	500
Portugal	4,250	Haiti	360
Yugoslavia	2,920	China	350
Hungary	2,590	India	340
South Africa	2,470	Bangladesh	180
Brazil	2,540	Mozambique	80

Source: World Bank, *World Development Report, 1991* (New York, 1991).

At the other extreme, per capita income in Mozambique, probably the world's poorest country, was less than $100, in Ethiopia about $120, in India about $340, and in Indonesia about $500. In the People's Republic of China, which contains more than one-fifth of the world's population, per capita income is believed to be about $350. Per capita income in Latin America ranged from $620 in Bolivia to $2,540 in Brazil, although the latter figure has subsequently declined. Table 1–1 shows recent per capita income for a representative sample of nations.

Altogether, in 1990 there were forty-one nations in which per capita incomes were lower than $500, and an additional fifty-three where the average income was between $500 and $5,000. The nations in these two latter categories are variously referred to as "poor," "low income," and "underdeveloped" (or, euphemistically, "less developed" or "developing"). It is obvious that, because of their low incomes, they are poor, but why underdeveloped?

Statistics of per capita income are, at best, crude measures of the level of economic development. In the first place, they are only approximations. Moreover, for a number of technical reasons, international comparisons of income are especially unreliable. But there are other measures of development and underdevelopment that, although less global or comprehensive, are more graphic. Table 1–2 presents a number of these. As a consequence of high death rates, life expectancy at birth ranges from about 40 years to 69 years in the underdeveloped nations of Asia, Africa, and Latin America, whereas it is well over 70 years in western Europe and North America. Infant mortality is especially high in poor countries. In the light of these figures it is not surprising that health care facilities are more plentiful in the wealthy nations: in the United States there are approximately 470 people per physician, in Switzerland about 700, and in Austria 323, compared with 1,540 in

TABLE 1-2. Indicators of Economic Development, Selected Countries

	Crude Birth Rate (1989)	Crude Death Rate (1989)	Life Expectancy at Birth (1989)	Population per Physician (1984)
High-Income Economies				
United States	15	9	76	470
Switzerland	12	10	77	700
United Kingdom	14	11	76	—
Spain	12	8	77	320
Japan	11	7	79	660
Middle-Income Economies				
Bolivia	42	13	54	1,540
Egypt	32	10	60	770
Hungary	12	13	70	310
Mexico	28	6	69	1,242
Low-Income Economies				
Indonesia	27	9	61	9,460
China	22	7	70	1,010
India	31	11	58	2,520
Haiti	36	13	55	7,130

6

	Energy Consumption per capita (kg oil equiv.)	Distribution of Gross Domestic Product (1988)			Urban Population % of Total 1985	Telephones per 1000	TVs per 1000
		Agriculture	Industry	Services			
High-Income Economies							
United States	7,655	2	43	54	75	760	813
Switzerland	4,193	*	*	*	60	855	411
United Kingdom	3,736	2	42	56	89	517	534
Spain	1,902	6	37	57	73	396	322
Japan	3,306	3	41	57	77	555	585
Middle-Income Economies							
Bolivia	249	24	27	49	51	29	76
Egypt	607	21	25	54	46	28	83
Indonesia	229	24	36	40	30	5	39
Mexico	1,305	8	35	56	72	96	117
Low-Income Economies							
Chad	18	47	18	35	29	*	*
China	580	32	46	21	53	7	10
Ethiopia	20	42	17	41	13	3	1.6
India	211	32	30	38	27	6	7

*Non-reporting member
Sources: World Bank, *World Development Report, 1990* (New York, 1990); *1991* (New York, 1991). *United Nations Demographic Yearbook 1989* (New York, 1990).

Bolivia, 20,000 in Ghana and 78,000 in Ethiopia! In even more materialistic terms, for every 1000 people in the United States there are 700 passenger automobiles, 450 in France, 41 in Ecuador, 3 in Tanzania, and 1.3 in Burma.

Growth, Development, and Progress

In ordinary discourse the terms *growth, development,* and *progress* are frequently used as though they were synonymous. For scientific purposes, however, it is necessary to distinguish among them, even if the distinctions are somewhat arbitrary. Economic *growth* is defined in this book as a sustained increase in the total output of goods and services produced by a given society. In recent decades this total output has been measured as national income, or gross national product (GNP). (Statistically, there is a small difference between gross national product and national income, which is slightly smaller than the former, but for most purposes in this book this difference can be ignored; the two aggregates almost always move together in the same direction. Another concept, gross domestic product (GDP), is also sometimes used; this value is normally intermediate between GNP and national income.) Although no national income data exist for earlier epochs, they can in some cases be estimated; in any case, even without specific quantitative data, one can usually determine, on the basis of indirect evidence, whether total product increased, decreased, or remained roughly constant during any given period.

Growth in total output may occur either because the inputs of the factors of production (land, labor, and capital) increase or because equivalent quantities of the inputs are being used more efficiently. If population is increasing, there may be growth in total output but not in output per capita; indeed, the latter may even decrease if the rate of population growth exceeds that of output, as has occurred in some underdeveloped countries in recent years. For welfare comparisons, economic growth is meaningful only if it is measured in terms of output per capita.

Difficulties also arise in comparing the outputs of two different societies or outputs within the same society at widely different points in time, for two main reasons. As a rule, national income and similar measures are given in monetary units, but values of monetary units are notoriously unstable and frequently difficult to compare. In principle, what is wanted is a measure of "real" income—that is, income measured in units of constant real value. We are not concerned here with the practical obstacles to obtaining such a measure, but assume that the reader will bear them in mind in evaluating the comparisons made hereafter.[2] A second difficulty is that of comparing the values of outputs of two different economies when the composition of the two outputs differs greatly—for example, when one consists mainly of vegetable products consumed directly or with negligible processing and the other consists largely of highly processed manufactured goods. This problem

---

[2] For a brief, only slightly technical introduction to the problems of comparisons of real income, see Dan Usher, *Rich and Poor Countries* (London, 1966). An excellent treatment of the history, uses, and construction of national income accounts can be found in Paul Studenski, *The Income of Nations* (New York, 1958).

has no clearcut, definitive solution, but normally its quantitative dimensions do not hinder fruitful analysis.

Economic *development,* as the term is used in this book, means economic growth accompanied by a substantial structural or organizational change in the economy, such as a shift from a local subsistence economy to markets and trade, or the growth of manufacturing and service outputs relative to agriculture. The structural or organizational change may be the ''cause'' of growth but not necessarily; sometimes the causal sequence moves in the opposite direction, or the two changes may be the joint product of still other changes within or outside the economy. The concepts of economic structure and structural change are discussed in somewhat greater detail later in this chapter.

Economic growth, as defined here, is a reversible process—that is, it may be followed by decline. Logically, economic development is equally reversible, although organizations or structures rarely revert to precisely the same forms as existed earlier. More often, during or following a prolonged period of economic decline some form of economic *retrogression* takes place—a reversion to simpler forms of organization, though not usually identical with those that formerly existed.

Although they are widely regarded as ''good things,'' both growth and development are, in principle, value-free terms in that they can be measured and described without reference to ethical norms. Such is clearly not the case with the term economic *progress,* unless it be given a highly restrictive definition. In modern secular ethics, growth and development are frequently equated with progress, but no connection necessarily exists between them. By some ethical standards an increase in material well-being might be regarded as harmful to the spiritual nature of human beings. Even by contemporary standards the increased production of the means of chemical, biological, and nuclear warfare and the use of production processes that poison the environment, although manifestations of economic growth, can scarcely be regarded as signs of progress.

Another reason economic growth and development cannot be automatically equated with progress is that an increase in per capita income tells us nothing about the distribution of that income. What constitutes a ''good'' or ''bad'' distribution of income is a normative question about which economics has very little to say. It can say what kind of income distribution is more favorable to growth in certain situations, but, from an ethical point of view, that amounts to circular reasoning. Given certain ethical assumptions, it is possible to argue that lower per capita incomes, more equally distributed, are preferable to high average incomes that are very unequally distributed. Such arguments, however, lie outside the range of this book. In the following, growth and development are described and analyzed without reference to the term *progress.*

Determinants of Economic Development

Classical economics evolved the tripartite classification of the ''factors of production''—land, labor, and capital. (Sometimes a fourth factor is included—entrepreneurship, the effort or talent involved in combining or organizing the other three.) At any given time, subject to certain assumptions that will be specified later, an economy's total output is determined by the quantity of the production factors employed. This classification and various formulas that can be derived from it, such as the famous law of diminishing returns (see more on this later), are indispensable to modern economic analysis and extremely useful in the study of economic history as well. As a framework for the analysis of economic development, however, the classification is much too limited. It assumes that tastes, technology, and social institutions (e.g., the forms of economic, social, and political organization, the legal system, and even religion) are given and fixed or, what amounts to the same thing, have no bearing on the process of production. In historical fact, of course, all of these bear strongly on the process of production, and all are subject to change. Indeed, changes in technology and in social institutions are the most dynamic sources of change in the whole economy. They are thus the deep wellsprings of economic development.

To put the matter another way, in analyzing an economy at a given time (economic statics), or even at successive points in time, provided the intervals are not great (comparative statics or dynamics), it is permissible to regard such factors as tastes, technology, and social institutions as parameters (i.e., constants) of a system within which the quantities and prices of conventional production factors are the principal variables. In moving from short-term economic analysis to the study of economic development, however, the parameters become the major variables. A broader classification of the determinants of output is therefore necessary for analyzing economic change in historical time.

One such classification envisages the total output at any point in time, and its rate of change through time, as functions of the "mix" of population, resources, technology, and social institutions.[3] Of course, these four factors are not single variables; each one is a cluster of variables. It is not sufficient to think of population in terms of numbers alone; the age and sex distribution, the biological characteristics (size, strength, health, etc., of the members), the level of acquired skills (see more, later, on the concept of "human capital"), and the rate of labor force participation, among other features, have a bearing on a population's economic performance.

Resources is the "land" of classical economics writ large. The term embraces not merely the amount of land, the fertility of the soil, and conventional natural resources, but also climate, topography, availability of water, and other features of the natural environment, including location.

In recent centuries technological innovation has been the most dynamic source of economic change and development. A century ago the automobile, airplane, radio, and television, not to mention computers and numerous instruments of destruction, did not exist; today, according to some social critics, they threaten to dominate our lives. But technological change has not always been so rapid. Stone-age technology endured for hundreds of thousands of years with little change. Even today methods of agricultural production in some parts of the world remain essentially unchanged from biblical times. With a given technology—whether

---

[3] See the appendix to this chapter for a simple mathematical model of this classification.

that of medieval Europe or of pre-Columbian America—the resources available to a society set the effective upper limits to its economic achievements. Technological change, however, allows those limits to be expanded, both through the discovery of new resources and by more efficient use of the conventional factors of production, especially human labor. The continental United States today supports a population of more than 250 million at one of the highest material standards of living ever achieved. Before the Europeans came the same area, whose inhabitants employed a stone-age technology, could support only a few million with difficulty. The population of medieval Europe, with a far more advanced technology than that of the pre-Columbian Americans, grew to a maximum of perhaps 80 million at the beginning of the fourteenth century before declining to 50 million or fewer as the result of a disastrous demographic crisis. Four hundred years later, after a long period of steady but undramatic technological and organizational change, the population had grown to approximately 150 million. Today, after a mere two centuries of economic growth based on modern technology, the population of Europe is more than 500 million, and its members are far more affluent than their ancestors of the fourteenth or even the nineteenth century could have imagined.

The interrelationship of population, resources, and technology in the economy is conditioned by social institutions, including values and attitudes. (This complex of variables is sometimes also called the "sociocultural context," or the "institutional matrix" of economic activity.) At the level of national economies and other similar aggregates, the most frequently relevant institutions are the social structure (number, relative size, economic basis, and fluidity of social classes), the nature of the state or other political regime, and the religious or ideological proclivities of the dominant groups or classes and (if different) of the masses. In addition, a host of lesser institutions may need to be taken into account, such as voluntary associations (business firms, labor unions, farmers' groups), the educational system, and even family structure (extended or nuclear) and other value-forming agencies.

One social function that institutions perform is to provide elements of continuity and stability, without which societies would disintegrate; but in performing this function they may serve as barriers to economic development by fettering human labor, withholding resources from rational exploitation (e.g., India's sacred cows), or inhibiting innovation and the diffusion of technology. But institutional innovation is also a possibility, with consequences not unlike those of technological innovation, permitting a more efficient or intensive use of both material resources and human energy and ingenuity. Some historical examples are the institutional innovations of organized markets, coined money, patents, insurance, and the various forms of business enterprise, such as the modern corporation. Many other innovations are highlighted in the chapters that follow.

A complete list of all the social institutions of relevance to the economy would cover many pages, and the analysis of their interactions with other relevant variables is the most difficult and frustrating aspect of the study of economic history. But any attempt to comprehend the nature and modalities of economic development without reference to them is foredoomed to failure. With the present state of knowledge there is no systematic a priori approach that can be used to study them in their relation to economic activity. Instead, the student or investigator must determine, in the context of a specific problem or episode, the relevant institutions and then seek to analyze the nature of their interactions with more purely economic variables.

Marxist scholars claim to have found the key to not only the process of economic development but the evolution of humanity. According to them, the "mode of production" (roughly equivalent to technology in the schema outlined previously) is the key element; all the rest—social structure, nature of the state, dominant ideology, and so on—is mere "superstructure." The dynamic element is furnished by the struggle between social classes for control of the means of production. While some aspects of the Marxist analysis are useful in understanding economic history, the system as a whole is oversimplified and, in the hands of its practitioners, overly dogmatic. One of its weakest points, in view of its emphasis on mode of production, is that it furnishes no satisfactory explanation of the process of technological change. It also errs in regarding social institutions as being determined exclusively by the economic substructure.

A somewhat similar but less ideological theory views economic development as the product of a permanent tension or struggle between technological change and social institutions. According to this theory, sometimes called the institutionalist theory, technology is the dynamic, progressive element, whereas institutions uniformly resist change.[4] This theory offers a number of brilliant insights into the process of historical change, but it, too, regards technological change as an automatic or quasi-automatic process and oversimplifies the relationship between institutions and technology. Like the Marxist theory, it also regards the ultimate outcome as predictable. In fact, as the following chapters demonstrate, the relationship between institutions, technology, resources, and population is complex, interdependent, and by no means wholly predictable.

Production and Productivity

Production is the process by which the factors of production are combined to produce the goods and services desired by human populations. Production can be measured in physical units (or units of identical services), or in value—that is, monetary—terms. One can compare the output of, say, two apple orchards in terms of the number of bushels produced by each; comparing the output of an apple orchard and an orange grove in these terms is much less meaningful. To get a useful comparison in that case, it is necessary to convert the physical measure into value terms, that is, to multiply the number of bushels of each by the respective prices to arrive at their aggregate values.

Productivity is the ratio of the useful output of a production process to the inputs of the factors of production. As in the case of production, it can be measured in physical units—x bushels of wheat per acre, y widgets per man-hour—or in value terms. To measure *total factor productivity*—that is, the combined productivity of all factors—value terms are necessary.

---

[4] For a forceful and comprehensive exposition, see Clarence Ayres, *The Theory of Economic Progress* (Chapel Hill, NC, 1944, 1978).

The productivity of the factors of production depends on a host of elements. Some land is naturally more fertile than other land. Some workers are stronger or more skillful than others. The productivity of capital is in part a function of the technology it embodies; a mechanical tractor (in proper working order) is more productive than an equivalent value of ox-drawn plow teams, and a hydroelectric generator is more productive than an equivalent value of simple waterwheels. Moreover, certain *combinations* of the factors of production increase productivity. For example, the fertility of the soil is increased by the addition of fertilizer—that is, capital. Workers furnished with appropriate machines are more productive than those who work with their bare hands or with simple tools. In most instances literate workers are more productive than illiterate ones.

That consideration brings us to a most important special combination of the factors of production, namely, the concept of *human capital*. Human capital (*not* slaves, although they were once regarded as capital) results from investments in knowledge and ability or skill. The investment can take the form of formal schooling or training (a college education is a considerable investment), an apprenticeship, or "learning on the job" (i.e., practice). However human capital is acquired, the differences in levels of human capital per capita between the most and least advanced economies are among the most striking and important to be observed.

Empirical measurements in recent decades show unambiguously that increased inputs of the conventional factors of production account only in small part for increased output in advanced economies. In other words, the productivity of *all* factors of production has increased greatly. What accounts for those increases? Various answers to that question have been advanced, it is clear that among the most important determinants are advances in technology, improvements in organization at both the macro and micro levels (including so-called "economies of scale"), and especially increased investments in human capital. The increases in productivity have been particularly striking in the last hundred years or so, but as later chapters show, they have been important throughout recorded history, and even before.

At this point it is useful to consider in somewhat greater detail the so-called law of diminishing returns, which might more accurately be stated as the law of diminishing marginal productivity. A simple hypothetical example will illustrate its significance. Imagine a cultivated field of 100 acres. (The exact size is irrelevant.) A single worker employing a given technology, whether simple or complex, is able to produce *some* output—let us say 10 bushels of grain. The addition of a second worker permits a simple division of labor, which more than doubles production to perhaps 25 bushels; that is, the marginal product is 15. A third worker may raise output still more, to 45 bushels, for a marginal product of 20; and so on. In other words, as more workers are added, up to a point, the marginal product increases. Eventually, however, as more and more workers are added they get in each other's way, trample on the crop, and so on, and the marginal product *declines*: that is the concept of the law of diminishing returns.

Now let us transfer the lesson of this simplistic example to the case of a whole society. Remember that the example assumed fixed resources (100 acres) and a given technology (no productivity-enhancing innovations). If, at some point in time, the society is underpopulated relative to its resources, its population and per capita income will be able to grow even without technical or institutional change—for a time. Eventually, however, as it fully uses its resources, the increase in numbers will result in diminishing marginal productivity, hence declining real incomes. In this situation only a significant productivity-enhancing innovation (technical or institutional, or both) could relieve the dilemma.

In 1798 the Reverend Thomas R. Malthus, an English clergyman turned economist, published his famous *Principle of Population*. In it he assumed that "the passion between the sexes" would cause populations to grow at a "geometric ratio" (2, 4, 8, . . .) but that food supply would grow in an "arithmetic ratio" (1, 2, 3, . . .). In the absence of "moral restraint" such as celibacy and late marriage (he did not foresee artificial contraception), he concluded, the law of diminishing returns and the "positive checks" on population of war, famine, and pestilence would condemn the great majority of people to a bare subsistence standard of living. Now, almost 200 years later, it would seem that Malthus was wrong—at least as far as the industrialized nations are concerned. The other thing that Malthus did not foresee, of course, was the host of productivity-enhancing technological and institutional innovations that have repeatedly postponed the operation of the law of diminishing returns. For many Third World nations, however—the poorest of the poor—the Malthusian specter is still a grim reality.

Economic Structure and Structural Change

Economic structure (not to be confused with social structure, although the two are related) deals with the relationships among the various sectors of the economy, especially the three major sectors known as primary, secondary, and tertiary.[5] The primary sector includes those activities in which products are obtained directly from nature: agriculture, forestry, and fishing. The secondary sector includes those activities in which the products of nature are transformed or processed: that is, manufacturing and construction. The tertiary, or service, sector deals not with products or material goods at all, but with services; these cover a wide range, from domestic and personal services (cooks, maids, barbers, etc.) to commercial and financial (retail clerks, merchants, bankers, brokers, etc.) to professional (doctors, lawyers, educators) to governmental (postal workers, bureaucrats, politicians, the military, etc.). (There are some ambiguities and anomalies. For example, mining logically belongs in the primary sector, but it is frequently regarded as secondary; similarly, transportation, a service, is also often treated as part of the secondary sector. Hunting, the most important primary activity of paleolithic times, is now regarded as a recreational activity—consumption rather than production.)

For thousands of years, from the earliest civilizations until less than a century ago, agriculture was the principal occupation of the vast majority of the human race.

---

[5] The pioneering work on economic structure is Colin Clark's *Conditions of Economic Progress* (London, 1940, 1957). Simon Kuznets made major contributions to the elaboration of the concept, notably in *Modern Economic Growth: Rate, Structure, and Spread* (New Haven, 1966), and *The Economic Growth of Nations: Total Output and Production Structure* (Cambridge, MA, 1971).

As a perusal of Table 1-2 will show, this is still the case for the low-income nations. This was true because productivity was so low that mere survival required concentrating on the production of foodstuffs. A few hundred years ago, for reasons explained in subsequent chapters, agricultural productivity began to rise, slowly at first and then more rapidly. As it rose, fewer workers were needed for the production of subsistence goods and could be spared for other productive activities. Thus began the process of industrialization, which extended from the late Middle Ages to the mid-twentieth century (in western Europe and North America; it is still continuing in much of the rest of the world). The proportion of the labor force engaged in agriculture fell from 80 or 90 percent of the total to less than 50 percent by the end of the nineteenth century in the most advanced industrial nations, and to less than 10 percent more recently. Concomitantly, the proportion of total income, or GNP, originating in agriculture also fell, even though in absolute terms the total value of agricultural production increased manyfold.

Meanwhile, as the percent of the labor force engaged in agriculture fell, that in the secondary sector rose, although not in proportion; typically, in highly industrialized nations, manufacturing and related occupations employ between 30 and 50 percent of the labor force, with the remainder divided between the primary and tertiary sectors. As the share of the labor force in the secondary sector rose, so did that of income originating in that sector.

The twin processes of shifts in the proportions of the labor force employed and of income originating in the two sectors are major examples of *structural change* in the economy. Since about 1950 the most advanced economies have experienced a further structural change, from the secondary to the tertiary sector.

How can these structural changes be explained? The shift from agriculture to secondary activities involved two major processes. On the supply side, increasing productivity, as already explained, made it possible to produce the same amount of output with less labor (or more output with the same amount of labor). On the demand side a regularity of human behavior called Engel's Law (named for Ernst Engel, a nineteenth-century German statistician, not Friederich Engels, Karl Marx's collaborator) came into play. Based on numerous family budget studies, Engel's Law states that as a consumer's income increases, the proportion of income spent on food declines. (This, in turn, may be related to the law of diminishing marginal utility, namely, the more one has of a given commodity, the less one values any single unit of it.)

The second structural change now underway, the relative shift from commodity production (and consumption) to services, involves a corollary of Engel's Law: as income increases, the demand for all commodities increases, but at a lower rate than income, with an increased demand for services and leisure partly replacing the demand for commodities.

Changes in technology, with increased productivity, and in tastes are basically responsible for such structural changes, but the immediate motivating force for the changes is usually change in relative prices (and wages). This is also true for many other economic changes, such as the rise of new industries and the decline of old ones, or the shift of production from one geographic area to another. The prices of commodities and services are determined by the interaction of supply and demand, as taught in elementary economics textbooks. A high relative price indicates that supply is scarce in relation to demand; a low relative price indicates the opposite. As a general rule, the factors of production move to uses for which they are best rewarded, that is, those for which their prices are highest. The importance of relative scarcity and relative prices as dynamic elements in economic change will become evident in the historical cases considered later.

The Logistics of Economic Growth

In ordinary usage the term *logistics* refers to the organization of supplies for a large group of people, such as an army. But *logistic* (singular) is also a mathematical formula. The logistic curve derived from it has the form of an elongated S and is sometimes called the S-curve (see Fig. 1-1). Biologists also call it the growth curve, because it describes rather accurately the growth of many subhuman populations, such as a colony of fruit flies in a closed container with a constant food supply. The curve has two phases, one of accelerating growth followed by a deceleration phase; mathematically, at its limit the curve asymptotically approaches a horizontal line that is parallel to the asymptote of origin.

FIGURE 1–1

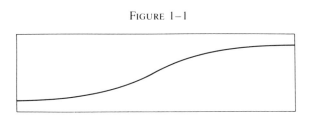

It has also been observed that logistic curves can also roughly describe many social phenomena, especially the growth of human populations. In the case of Europe, long-period surges of population growth have been identified, each being followed by a period of relative stagnation or even decline. The first of these began in the ninth or tenth century, probably reached peak rates in the twelfth century, began to slow down in the thirteenth, and was abruptly terminated by the Great Plague of 1348, when Europe lost a third or more of its total population. After a century of relative stagnation, the population began to grow again around the middle of the fifteenth century, reached peak rates in the sixteenth century, and again leveled off or possibly even declined in the seventeenth century. Toward the middle of the eighteenth century the process got underway again, this time much more powerfully, and continued at unprecedented rates until interrupted by the world wars and related misfortunes of the first half of the twentieth century. There is evidence of a fourth logistic, this time on a world scale, beginning after World War II.

Although precise quantitative data are lacking, it seems likely that the growth of the Greek population between the ninth and fifth century B.C. followed a logistic pattern, as did the population of the Mediterranean basin in the era of the *pax romana* (ca. 50 B.C.–200 A.D.). Some scholars believe the three identifiable European logistics were, in fact, worldwide and related to climatic variations. The population of China, for example, seems to have kept pace with that of Europe. We are even more ignorant about the pattern of population growth in earlier epochs, but, as Chapter 2 shows, the population of the present-day Near and Middle East definitely grew after the advent of agriculture in the neolithic period, and the populations of the great river valleys (the Nile, Tigris-Euphrates, Indus, and Yellow River in China) likewise grew rapidly after the introduction of irrigation agriculture.

Whether or not the growth of population actually conforms to the logistic curve, other related aspects of the phenomenon intrigue the scientific imagination. It is virtually certain that each accelerating phase of population growth in Europe was accompanied by economic growth, in the sense that both total and per capita output were increasing. (If per capita output merely remained constant as the population grew, the total output, of course, would increase; but we are warranted in making the stronger statement.) This is most clearly attested for the third logistic (and the incipient fourth), for which statistical evidence is relatively plentiful; but there is also much indirect evidence for similar behavior during the first and second logistics.

The hypothesis of economic growth accompanying the growth of population is strongly supported by the unquestioned evidence of both physical and economic expansion of European civilization during each of the accelerating phases of population growth. During the eleventh, twelfth, and thirteenth centuries European civilization expanded from its heartland between the Loire and Rhine rivers to the British Isles, the Iberian peninsula, Sicily and southern Italy, into central and eastern Europe, and even temporarily, during the Crusades, to Palestine and the eastern Mediterranean. In the late fifteenth and sixteenth centuries maritime exploration, discovery, and conquest took Europeans to Africa, the Indian Ocean, and the Western Hemisphere. Finally, in the nineteenth century, through migration, conquest, and annexation, Europeans established their political and economic hegemony throughout the world.

There is also evidence that conditions of life for ordinary men and women were becoming increasingly difficult in the decelerating phases of the first two logistics (the first half of the fourteenth and seventeenth centuries, respectively), suggesting a decline in or at least stagnation of per capita incomes. In the third logistic the opportunity for large-scale emigration from Europe in the late nineteenth and early twentieth centuries palliated the condition of the masses; even so, a number of countries experienced localized subsistence crises, of which the Irish famine of the 1840s was the most dramatic. In the light of these observations Adam Smith's remark, written in the accelerating phase of the third logistic, to the effect that the position of the laborer was happiest in a ''progressive'' society, dreary in a stationary one, and miserable in a declining one takes on a new significance.

Another noteworthy similarity is that the final phases of all the logistics, and the intervals of stagnation or depression that followed, witnessed the spread of social tension, civil unrest and disorder, and the outbreak of unusually fierce and destructive wars. To be sure, wars and civil strife occurred at other times as well; and there is no obvious theoretical reason that the decline of population growth should have resulted in the breakdown of international relations. Possibly the wars were simply fortuitous occurrences that ended periods of growth that were already waning. But the question is worthy of further study.

It would no doubt strain credulity to suggest that notable periods of intellectual and cultural ferment were also related somehow to the logistic. It is nevertheless remarkable that the accelerating phases of each period of population growth in Europe witnessed outbursts of intellectual and artistic creativity followed by a proliferation of monumental architecture—medieval cathedrals, baroque palaces, and the nineteenth-century Gothic revival. Earlier, the "Golden Ages" of Greece and Rome—and still earlier, those of Mesopotamia and Egypt—were periods of population growth and ended with civil strife and internecine warfare (the Peloponnesian War, the decline of Rome).

Of course, human creative efforts are not confined to specific historical periods any more than our destructive tendencies. The origins of the Renaissance were the great depression of the late Middle Ages, and the century of genius that included Galileo, Descartes, Newton, Leibnitz, and Locke spanned the interval of stagnation and upheaval between the second and third European logistics. Still, it is possible that periods of crisis in human affairs, when the established order appears to be breaking down, may stimulate the best intellects in a variety of fields to reexamine accepted doctrines. Such lofty considerations lie outside the scope of this volume, however.

A possible explanation for the correlation of population growth/stagnation/decline with income movements can be fashioned by analyzing the interaction of the fundamental determinants of economic development introduced previously (p. 9). As indicated, with a given technology the resources available to a society set the upper limits to its economic achievements, including the size of its population. Technological change, by increasing productivity and opening up new resources, has the effect of raising the ceiling, as it were, thereby permitting further growth in population. Eventually, however, without further technological change, the phenomenon of diminishing returns sets in, the society encounters a new ceiling on production, and population again levels off (or declines) until a new "epochal innovation" (the phase is that of Simon Kuznets, a Nobel Prize winner in economics; see Chapter 8) again increases productivity and opens up still newer resources. Figure 1-2 presents a simplified representation of the relationship between population and epochal innovations.

The chapters that follow provide an empirical test for this hypothesis as they attempt to explain economic development in history.

FIGURE 1–2

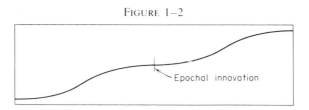

APPENDIX

Let Y stand for national income (or output) and P, R, T, and X for population, resources, technology, and social institutions (the "great unknown"), respectively. Then

$$Y = f\,(P,\,R,\,T,\,X)$$

and the rate of change through time is

$$\frac{dy}{dt} = \frac{df}{dt}\,.$$

For reasons noted in the text, the equation cannot be written in operational form.

2

Economic Development in Ancient Times

Humans, the tool-using animals, may have appeared on earth as long ago as 2 million years,[1] but, if so, their only tools for the first 1,990,000 years or so of their existence were rough hand tools—clubs, fist hatchets, scrapers, and such—made of wood, bone, and stone. Although we have little definite, detailed knowledge of this long period in our evolution, scholars have pieced together the widely scattered fragmentary evidence with much ingenious speculation to construct a plausible account of it.

The earliest humans, forerunners of *Homo sapiens,* were probably omnivorous creatures who supplemented their basic diet of tubers, berries, and nuts with insects, fish, mollusks (where available), the flesh of small game, and possibly carrion. Their crude tools, either appropriated directly from nature or subjected to minimal refinement, would have been used mainly for digging, scraping, and pounding—that is, as extensions or modifications of the human hand. In successive millennia biological evolution was accompanied, and eventually outstripped, by social and technological evolution. Stones formerly used for pounding were chipped or flaked to make rough cutting edges; straight sticks were given pointed ends to serve as primitive spears. Special types of stones, such as flint and obsidian, were discovered to be especially suitable for toolmaking, and bones, horns, and ivory entered the toolmakers' list of materials. In the beginning this technological evolution was probably as slow as biological evolution itself, but it must have been accelerated in the last 50,000 years or so. Toward the end of the last (Würm) glaciation, some 20,000 to 30,000 years ago, late paleolithic humans had reached a relatively advanced state of technological, and probably also social, development. They made a great variety of chipped and flaked stone tools, including knives, awls, and chisels, and used bones, horns, and shells for fishhooks and needles (Fig. 2-1). For weapons they had lances, spears, harpoons, slings, and bows and arrows. By this time humans were primarily carnivorous hunters, at least in Eurasia, North

[1] The issue is clouded not only by a paucity of evidence but by differing definitions of humans. *Homo sapiens,* the species to which all existing races belong, is thought to be only about 250,000 years old, but it was preceded by *Homo erectus* and *Homo habilis.* Hominoid remains recently found in Kenya, in proximity to what may have been very crude stone tools, have been estimated to be almost 20 million years old.

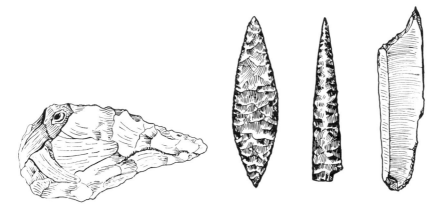

FIGURE 2–1.   Paleolithic tools. These stone tools had many uses. Most were fitted with wooden hafts or handles to facilitate their use. (From *Art Through the Ages,* 3rd ed., by Helen Gardner, Copyright 1948, 1975 by Harcourt Brace Jovanovich, Inc. Reprinted with permission.)

America, and North Africa, and among their favorite prey were the wild horses, bison, reindeer, and mammoths that abounded at that time. They had long known and used fire.

The unit of social organization was the band or tribe, consisting of about half a dozen families. It was essentially migratory, following the game, but usually limited its migrations to a given geographical area, and might return at periodic intervals to a ceremonial center such as a sacred grove or cave. Contact between bands or tribes was probably rare, but not so rare as to prevent the diffusion of social traits and techniques, and perhaps some primitive barter trade including the exchange of women. Marriage and kinship rules had evolved, and incest was universally tabooed. Animistic beliefs presaged religion, just as a primitive calendar portended science. Some indication of the level of cultural development is given by the magnificent cave paintings of northern Spain and southwestern France, dating from about 20,000 years ago (Fig. 2-2). They not only show a high level of artistic skill, but reflect aspects of their creators' economic activities and probably their religious conceptions. The most common subjects are the animals they hunted; the paintings may have been intended as memorials to particularly successful hunts, or they may have been evocations to the spirits to supply them with plentiful game.

In material terms, life was, as the seventeenth-century philosopher Thomas Hobbes described it in the state of nature, "nasty, brutish, and short." From skeletal remains it has been estimated that the average length of life was no more than about twenty years; infant and childhood mortality was especially high, with fewer than 50 percent of children surviving to the age of ten; and survivors beyond the age of fifty were extremely rare. Given the nature of their economy, paleolithic humans were subject to recurrent rounds of feast and famine, depending on the movement of game and the luck of the hunt. In periods of famine all but the strongest perished, and in prolonged famines entire communities did so.

FIGURE 2–2. Cave paintings. Paleolithic people—or some of them—had fine artistic sensibilities and talents. The hand prints above the painting may be the artist's "signature." (Studio Laborie, Bergerac, France.)

In spite of these hazards, paleolithic humans were distributed widely over the face of the earth. By the end of paleolithic times, some 10,000 or 12,000 years ago, virtually every inhabitable part of the earth, from the Arctic to South Africa, Australia, and the Tierra del Fuego, had been occupied, however thinly or tentatively. Population densities no doubt varied in proportion to the flora and fauna that served as their means of subsistence, with the higher densities in tropical and subtropical areas; but nowhere were densities high by modern standards. Modern authorities have estimated, largely on the basis of deductive reasoning, that the world population of *Homo sapiens* at the end of the paleolithic era could not have exceeded 20 million and more likely was in the range of 10 million inhabitants.

Economics and the Emergence of Civilization

The retreat of the last continental glaciers about 10,000 or 12,000 years ago ushered in a period of significant geographic and climatic change, especially in the northern hemisphere, with correspondingly significant consequences for human history. The amelioration of the climate of Eurasia and North America was offset by the disappearance of many of the mammals that made up the basic food supply of the late paleolithic hunters. The hairy mammoth and the woolly rhinoceros became extinct, and the reindeer migrated northward to its present habitat. North Africa and central Asia became more arid, forcing their inhabitants to migrate and adopt new ways of life, while huge forests grew north of the Alps and great grasses covered the highland areas at the eastern end of the Mediterranean.

Whether or not they were directly related to the climatic changes, important technological changes also occurred in the four or five millennia following the retreat of the glaciers, especially in the Near and Middle East. Stone tools (and art and religious objects as well) became more intricate and refined. Grinding and polishing of stone replaced the older methods of chipping and flaking. The neolithic, or new stone, age had arrived. (Some scholars assert that a rather nebulous mesolithic, or transitional, era occurred between the end of the ice age and the full establishment of neolithic cultures in the Near and Middle East by the beginning of the sixth millennium B.C.) The most significant new departures, however, were the invention of agriculture and the domestication of animals.

The exact time and location of these latter achievements are open to dispute. It is not even certain that they occurred in conjunction with one another, although it seems likely that they did, at least for some animals. The most probable site is somewhere on the so-called Fertile Crescent, the belt of land (perhaps more fertile then than now) extending along the eastern end of the Mediterranean, arching across the hills of northern Syria and Iraq, and down the valleys of the Tigris and Euphrates to the Persian Gulf. One hypothesis, as plausible as any, is that the domestication of plants was the work of women in the hills of northern Iraq or Kurdistan. The wild ancestors of wheat and barley grew naturally in the area. The women, left behind in temporary camps while their men hunted sheep and goats in the nearby mountains, harvested the wild seeds and eventually began to cultivate them. This hypothesis is reinforced by the fact that sheep and goats were probably the first animals to be domesticated (except for the dog, which may have been associated with paleolithic hunters). The process (for it almost certainly was not a unique event) may have begun as early as 8000 B.C. or even earlier. What is certain is that by 6000 B.C. settled agriculture, involving cultivation of wheat and barley and tending of sheep, goats, pigs, and possibly cattle, was well established throughout the area from western Iran to the Mediterranean and across the Anatolian highlands to both sides of the Aegean Sea. From that area it spread gradually to Egypt, India, China, western Europe, and other parts of the Old World. (Arguments for the independent origins of agriculture in China and Southeast Asia, though interesting, have not been proven.)

The significance of these developments for human history was momentous. For the first time people were able to establish relatively permanent settlements. This, together with the greater productivity of their efforts, enabled them simultaneously to accumulate greater stores of material goods, or wealth, and to devote more time to nonsubsistence activities, such as art and religion. The greater reliability of their food supply (fluctuations were at least annual rather than daily) no doubt introduced an element of psychological as well as physical stability into personal and social relationships. The entire basis of existence was thereby revolutionized, with consequences that are very much a part of our lives in the twentieth century.

One should not, of course, exaggerate the revolutionary nature of changes that were accomplished over a period of hundreds, even thousands, of years. The changes were so gradual that the people experiencing them were probably unaware, or at best dimly aware, of them, and without written records they could have no notion of the significance of the transition. Hunting and farming were complementary activities for many generations, with pastoralism as a transitional stage. As the techniques of agriculture were mastered and it became more efficient and productive, the economic importance of hunting receded, but it never lost its symbolic significance: the transition from hunter to warrior to ruler was a natural one. Insofar as one can speak of motivation, the changes were simply a process of adaptation to a mostly hostile environment. Custom and tradition governed both social relationships and methods of production, and the idea of deliberate invention in either area could scarcely have entered the mind of neolithic humans.

The implements used by the earliest agriculturists were simple in the extreme. The first was a primitive sickle or reaping knife—typically a blade of flint chips or teeth attached to a wood or bone handle—used in harvesting the seeds of wild grasses and eventually those of cultivated grains. The first instruments of cultivation were plain digging sticks and simple hoes made by attaching a stone blade to a wooden handle. This type of agriculture, which subsequently spread to many parts of the world and still persists in some remote areas, is frequently referred to as "hoe culture." Ploughs pulled by oxen or donkeys belonged to a later stage of development, first making their appearance in the great river valleys in the third or fourth millennium B.C.

To this basic equipment new tools, new techniques, new crops, and new livestock were gradually added. Cattle, if not domesticated before 6000 B.C., came into the fold soon afterward. Lentils and peas, as well as various root crops, were cultivated in Anatolia well before that date. Grain was probably first consumed in the form of gruel or porridge, but primitive querns and mortars for grinding grain into meal or flour have been found in some of the earliest archeological sites, evidence that the art of baking was discovered almost as early as the invention of agriculture. By the sixth millennium grain was also fermented to make a kind of mead or ale. Pottery, more fragile but requiring less labor to produce than stone containers, was invented about the same time; pottery also provided a new esthetic outlet and was widely used for ornamental and ceremonial as well as utilitarian purposes. Although no evidence survives, it seems likely that basketry preceded pottery. It almost certainly preceded the manufacture of textiles (spinning and weaving), and there is evidence that linen cloth was being made by the beginning of the fifth millennium (which also suggests the domestication of flax). There is no clear evidence that woolen cloth was manufactured before the middle of the third millennium but, given the early domestication of sheep and goats and the fact that the technique of making woolen yarn is simpler than that for linen, wool was probably the first substitute for the skins and furs that clothed paleolithic humans.

The sedentary existence of the peasant village permitted a finer division of labor than that determined by age and sex. As Adam Smith pointed out more than two centuries ago, division of labor involves specialization, and specialization leads to greater efficiency and technological progress. Precisely how and when specific innovations took place is a matter for speculation, since the surviving evidence is rarely explicit. It seems logical, nevertheless, that advances in one area would stimulate advance in others ("spin off" or "fall out" in modern research and development jargon). For example, as migratory bands settled in one location they would replace such temporary shelters as tents of skins or windbreaks made of boughs with more permanent and comfortable abodes: dugouts or pit houses at first, followed by sod houses, and eventually (the typical dwelling of the peasant villagers in the Near and Middle East) houses of sun-dried mud brick. Experience in making bricks needed for dwellings may have led to the use of clay for pots, and thus to pottery. As potters refined their art they invented the potter's wheel, which almost certainly preceded the use of wheels for transport.

Metallurgy may have arisen in an analogous fashion. Although some gold and copper objects have been found that date from the sixth millennium, regular production of copper did not begin until the fifth or perhaps even the fourth millennium, and bronze (an alloy of copper and tin) came even later. Copper ore occurs in the mountains of Anatolia, the southern Caucasus, and northern Iran. It was in precisely these areas that some neolithic cultivators obtained the flint, obsidian, and other stones required for their tools. A common method of extraction of fresh flintstone from its mother lode was to heat it to high temperatures with fire, then cool it rapidly by dashing water on it. It is not unlikely that copper ore was accidentially smelted in this fashion. Whatever the method of its discovery, however, copper smelting was widely practiced in the Near and Middle East by the middle of the fourth millennium, and tools, weapons, and ornaments of copper and bronze were added to (but did not wholly replace) those of stone, clay, and other materials.

The division of labor and evolution of new crafts, such as pottery and metallurgy, required some form of exchange or commerce. The nature of the exchange varied with the distance over which commodities had to be transported. Within individual communities the terms of exchange were probably determined by custom, but this would not do for long-distance trade in highly localized commodities, such as metal. Some form of organized exchange was necessary. Indeed, trade in the form of barter had been practiced in late paleolithic and early neolithic times; the mining of flintstones and the manufacture of stone axes and other weapons had become specialized crafts by the eighth millennium, as evidenced by the widespread distribution of implements that can be identified as coming from particular mines or mining areas. Unfortunately, we do not know who the agents of this commerce were. The trade in stone implements may have been carried on by migratory hunters, that in metals by nomadic pastoral tribes, but this is just speculation. After the rise of city-states and empires, organized expeditions were sent out for trading and raiding.

One of the principal consequences of the invention of agriculture was the enhanced ability of given areas to support their populations. Thus, the population grew wherever neolithic agriculture was diffused. Agriculture reached the Nile valley before 4000 B.C. and the Indus valley within the next millennium. By about 2500 B.C. it had penetrated the Danube valley, the western Mediterranean, southern Russia, and possibly China. Modifications were sometimes introduced in the course of diffusion, because of differences in resources and climate. In northern China, for example, millet and soy beans became the staple food crops. In Southeast Asia the basis of farming was first the taro root and later (after about 1500 B.C.) rice. In the latter area the water buffalo was the most important domesticated animal. In the arid steppe lands of southern Russia and central Asia, neolithic hoe cultures did not take root, but the inhabitants developed a pastoral way of life; it was probably in this area that the horse was domesticated sometime in the third millennium.

The basic unit of economic and social organization in the early agricultural communities was the peasant village, consisting of from 10 to 50 families, with a total population between 50 and 300 persons. The peasant villages may be regarded as the logical, and perhaps in some cases the actual, successors of the late paleolithic hunting bands, although on the average they were substantially larger because of their better adaptation to the environment. Living conditions were slightly improved over those of hunting and gathering communities. The food supply was somewhat more regular and dependable, and dwellings were no doubt more comfortable; but because population tended to grow along with the means to support it, the peasants still lived at the margin of subsistence. A natural disaster such as a drought, flood, or plague of insects could wreak havoc on an entire village or group of villages; and their settled existence and denser populations than those of hunting tribes made them more subject to epidemic disease. The average length of life probably did not exceed twenty-five years.

It was formerly believed that the neolithic peasant villages were relatively uniform and undifferentiated until the rise of more powerful city-states, in the middle of the fourth millennium. Recent archeological discoveries, however, have demonstrated the existence of communities with a fundamentally different structure from the peasant villages, which can rightfully be called cities (Fig. 2-3). A city discovered at Catal Hüyük in Anatolia, dating from the middle of the seventh millennium, had closely set houses of uniform structure and dimensions built of

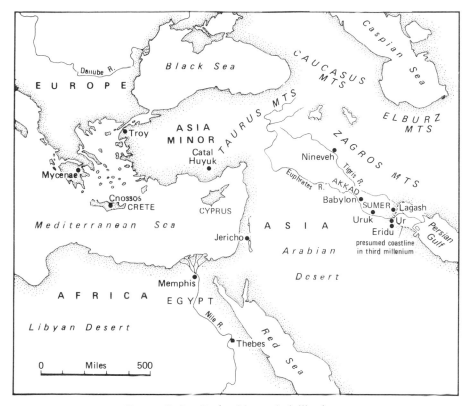

FIGURE 2–3. Early cities and civilizations.

clay and bricks, also of uniform dimensions, implying a well-organized division of labor. Obsidian, the raw material for most of its tools, was imported from volcanic deposits more than seventy miles distant. At Jericho, perhaps the oldest continuously occupied site in the world, with traces of neolithic settlement as early as 8000 B.C., a huge stone wall was erected by 7000 B.C. This achievement was certainly beyond the capacity of a simple agricultural village. There are traces of other such cities in the Aegean area and elsewhere in the Near East, and undoubtedly other yet undiscovered urban sites must have existed before the rise of the great river valley civilizations in Mesopotamia and Egypt. The exact function and basis for the existence of these proto-cities has not yet been discovered. They probably served, however, as primitive manufacturing centers and commercial entrepôts for the surrounding agricultural communities. If so, their existence is evidence of a far more complex organization of the economy—with no central organizational authority—than was formerly believed possible for that time.

Before about 4500 B.C. Lower Mesopotamia, the region between the Tigris and Euphrates rivers just north of the Persian Gulf, was much less densely populated than other inhabited regions of the Near and Middle East. Its marshy soil, subject to annual inundations from the rivers, was not suited to the primitive hoe culture of neolithic agriculture. Moreover, the land was virtually treeless and lacked building stone and mineral resources. During the next thousand years, however, this unpromising area became the seat of the first great civilization known to history, that of Sumer, with large concentrations of people, bustling cities, monumental architecture, and a wealth of religious, artistic, and literary traditions that influenced other ancient civilizations for thousands of years. The exact sequence of events that led to this culmination is unknown, but it is clear that the economic basis of this first civilization lay in its highly productive agriculture.

The natural fertility of the black alluvial soil was renewed annually by the silt left from the spring floods of the Tigris and Euphrates rivers. Harnessing its full productive powers, however, required an elaborate system of drainage and irrigation, which in turn required a large and well-disciplined work force as well as skilled management and supervision. The latter were provided by a class of priests and warriors who ruled a large, servile population of peasants and artisans. Through tribute, taxation, and slavery the rulers extracted the wealth that went into the construction of temples and other public buildings, the creation of works of art, and that gave them (or some of them) the leisure time to perfect the other refinements of civilization.

The rise of civilization brought with it a far more complex division of labor and system of economic organization. Full-time artisans specialized in the manufacture of textiles and pottery, metal working, and other crafts. The professions of architecture, engineering, and medicine, among others, were born. Weights and measures were systematized, mathematics was invented, and primitive forms of science emerged. Since Sumer was virtually devoid of natural resources other than its rich soil, it traded with other, less advanced people, thereby contributing to the diffusion of Sumerian civilization. The scarcity of stone, for tools as well as for buildings, probably hastened the adoption of copper and bronze. Copper, at least, was already known before the rise of Sumerian civilization, but lack of demand for it among the neolithic peasant villages inhibited its widespread use. In the Sumerian cities, on the other hand, imported stone had to compete with imported copper, and the latter proved more economical as well as more effective in a variety of uses. It was imported by sea through the Persian Gulf from Oman, and downriver from the mountains of Anatolia and the Caucasus. Thereafter metallurgy was regarded as one of the hallmarks of civilization.

Sumer's greatest contribution to subsequent civilizations, the invention of writing, likewise grew out of economic necessity. The early cities such as Eridu, Ur, Uruk, and Lagash were temple cities; that is, both economic and religious organization centered on the temple of the local patron deity, represented by a priestly hierarchy. Members of the hierarchy directed the labor of irrigation, drainage, and agriculture generally, and supervised the collection of the produce as tribute or taxation. The need to keep records of the sources and uses of this tribute led to the use of simple pictographs on clay tablets, sometime before 3000 B.C. By about 2800 B.C. the pictographs had been stylized into the cuneiform system of writing, a distinctive characteristic of Mesopotamian civilization. It is one of the few examples in history of a significant innovation issuing from a bureaucratic organization.

Although writing originated in response to the need for administrative recordkeeping, it soon found many other religious, literary, and economic uses. In a later phase of development, after the strict temple-centered organization of the economy had given way to greater freedom of enterprise, clay tablets recorded the details of contracts, debts, and other commercial and financial transactions.

From its earliest seat at the head of the Persian Gulf, Mesopotamian civilization spread northward into Akkad, whose principal center was the city of Babylon, and subsequently into the upper reaches of the Tigris and Euphrates valleys. Through their trading expeditions in search of raw materials, especially metal and perhaps other commodities, the Mesopotamian city-states stimulated the nascent civilizations of Egypt, the eastern Mediterranean and Aegean area, Anatolia, and the Indus valley. Of these, Egypt and the Indus valley were, like Mesopotamia itself, riverine civilizations that owed their existence to the control and use of the flood waters of the great rivers along which they lay. Little is known about the early development of the Indus valley civilization, although it apparently had contacts with Mesopotamia by both land and sea.

Egypt, near the end of the fourth millennium, was still in a neolithic stage of development, but its contacts with Mesopotamia—especially those of Upper Egypt via the Persian Gulf, Indian Ocean, and Red Sea route—stimulated a rapid development in all aspects of civilization. By the middle of the third millennium Egyptian civilization had reached a stage of maturity in its government, art, religion, and economy that remained virtually unaltered until the beginning of the Christian era, in spite of foreign conquests and domestic upheavals.

The Economic Foundations of Empire

One of the notable features of ancient history, reflecting the interests of the annalists who chronicled it as well as those of subsequent historians, is the rise and fall of empires. From the rise of the first great world empire of Sargon of Akkad (ca. 2350–2300 B.C.) to the fall of the Roman Empire in the West (the traditional date is 476 A.D.), the historical record is filled with a bewildering profusion of empires and their rulers: Babylonia, Assyria, the Hittites, the Persians, Alexander the Great and his successors are but a few. (The process continued in the European Middle Ages, with the explosive exfoliation of various Islamic empires in the seventh and succeeding centuries, the tribulations of the venerable Byzantine Empire, and its final extinction by the Ottoman Turkish Empire in 1453 A.D.) Much less has been written about the economic foundations of these empires. What were the economic bases of their military exploits and political power? What did they contribute to the material progress of civilization? What was the daily existence, the standard of living of ordinary men and women? The historical records relating to these questions have not yet been analyzed extensively, but through indirect evidence (much of it archeological) and with judicious inference and deduction, it is possible to fashion at least tentative answers.

Before the rise of the first great urban civilizations the social structure of the neolithic peasant villages appears to have been relatively simple and uniform. Custom and tradition, as interpreted by a council of elders, governed relations among members of the community. The concept of property would have been vague, at best. Private ownership of tools, weapons, and ornaments was no doubt recognized, but land and livestock were probably owned collectively. (In economic terminology, land was not scarce, and hence would not command a premium, or rent.) Although some individual or individuals in each village may have been accorded special status because of wisdom, strength, courage, or other leadership qualities, it does not appear that there were any privileged or leisured classes; the obligation of all to work was dictated by both technology and resources.

In the early temple cities of Sumer, in contrast, the social structure was definite-ly heirarchical. The masses of peasants and unskilled workers, probably amounting to 90 percent or more of the total population, lived in a state of servitude, if not outright slavery; they had no rights, property or other. The land belonged to the temple (or to its deity) and was administered by the deity's representatives, the priests. At a somewhat later date—but not later than the beginning of the third millennium—a warrior class, led by chiefs or kings, asserted its authority in con-junction with or over that of the priests. Unfortunately, the details of this transition from a relatively undifferentiated society to a stratified society are unknown. Ac-cording to Marxist theory, it resulted from the creation of the institution of private property out of the former communal property, which allowed one segment of society to live from the labor of the others—"the exploitation of man by man." Although it is true that the priest and warrior classes did not engage in economically productive activities (except to the extent that their directive and supervisory func-tions were necessary), and in that sense exploited the peasants and workers, one may fairly doubt that the institution of private property was closely associated with the phenomenon. Property relationships varied considerably from one area to an-other, and over time within the same area; but nowhere in ancient civilization did private property, in the modern sense, constitute the legal foundation of society or state. Some form of collective or state ownership of land was the general rule. Certain parcels of land, or a portion of its produce, were frequently designated for the support of particular officials and warriors, and private ownership of tools, weapons, and other personal possessions was no doubt recognized, but private property was not an absolute right.

More likely, the root of class differentiation and formal political organization was ethnic or tribal differences. Significantly, Sumerian, the first written language, was unrelated to any of the neighboring Semitic languages—unrelated, in fact, to any other known language. Possibly the organizers of the earliest Sumerian city-states were alien conquerors who imposed themselves on a preexisting neolithic population. In any event, it is clear from subsequent developments that the riches of the riverine city-states were tempting prizes that repeatedly drew their more primi-tive neighbors from the surrounding hills and deserts to invade and conquer or pillage the Sumerian cities. In some instances the invaders merely seized what they could conveniently carry away and departed; in others they slaughtered or subjugat-ed the existing ruling class and established themselves as rulers over the servile population. The numerous references in ancient mythology to conflicts among the gods probably reflect the struggle for mastery among the various warrior tribes, each with its own deity. Such successions of ruling classes mattered little to the peasant populations, except when they became accidental victims of violence or one group of rulers was more ruthless and efficient than another in extracting tribute and taxation.

As the early city-states expanded in proximity to one another, disputes over boundaries and water rights became additional sources of conflict and conquest. The earliest written records from the classical Sumerian civilization of the third millennium contain numerous references to the succession of dynasties that ruled the various cities. Economic considerations were not, of course, the only motivating influences in these struggles. The lust for power, dominion, and magnificence soon overtook mere economic motivations. Sargon the Great not only brought all of the city-states of Sumer and Akkad under one central administration but extended his conquests to Iran, northern Mesopotamia, and Syria, thus ruling virtually all of the civilized world of his time except Egypt. Similar ambitions stirred other conquerors, great and small, including Cyrus of Persia, Alexander of Macedonia, and Julius Caesar and his successors, the Roman emperors. Whatever the motives, however, the economic bases of these ancient empires lay in the booty, tribute, and taxation that the conquerors could wring from the conquered and from the peasant masses.

Given the predatory character of ancient empires, did they make any positive contributions to economic development? In terms of technological development the record is extremely sparse. Almost all of the major elements of technology that served ancient civilizations—domesticated plants and animals, textiles, pottery, metallurgy, monumental architecture, the wheel, sailing ships, and so on—had been invented or discovered before the dawn of recorded history. The most notable technological achievement of the second millennium (ca. 1400–1200 B.C.), the discovery of a process for smelting iron ore, was probably made by a barbarian or semibarbarian tribe in Anatolia or the Caucasus Mountains. Significantly, the principal use of iron in ancient times was for weapons, not tools. Other innovations, such as chariots and specialized fighting ships, were even more directly related to the art of war and conquest.

Although there were few major breakthroughs, many minor technical improvements were made, especially in agriculture, but these can rarely be credited to government actions or policy. During Hellenistic times and under the Roman Empire scores of treatises were written on various aspects of agriculture and related occupations (the famous library in Alexandria contained fifty manuscripts devoted solely to the art of baking bread!), designed to inform wealthy landowners and their stewards how to increase the yields of their estates. The peculiarities of climate, topography, and soil in the Mediterranean basin determined optimum agricultural methods, which evolved gradually and imperfectly through many centuries of trial and error. The wealth of the great riverine civilizations was based on irrigation agriculture, which required a high degree of organization and discipline of the labor force. Elsewhere (for example, in North Africa and southern Spain) irrigation sometimes supplemented other methods, but for the most part it was uneconomic if not impossible for generalized use. Instead, the technique of "dry farming" (as it came to be known in nineteenth-century America) evolved. Given the light, shallow soils and the long, dry summers that characterize most of the area, arable land had to be plowed frequently but lightly to hold and use the moisture that collected during the rainy winter season. To maintain the fertility of the soil, with no artificial fertilizers and scarce natural manure, fields were planted only every other year (biennial rotation with fallow); moreover, to reduce unwanted growth that would rob the fallow of its nutrients, it too had to be plowed, usually three or four times but, optimally, up to nine times per season. Numerous variations of this basic pattern occurred, especially in the areas where horticulture, arboriculture, and viticulture flourished. In general, however, all of them were highly labor intensive, that is, requiring much labor per unit of land. This severely limited the size of units that could be exploited by independent proprietors or single tenants, and according-ly left little surplus for taxation. On the other hand, where the terrain was suitable and the supply of labor adequate, large estates using gang labor of either cheap servile workers (an agricultural proletariat) or slaves could be profitable to both the owner and the government. From earliest times to the later Roman Empire, the latter system gained ground at the expense of the former, especially in the most fertile regions.

In spite of the near-stagnation of technology, the economic achievements of the ancient empires were considerable. Organized expeditions, whether for trade or conquest, diffused the existing elements of technology more widely and brought new resources into the ambit of the economy. Explicit formulation of civil law, even if drawn up for the enlightened self-interest of the ruler or the ruling class, contributed to smoother functioning of the economy and society. Most important of all, perhaps, establishing order and common laws over larger and larger areas facilitated the growth of trade and, with it, regional specialization and division of labor. The outstanding example of this tendency is, of course, the Roman Empire.

Trade and Development in the Mediterranean World

In the millennium extending roughly from 800 B.C. to 200 A.D., the classical civilization of the Mediterranean world achieved a level of economic development that was not surpassed, at least in Europe, until the twelfth or thirteenth century. Given the absence of notable technological progress in the era, the explanation for this achievement should be sought in the extensive division of labor made possible by a highly developed network of trade and markets. Trade was not a new phe-nomenon, of course; allusion has been made to the traffic in stone tools and weap-ons in neolithic times and to the expeditions of the Mesopotamian city-states and empires. The latter were usually state-sponsored, and it was not always easy to distinguish a trading from a raiding mission. Rulers of neighboring states also engaged in the ritual exchange of gifts, a disguised form of barter. Given the high cost of land transport, however—goods were carried by pack animals and human porters—such commerce was limited to commodities of high value in relation to their bulk, such as gold, silver, and precious stones, luxury cloth, spices and perfumes, and art and religious objects. (The only apparent exception to this rule, the traffic in copper and bronze, was not really an exception because metals des-tined primarily for the weapons and ornaments of the ruling classes commanded a much higher relative price than they do today.) Mesopotamian civilizations established contacts through the Indian Ocean with both Egypt and the Indus valley at a very early date, but these routes do not appear to have carried large-scale or sustained traffic, because of both the lack of suitable complementary trading goods and the hazards of navigation in the monsoon region.

Navigation on the Mediterranean was another matter. Already at the beginning of recorded history (ca. 3000 B.C.), a seafaring people had established themselves at the eastern end of the Mediterranean, serving as intermediaries between the developing civilizations of Mesopotamia and Egypt (Fig. 2-4). The Phoenicians were the first specialized sailors and merchants; according to their own traditions, they came to the Mediterranean from either the Persian Gulf or the Red Sea, which raises the possibility that they (or their forerunners) may have been the early intermediaries between Sumer and Upper Egypt via the Indian Ocean. In any case, they virtually monopolized the commerce of Egypt for long periods, in a sense serving as the pharoahs' agents or contract merchants. Among their commercial articles were copper from Cyprus and the fabled cedars of Lebanon. In connection with their commerce the Phoenicians also developed a number of processing industries, including the manufacture of their famous purple dye. The word *Phoenicia*, in fact, comes from the Greek, meaning "land of the purple [dye]."

The Phoenicians organized themselves politically into autonomous city-states, of which the most famous were Sidon and Tyre. Dependent, to a large extent, on the goodwill or tolerance of their more powerful neighbors, they experienced fluctuations of fortune, but for almost three millennia, until their cities were overrun by the armies of Alexander the Great, they were among the foremost mercantile peoples in ancient civilization. Their commercial activities led them to develop the alphabet as a more efficient substitute for hieroglyphic or cuneiform writing, which the Greeks and the Romans adopted, along with other of their commercial techniques. To foster trade as well as to relieve population pressure in their narrow homeland, they established colonies along the North African coast and in the western Mediterranean on Sicily, Sardinia, the Balearic Isles, and the coast of Spain. One of the Phoenician colonies, Carthage, later founded an empire of its own and struggled with Rome for hegemony in the western Mediterranean. Daring sailors as well as skillful merchants, the Phoenicians sailed in the Atlantic to obtain tin from Cornwall, and may have circumnavigated the continent of Africa.

The other great maritime traders of the Mediterranean were the Greeks. Unlike the Phoenicians, the Greeks were originally cultivators, but the rocky, mountainous character of their adopted homeland (they had come from the north) soon drove them to the sea to supplement the meager produce of their agriculture. Their excellent natural harbors and the numerous islands of the adjacent Aegean Sea also encouraged this departure. As early as the Mycenaean period (from the fourteenth to the twelfth centuries B.C.), Greek merchants could be found throughout the Aegean and eastern Mediterranean and as far west as Sicily; the Homeric epic of the Trojan War likely reflects an episode of commercial rivalry between the Greeks and the city of Troy, which commanded the entrance to the Black Sea, just as the legend of Jason and the Golden Fleece probably reflects a pioneering venture to the Black Sea in search of wool. After a "dark age" occasioned by a new wave of invasions from the north,

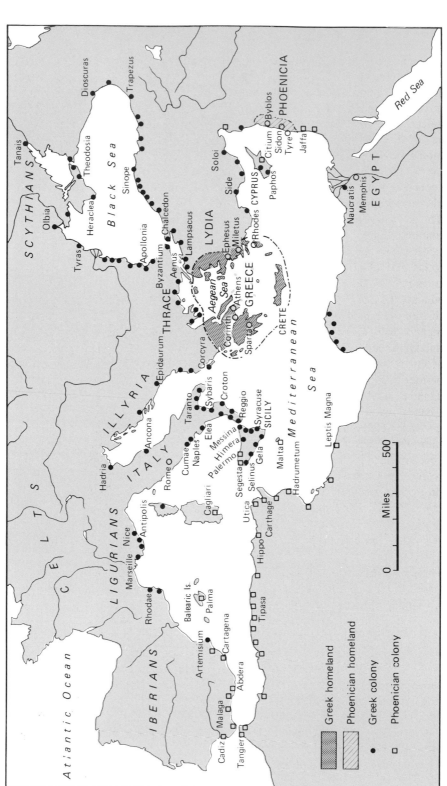

FIGURE 2-4. Greek and Phoenician colonization.

Greek commerce and civilization revived around the beginning of the eighth century B.C. By that time the Aegean was already a Greek lake, with Greek settlements on the coast of Asia Minor as well as on the islands. The pressure of population on limited resources was probably at least partly responsible for the settlement of the islands and the adjacent coast of Asia Minor; but even these measures did not relieve the pressure. In the middle of the eighth century the Greeks undertook massive organized ventures in colonization that resulted in the foundation of Greek cities throughout the Mediterranean, as far west as present-day Marseilles, and on the Black Sea coasts as well. The concentration of Greek cities in southern Italy and Sicily was so great that the area became known as Magna Graecia ("Greater Greece").

The colonization movement served economic purposes other than relieving population pressure at home (and incidentally relocating political dissidents). Many new cities were located in fertile agricultural regions and could thus supply grain and other agricultural products to the mother city. They also served as markets or trade centers for the processed and manufactured wares of the mother city, thus introducing the indigenous neighboring populations (mostly neolithic cultivators) to civilization by means of the market system. The founding cities did not generally attempt to maintain political control over their colonies, but ties of kinship and commercial relations kept them closely affiliated. In these circumstances the cities of the Greek mainland (and those of Asia Minor as well) became more specialized in commerce and industry. Grain made way for grapes and olives, which, by nature, were better suited to the Greek soil and climate, and their final products—wine and oil—had a much higher value per unit of weight. Greek craftsmen, especially potters and metal workers, became highly skilled, and their wares commanded a premium throughout the area of classical civilization. Greek sailors and merchants also became carriers for other, nonseafaring peoples such as the Egyptians. Some cities, such as Athens, concentrated a number of commercial and financial functions within their boundaries in much the same way as Antwerp, Amsterdam, London, and New York did in subsequent eras. Banking, insurance, joint-stock ventures, and a number of other economic institutions that are associated with later epochs already existed in embryonic form in classical Greece; indeed, they had roots in ancient Babylon.

These commercial and financial developments were facilitated by an innovation of minor technical significance but great economic importance—the introduction of coined money. Money and coinage, of course, are not identical. Before the invention of metallic coins many other commodities had served as standards of value, the most fundamental function of money, and also as media of exchange. In an actual exchange it was not necessary for the standard of value to be physically present or to be a part of the exchange, as long as the commodities involved could be valued in relation to it. On this basis barter and even credit transactions had long preceded the use of coined money. The latter, nevertheless, greatly simplified commercial transactions and permitted the extension of the market system to many individuals and groups who would otherwise have remained isolated in a closed subsistence economy.

As with most inventions of ancient times, the inventor of coins is unknown to history. The earliest surviving coins, dating from the seventh century B.C., came from Asia Minor. Moralistic legends ascribe the invention to both Midas, a king of Phrygia who had the "golden touch," and Croesus, a fabulously wealthy king of Lydia who was executed by Cyrus the Great by being forced to swallow molten gold; but more likely the first coins were struck by some enterprising merchant or banker of one of the Greek cities on the coast as a form of advertising. In any event, their potential for both profit and prestige was quickly recognized by governments, which arrogated the coining of money as a state monopoly. The effigy of a ruler or the symbol of a city (the owl of Athens, for example) stamped on a coin testified not only to the purity of the metal but also to the glory of its issuer.

The earliest coins were apparently made of electrum, a natural alloy of gold and silver that was found in the alluvial valleys of Anatolia, but because of the vari-

FIGURE 2–5.   Greek coins. The coin on the left, with simple punch marks on the face (A) and a striated surface on the back (B), is of electrum, dating from about 600 B.C. The silver coin on the right, with the face of Athena (C) and the owl of Athens on the reverse (D), dates from about 480 B.C.; it shows how far the technology of minting advanced in little more than a century. (Hirmer Fotoarchiv München.)

ability in proportions of the two metals in electrum, the pure metals were preferred (Fig. 2-5). Although both gold and silver coins were struck, silver was both more plentiful and more practical for commerce. The leading role of Athens in fifth-century commerce as well as culture also contributed to the predominance of silver, at least among the Greeks; in fact, the two phenomena were intimately related. Athens' state-owned silver mines at Laurium, on the Attic peninsula, provided the resources for the construction of the triremes. This new type of warship was decisive in the struggle of the Greeks against Persian encroachment, and it subsequently allowed Athens to dominate the Delian League to such an extent that the Aegean and surrounding territories effectively became an Athenian empire. Silver from Laurium also helped finance Athens' persistently unfavorable balance of trade (shipping and financial services were also important sources of earnings), and thus indirectly aided in the construction of the great public buildings and monuments for which Athens became famous. The Athenian Golden Age, in fact, was made possible by the silver of Laurium.

The Greek cities exhausted themselves in internecine struggles, but the conquests of Alexander the Great spread Greek (or Hellenistic) culture throughout the Near and Middle East. Although Alexander's empire disintegrated after his death, the cultural and economic unity remained. The Greek language was spoken from Magna Graecia to the Indus River. Greeks manned the civil services of the successor states, and Greek merchants established their precincts in all important cities. Alexandria—probably the largest city in the world before the rise of Rome, with a population of over 500,000—was virtually a Greek city, and the most important emporium of the age. Through its markets passed not only the traditional exports of Egypt (wheat, papyrus, linen cloth, glass, etc.) but hundreds of staple and exotic products from many parts of the world, including elephants, ivory, and ostrich feathers from Africa, carpets from Arabia and Persia, amber from the Baltic, cotton from India, and silk from China. The mere number of these commodities testifies to the scale and extent of commercial organization.

Economic Achievements and Limits of Ancient Civilization

The apogee of classical civilization, at least in its economic aspects, occurred during the first and second centuries of the Christian era, under the dominance of Rome (Fig. 2-6). Rome had already absorbed Hellenistic culture before mastering the Mediterranean, and with the latter feat it inherited—or appropriated—Hellenistic economic achievements and institutions as well.

The Romans were originally an agricultural people, mostly small farmers with a high regard for property rights. In the course of their expansion they became increasingly concerned with military and administrative affairs, but their traditional attachment to the soil lingered. Commerce, on the other hand, did not rate highly in the Roman system of values; it was left in the hands of inferior social classes, foreigners, and even slaves. Nevertheless, the Roman legal system, initially adapted to an agrarian regime but gradually modified by the incorporation of Greek elements, allowed considerable freedom of enterprise and did not penalize commer-

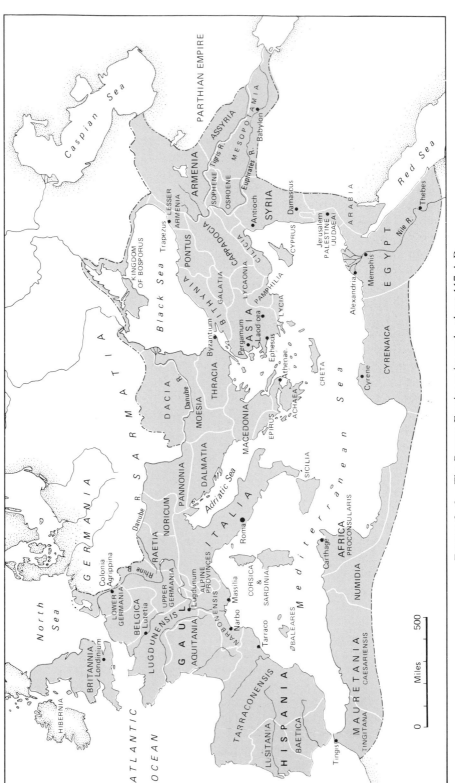

FIGURE 2–6. The Roman Empire at its peak, about 117 A.D.

cial activities. In particular, it provided for strict enforcement of contracts and property rights and prompt (and usually equitable) settlement of disputes. As Roman law spread, in the wake of the conquering legions, it provided a uniform, coherent legal framework for economic activity throughout the empire. (Some regions, notably Egypt, were subject to special regimes in which their traditional customs and usages were retained.)

The urban character of the Roman Empire both stimulated and was made possible by the highly developed commercial network and the fine division of labor that supported it. The city of Rome alone may have had a population in excess of 1 million people at its height. Since feeding such a concentrated population from local resources was manifestly impossible, great fleets were organized to bring wheat from Sicily, North Africa, and Egypt. (These shipments also provided the occasion for one of the major exceptions to the rule of free enterprise; grain was distributed free to as many as 200,000 families of the Roman proletariat. To guard against nondeliveries, which might provoke riots, the government granted special privileges to the agents charged with providing grain, and at times even undertook the task itself.) Although no other city could compare in size or magnificence with Rome at its zenith, many ranged in size from 5,000 to 100,000, and a few, such as Alexandria, were much larger. Probably no sizable area of the world was so highly urbanized again until the nineteenth century.

Rome's greatest contribution to economic development was the *pax romana*, the long period of peace and order in the Mediterranean basin that allowed commerce to develop under the most favorable conditions. Although Roman legions were almost constantly involved in conquering new territory, punishing an upstart neighbor, or suppressing a native rebellion, before the third century these disturbances normally took place on the periphery of the empire and rarely disturbed the most active commercial routes. Piracy and brigandage, which had been serious threats to commerce even in the Hellenistic era, were almost completely eliminated. The famed Roman roads were designed for strategic rather than commercial use; wheeled vehicles other than chariots were rarely used, and not at all for long-distance transportation. Yet the roads did facilitate communications and the transportation of light merchandise. The major artery of transportation, however, was the Mediterranean, which flourished as never before—and rarely since—as a highway for commercial traffic.

One major consequence of the *pax romana* was population growth. Estimates of the population of the empire at its height range from 60 million to more than 100 million, with the more recent estimates favoring the latter figure. Unfortunately there are no reliable estimates of the population of the same area at an earlier date, such as the time of Alexander or that of Greek colonization in the eighth century B.C. It is likely, though, that the population of the empire at the death of Marcus Aurelius (180 A.D.) was at least double that of the same area at the death of Julius Caesar (44 B.C.). Growth was most marked in the western Mediterranean, including Italy, because the East was already well populated. (Egypt, for example, probably had 5 million inhabitants as early as 2500 B.C.; in the first century A.D. it had about 7.5 million.) In the era of Phoenician and Greek colonization much arable land in the West was totally uninhabited; even during the period of Roman expansion in

Italy many areas of the peninsula were sparsely populated. Gaul, which later became one of the largest Roman provinces, with more than 10 million inhabitants, probably had fewer then half that number at the time of the Roman conquest. North Africa and Spain also experienced both prosperity and population growth in the first half of the imperial period.

To what extent the average standard of living improved concurrently with demographic growth is a much more difficult question. Undoubtedly there was some improvement, which both permitted and encouraged population growth. A distinguished economist, Colin Clark, has estimated that the real earnings of a typical free artisan in Rome in the first century A.D. were approximately equivalent to those of a typical British factory worker in 1850 and to those of an Italian worker in 1929. By extrapolation, this implies that Roman artisans were substantially better off economically than millions of peasants and urban dwellers in Asia, Africa, and Latin America today. Such comparisons, however, contain difficult conceptual problems as well as statistical pitfalls. Granted, one can (with adequate statistical data) compare the purchasing power of the wages of distinct populations in terms of grain or bread, for example, or perhaps the average caloric intake of foodstuffs. But how does one evaluate the relative contributions to material or psychic well-being of Roman circuses and modern transistor radios and television, of travel by foot (even on Roman roads!) and travel by subway, private automobile, or jet aircraft, or of different types of housing, which vary in comfort and convenience with the climatic conditions as well as with their construction characteristics? Moreover, statistics (even if accurate) about ''average'' or ''typical'' peasants or urban workers tell us nothing about the relative distribution of income.

The prevalence of slavery in ancient times is an especially vexing problem for statistical comparisons. The absolute and relative numbers of slaves varied considerably over time; slaves were numerous in the expansive phase of the empire, when war captives and hostages were plentiful, but much less so in later periods when the empire was on the defensive. (The ratio was also affected by the rate of manumission and the relative birth rates of the slave and free populations; generally, birth rates among slaves are not as high as those of free people.) Some slaves were no doubt well treated by their masters, especially literate Greeks and others who served as tutors, scribes, household servants, and business agents; but the great majority were employed in agriculture and as common laborers, and received little more than bare subsistence. The relative numbers of slaves also affected the price of free labor; freemen rarely worked in such unpleasant and unsafe occupations as mining, but in other areas they might have to compete with the subsistence standards of slaves.

Another possible measure of material well-being is the average length of life. Again, one must be wary of incomplete and inconclusive statistics, especially since they reveal little about the relative incidence of disease and other causes of death among different social classes. In general, however, the average length of life in the best years of the empire appears to have been about twenty-five years—a slight improvement over earlier societies but still considerably below all but the very poorest societies in recent times.

The ''best years of the empire'' constituted a transitory period. Even before the death of Marcus Aurelius a number of problems foreshadowed the decline of the empire and the economy on which it rested. Among these were Germanic incursions from the north, localized labor shortages, and gradual monetary inflation. All of these problems increased in severity in the third century, especially the inflation resulting from the continual debasement of the coinage by a treasury whose expenses always exceeded its revenue. The inflation, however, was symptomatic of more fundamental economic problems. Diocletian attempted to deal with it at the beginning of the fourth century by decreeing legal controls on prices and wages and by reorganizing the bureaucracy and fiscal system. His reforms and those of his successor Constantine shored up the imperial structure for a time, but they did not deal with the fundamental problems; in fact, they exacerbated them.

Economically, the twin pillars of the Roman Empire were agriculture and commerce. Agricultural surpluses (production in excess of that required to maintain the cultivator and his family), though small in terms of the individual cultivator, bulked large when collected and concentrated through taxation. They provided the resources that supported the army, the imperial bureaucracy, and the urban population. Effective marshaling of these surpluses, however, depended on the unimpeded flow of commerce throughout the empire. Barbarian invasions and depredations interfered with this commerce, but perhaps even greater problems were the inefficiency and corruption of the imperial government itself. Pirates again infested the Mediterranean, and robber bands controlled the mountain passes. On occasion the army itself preyed on peaceful commerce.

Taxation grew steadily heavier, but its burden varied inversely with the benefits government conferred. Many great estates, the property of the nobility, were exempt from taxation, leaving the burden increasingly on those least able to bear it. During the inflation of the third century, when tax revenues fell consistently below the expenditures of the army and the bureaucracy, the government resorted to levies in kind, which Diocletian transformed into a regular system of contribution. Although this drastic measure achieved its purpose in the short run, it subverted the very nature of the economic system of the empire. Production for the market declined. Cultivators, even small proprietors, fled the land and placed themselves under the protection of the great lords, whose tax-exempt estates grew accordingly. Moreover, as trade declined and populations of towns and cities dwindled for lack of provisions, the great estates became more self-sufficient, not only retaining their own food production but instituting metalworking, clothmaking, and other trades as well, thus depriving the towns of their function. It was a vicious spiral of contraction.

Diocletian's attempt to fix wages and prices by imperial edict failed almost completely, in spite of the severe penalties it imposed for infractions. In 332 the government resorted to an even more drastic measure, binding cultivators to the soil they tilled and making all occupations and offices—those of farmers, artisans, tradesmen, even municipal officials—compulsorily hereditary. As with the requisitioning of supplies in kind, the measure had some short-run success, but it was even more subversive of the economic system. The economy reverted to a primitive subsistence basis as population declined, towns and cities were deserted, and the villas of the great estates came more and more to resemble fortified castles. By the end of the fourth century the empire in the west was a hollow shell that gradually collapsed under its own weight.

The fall of the Roman Empire and the decline (or retrogression) of classical economy were not identical, in spite of their intimate relationship. Had the economy been able to meet the demands made on it by the increasingly parasitic imperial bureaucracy and army, the empire might have lasted for another thousand years—as, indeed, the Eastern or Byzantine Empire did. Conversely, if the empire, the institutional framework within which the economy functioned, had continued to provide efficient protection from both internal and external threats to peaceful productive activities and an effective administration of justice, there is no obvious reason the economy could not have performed as well under the Severi or Diocletian as under the Antonines. In fact, neither of these conditions prevailed.

A still more fundamental reason for the limits to, and ultimate failure of, the classical economy transcends the immediate causes of the decline of Rome, however: the lack of technological creativity. This technological sterility stands in sharp contrast to the cultural brilliance of at least some periods of ancient civilization. Even today classical art and literature provide standards against which to measure contemporary works, and notable progress was also achieved in philosophy, mathematics, and some branches of science. Some of the properties of steam were known to the ancients, although the only applications were in producing toys and devices to mystify the credulous; the waterwheel and windmill were invented at least as early as the first century B.C., but were not widely adopted until the European Middle Ages. Roman engineering ingenuity manifested itself in roads, aqueducts (Fig. 2-7), and domed buildings, but not in labor-saving machinery. Clearly, it was not

FIGURE 2–7.   This Roman aqueduct, in Segovia, Spain, still standing today, testifies to the Romans' engineering genius, but they did not use it to create labor-saving devices. (Arlene DeBevoise.)

lack of intelligence that prevented the ancients from contributing more to the progress of technology.

The explanation appears to lie in the socioeconomic structure and the nature of the attitudes and incentives that it generated. Most productive work was done either by slaves or by servile peasants whose status differed little from that of slaves. Even if they had had an opportunity to improve technology, they would have reaped few if any benefits, either in terms of higher incomes or reduced labor. Members of the small privileged classes devoted themselves to war, government, the cultivation of the fine arts and sciences, and conspicuous consumption. They lacked both the experience and the inclination to experiment with the means of production, since labor carried the stigma of menial status. Archimedes was a scientific genius who frankly disdained practical application of science; his one concession to practicality was to design a mechanical catapult for the (unsuccessful) defense of his native Syracuse against the Romans. Aristotle, who had perhaps the most encyclopedic knowledge of any ancient philosopher or scientist, believed that the distinction between masters and slaves was biologically determined. For him, it was a part of the natural order of the universe that slaves should labor to provide their masters with the leisure to develop the arts of civilization. And St. Paul wrote that "masters and slaves must accept their present stations, for the earthly kingdom could not survive unless some men were free and some were slaves." In view of such attitudes, it is scarcely surprising that little serious thought should have been given to devising methods for lightening the burden of labor or improving the status of the servile masses. A society based on slavery may produce great masterworks of art and literature, but it cannot produce sustained economic growth.

3

Economic Development in Medieval Europe

To an earlier generation the phrase "medieval economic growth" would have seemed a contradiction in terms. Under the influence of Renaissance authors, who belittled their immediate predecessors in their praise of the rediscovered glories of classical civilization, the Middle Ages have long been regarded as a period of both economic and cultural stagnation. In fact, medieval Europe experienced a flowering of technological creativity and economic dynamism that contrasts strongly with the routine of the ancient Mediterranean world. Moreover, the distinctive institutions created in the Middle Ages served as a framework for economic activity until recent times; medieval survivals in rural areas are still prominent features of the landscape, even in the formerly socialized economies of eastern Europe.

The Agrarian Basis

Until the advent of industrialism in the nineteenth century, agriculture everywhere constituted the most important sector of economic activity, both in terms of the value and volume of output and of the proportion of the labor force engaged in it. Medieval Europe was unique among developed civilizations, however, in its agrarian orientation. From the ancient city-states of Sumer to the Roman Empire, urban institutions determined the character of the economy and society, even though most of the population was engaged in agricultural labor. In medieval Europe, on the other hand, although the urban population grew in size and importance, especially in Italy and Flanders, agrarian and rural institutions set the tone.

To understand the distinctive character of the medieval economy, one should recall the political and social conditions that surrounded its origins—the growing burden of taxation and the increasing inefficiency and corruption of the Roman Empire, the final breakdown of central authority and the resulting anarchy, the growth of large self-sufficient estates and the decline of towns and interregional trade. After the collapse of the empire barbarian tribes continued to roam and ravage; petty kingdoms rose and fell but were unable to maintain effective order for more than brief periods or to establish regular systems of taxation. The Frankish kingdom, based in the heartland of medieval Europe between the Loire and Rhine rivers, maintained its existence longer than the others; but without a regular system of taxation or permanent bureaucracy, it, too, depended on the uncertain loyalty of the great nobles and their retainers for the preservation of order and unity.

Beginning in the eighth century, new hordes of invaders threatened the Franks and other Europeans for more than two centuries. In 711 Muslims from North Africa invaded Spain and quickly overthrew its Visigothic kingdom; by 732 they had penetrated as far as central France before being turned back. Although the Franks drove the Muslims back across the Pyrenees, the latter conquered Sicily, Corsica, and Sardinia and turned the Mediterranean into virtually a Muslim lake.

Later in the century the Vikings poured out of Scandinavia, dominated the British Isles, conquered Normandy, raided coastal and riverine sites as far inland as Paris, and even penetrated the Mediterranean. In the ninth century fierce Magyar tribesmen crossed the Carpathians into central Europe and raided, pillaged, and extracted tribute in northern Italy, southern Germany, and eastern France before settling down in the following century in their newly chosen homeland in the Hungarian plain.

To meet these threats the Frankish kings devised a system of military and political relationships, subsequently called feudalism, which they grafted onto the evolving economic system. Military considerations required troops of mounted warriors, since the recent introduction of the stirrup (probably from central Asia) had made foot soldiers almost obsolete. Directly supporting such troops was impossible in the absence of an effective system of taxation and the virtual disappearance of a money economy. Moreover, considerations of domestic order and administration called for numerous local officials who, again, could not be paid directly by the state. The solution was to grant the warriors the income from great estates, many of which were confiscated from the church, in return for military service; the warriors—lords and knights—were also charged with maintaining order and administering justice on their estates. Great nobles—dukes, counts, and marquises—had many estates encompassing many villages; some of these they granted to lesser lords or knights, their vassals, in return for an oath of homage and fealty, similar to that which they gave the king; this procedure was called subinfeudation.

Underlying the feudal system, but with older and quite different origins, was the form of economic and social organization called (in English) manorialism.[1] Manorialism began to take shape under the later Roman Empire, when the *latifundia* (large farms) of Roman nobles were transformed into self-sufficient estates, and cultivators were bound to the soil either by legislation or by more direct and immediate economic and social pressures. The barbarian invasions modified the system, mainly by introducing tribal chieftains and warriors into the ruling class, and manorialism received its ''definitive'' stamp in the eighth and ninth centuries, during the Saracen, Viking, and Magyar invasions, when it became the economic basis of the feudal system.

The earliest documentary evidence that provides direct information on the oper-

[1] Since France was the classic home of the manor, the French terms *seigneurie* and *seigneurialisme* (or, in bastardized anglicization, seignorialism) are frequently used. Other languages have similar but not identical terms, because of regional variations in the nature of the manor.

ation of the manorial system dates from the ninth century. By that time it was already well established in the areas between the Loire and Rhine rivers (northern France, the southern Low Countries, and western Germany) and in the Po valley of northern Italy. Subsequently it spread, with modifications, to England with the Norman Conquest, to reconquered Spain and Portugal, to Denmark, and to central and eastern Europe. Some areas, such as Scotland, Norway, and the Balkans, were never effectively manorialized; even within the areas of manorial economy some regions, usually hilly or mountainous, maintained different forms of organization.

There was no such thing as a typical manor. Variations, both geographical and chronological, were far too numerous. It is useful, nevertheless, to construct a hypothetical, idealized manor for purposes of comparison (Fig. 3-1 shows an actual manor). As an organizational and administrative unit, the manor consisted of land, buildings, and the people who cultivated the former and inhabited the latter. Functionally, the land was divided into arable, pasture and meadow, and woodland, forest, or waste. Legally, it was divided into the lord's demesne (since the English word *domain* has a more general meaning, the anglicized French is preferred for this special meaning), peasant holdings, and common land. The lord's demesne, some-

FIGURE 3–1.   A medieval manor. This map of the village of Shilbottle in Northumberland, England, dates from the early seventeenth century, but is representative of medieval times. Note the crofts (cottages with gardens) of the peasants surrounded by open fields and common (waste) land. The manor house is not shown, but the enclosed lord's demesne is located in the lower right portion of the map. (From *Studies of Field Systems in the British Isles*, edited by A. R. H. Baker and R. A. Butlin. Copyright 1973 by Cambridge University Press. Reprinted with permission.)

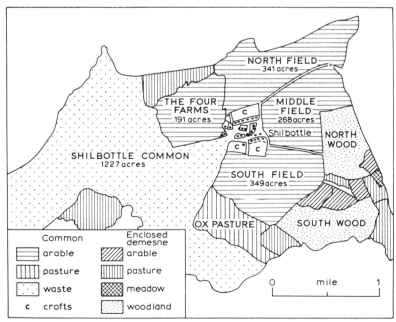

times though not necessarily enclosed or separated from peasant land, might account for 25 or 30 percent of the total arable land of the manor; it also included the manor house, barns, stables, workshops, gardens, and perhaps vineyards and orchards. The land the peasants tilled for themselves lay in large open fields surrounding the manor house and village; the fields were divided into strips, with the holdings of a single peasant household consisting of possibly two dozen or more strips scattered throughout the fields of the manor. Meadows, pastures (including *vaine pâture*, fallow fields used for grazing), and woodland or forests were normally held in common, although the lord supervised their use and maintained special privileges in the forests.

The manor house, frequently fortified, served as the residence of the lord or his agent. In the case of very great lords owning many manors, the manor might be let to a lesser lord, or vassal, in fee; that is, the vassal was entitled to the benefits of lordship of the manor in return for military service. Religious establishments such as cathedrals and abbeys also owned manors, which might be let to vassals, managed directly by clerics, or entrusted to lay stewards or managers. The feudal ideal was "no land without a lord, no lord without land," but it was not universally realized. In principle the function of the lord was defense and the administration of justice; he might take a direct interest in supervising the exploitation of his demesne, but more often he left this to a steward or bailiff. In addition, he frequently had other perquisites, such as ownership of the local mill, oven, and winepress.

The peasants lived in compact villages under the walls of the manor house or in its vicinity. Their cottages were simple one- or two-room affairs, sometimes with a loft that served as sleeping quarters. Construction might be of wood or stone, but was more often of mud and wattle, with earthen floor, no windows, and a thatched roof with a hole in it to serve as a chimney. There might be auxilliary buildings for livestock and equipment, but in winter the livestock frequently shared the living quarters with the family. Villages were normally located in the vicinity of a stream, which provided a water supply and actuated the mill and perhaps a bellows for a forge or smithy. Unless the manor house contained a chapel (or sometimes even if it did), a simple church would complete the village scene.

So much for the hypothetical manor. In fact, variations were endless. Although the ideal might have been one manor, one village, frequently one manor encompassed several villages, or less frequently a single village was divided among two or more manors. Sometimes the peasant subjects of the manor did not live in villages at all, but in scattered hamlets or even isolated farmsteads. The two latter types of settlement were most often found in regions of infertile soil or hilly land, where the manorial form of organization existed either in a diluted form or not at all; but in the Mediterranean basin, especially southern France and most of Italy, the small, square enclosed fields with isolated dwellings, typical of Roman times, persisted throughout the Middle Ages. In areas where manorialism was introduced from the outside, as it were, such as the Iberian peninsula, eastern Germany, and even England, its features were modified to take the soil, climate, terrain, and existing institutions into account. Last but not least, manorialism was nowhere the static institution sometimes pictured but was in a state of constant flux or evolution, usually gradual, almost imperceptible, but ineluctable.

Rural Society

There were various gradations of social status within the rural population. In the fully developed theory of feudalism—which, characteristically, was not elaborated until the institution itself was on the verge of decline—society consisted of three "orders," each with its assigned duty. The lords provided protection and maintained order; the clergy looked after the spiritual welfare of society; and the peasants labored to support the two higher orders. Stated more pithily, the lords fought, the clergy prayed, and the peasants worked. Significantly, town dwellers did not even figure in this hierarchy, although by the eleventh century at least they constituted a sizable category, certainly more numerous than either lords or clergy.

The ruling class—that is, the feudal order in the strict sense—which probably accounted for less than 5 percent of the total population, in principle formed a social pyramid ranging from the king at the apex down through the great nobles to the lowliest knights at the base. In fact, the situation was even more complicated, as many nobles held several manors (also called *benefices*), and thus were technically vassals of more than one lord. In extreme cases two nobles, even kings, might be vassals of one another with respect to particular estates. Not surprisingly, such complexities frequently led to quarrels and strife, which have given the feudal age a somewhat unjustified reputation for lawlessness and violence.

The clerical order, the only one that was not biologically self-perpetuating (in principle, at least, although in practice it was sometimes a different matter), also contained numerous social gradations. In the first place, there was the distinction between the regular clergy (i.e., the monastic orders), who withdrew from ordinary life into separate communities, and the secular clergy (priests and bishops), who participated more directly in the life of the community. In the early Middle Ages the regular clergy had greater prestige, but the status of the secular clergy improved with the revival of town life and the economic upswing from the tenth century onward, when bishops and archbishops played important roles in lay as well as religious life. Second, within both regular and secular clergy distinctions existed, based on the social status of the individuals entering the clergy. The younger sons of noble families were often destined, with or without the appropriate training, to become bishops or abbots from the time they took holy orders, whereas humbler folk could rarely aspire to more than a parish priesthood or a clerical office in a monastery. The opportunity for vertical mobility was somewhat greater within the church than in rural society generally, but much less than that offered by the new towns.

Even within the peasant population differences in status existed. Broadly speaking there were two categories, free and servile; but these categories were not always distinct, and degrees of servility and freedom existed within them. Chattel slavery, such as existed under the Roman Empire, gradually died out until, by the ninth century, almost the only remaining slaves were the household slaves of great nobles. On the other hand, the class of freemen—peasant proprietors and tenant farmers—who also existed under the Roman Empire, was depressed almost to the status of servile workers. Truly free men—free to move from one village to another, to acquire or dispose of land on their own initiative, to marry without their lord's permission—were rarities among the medieval peasantry. At the same time the power of the lords was limited. Serfs were not the property of their masters, but *adscripti glebae,* that is, bound to the soil. Lords might come and go, but, except in periods of great stress, the peasant cultivators whether nominally free or servile would remain secure in their tenures, protected by the "custom of the manor" and occasionally by documentary evidence (e.g., the English copyholders).

Two general tendencies in the social status of the peasantry are perceptible through the Middle Ages and early modern times—tendencies closely associated with the evolution of the manor. From the later Roman Empire to about the tenth or eleventh century, rights and obligations of the two extremes—freemen and slaves—were pressed closer and closer together. Then, from about the twelfth century to the French Revolution, a progressive relaxation of the servile restrictions (not necessarily economic exactions) occurred, resulting in the withering away of the institution of serfdom in some areas of western Europe (much less in central Europe, and not at all in eastern Europe, where a contrary evolution occurred).

Patterns of Stability

The organization of work on the manor involved a mixture of customary cooperation and coercion, with very little scope for individual initiative. The most important operations were plowing, sowing, and harvesting, which would involve almost all inhabitants of the village. Because of the open field system, and the fact that an individual peasant's strips were scattered throughout the fields, the work had to be undertaken in common. Moreover, in the heavier soils, which were also normally the most fertile, a plow team required four, six, or even eight oxen; since peasants rarely owned more than one or two (many none at all), cooperation was required. Harvesting was also done in common, to allow the livestock to graze on the stubble.

The place of livestock in the medieval agrarian economy varied considerably from one region to another. Their most important function was as draft animals, and oxen, the most common of these, were found in every part of Europe. Other draft animals included horses, used in northwestern Europe and in Russia from about the tenth century, donkeys and mules, used principally in southwestern France and Spain, and water buffaloes, used in parts of Italy. Oxen, unlike horses and mules, consumed mainly grass and hay, were docile, and were easy to raise, which accounts for their prevalence. Milk cows were, of course, necessary to breed oxen; in addition, they provided the raw material for butter and cheese, and in the poorest regions were also used as draft animals. In the "Celtic fringe" of Europe (Brittany, Wales, Ireland, and Scotland), outside the area of manorial economy where agriculture was little practiced, seminomadic tribes lived almost exclusively by their herds of cattle. In Scandinavia also, especially Norway and Sweden, stock raising was more important than tillage. In the principal manorialized areas cattle, sheep, and swine were raised for their meat products (and sheep for wool), and incidentally for the fertilizer they produced, but stock raising was definitely secondary to field agriculture. It was most often practiced in northwestern Europe, whose moister climate provided better natural pastures. The large forests of that region also provided forage for cattle and horses, as well as for swine. In the south, in areas with a Mediterranean climate, stock raising was much less important, and frequently took the form of transhumance pasturing for sheep and goats: flocks wintered in the lowland areas and were driven to mountain pastures for the spring and summer. Sometimes the passage of the flocks damaged agricultural fields, and overgrazing in the mountains contributed to deforestation and soil erosion.

Most peasants were obliged to perform some labor services on the lord's demesne, which (in principle) took precedence over work on their own strips. The extent and nature of the services varied from region to region (even from manor to manor), over time, and according to the social status of the peasant or the nature of his tenure. It was not uncommon for nominally free men to hold servile tenures, and occasionally a nominal serf might own a copyhold or leasehold. In general, those with servile tenures would be called on for more work, perhaps three or four days a week on the average, and those with freeholds did less. Women spun yarn and wove cloth, either in the lord's workshop or in their cottages, and children were used as servants in the lord's household. Beginning in the tenth century a progressive movement developed, faster in some areas than others, to suppress labor services or commute them into money rents.

In addition to labor services, most peasants normally owed their lord other dues, rents, and fees, in money and in kind. Some of these were collected on a regular basis—a sheep or a few chickens at Christmas, for example, in addition to annual money rents—whereas others were due on special occasions, such as the assumption of a deceased peasant's tenure by his heir or at his marriage. The nature and value of these exactions varied enormously. For thirteenth-century England the total of peasant dues and rents has been estimated at 50 percent of peasant income, but in some times and places it may have exceeded even that figure. Peasants were also obliged to use the lord's mill, winepress, and oven, for which they paid a fee, and were subject to the lord's justice in the manorial court, which often involved payment of fines. They also paid a tithe (not necessarily a tenth) to the church, and were sometimes subjected to royal taxation as well. Peasants whose tenures were too small to support a family, as they often were, performed additional labor on the lord's demesne (or, less frequently, for a more prosperous fellow peasant), for which they received wages denominated in money, although the actual payment was frequently in kind.

The manorial system developed gradually over a period of several centuries, a period characterized by political uncertainty, frequent outbreaks of violence, declining commercial activity and occupational specialization, and primitive production techniques. Although not consciously designed, it maintained social stability and continuity and supported a sparse population at a low but tolerable level of living. Apparently antithetical to individual initiative and hence to innovation, the system nevertheless evolved in response to the interplay of institutions and resources, giving rise to technological changes that increased productivity and stimulated population growth, thus altering the bases of its own existence.

Forces of Change

The most important innovation in medieval agricultural practice was the substitution of a three-course crop rotation for the classical two-course rotation of Mediterranean agriculture. It was closely associated with two other significant innovations, the introduction of a heavy wheeled plow (Fig. 3-2) and the use of horses as draft animals. The latter innovation, in turn, depended on other innovations in the harness and equipment for horses.

The classical two-course rotation, in which fields were planted and left fallow in alternate years to maintain soil fertility and accumulate moisture, was adapted to the light soils and long dry summers of the Mediterranean basin. Before the power of Rome extended to northwestern Europe, settled agriculture was seldom practiced there. The Gauls and various Germanic tribes depended primarily on their herds of cattle; when they planted field crops, they used a slash-and-burn technique to clear the ground, moving to a new location as soon as the fertility of the soil declined. The Romans brought with them their two-course rotation, but their plows were unable to penetrate the heavy soils characteristic of northwestern Europe; consequently, they cultivated sandy or chalky hills with adequate natural drainage and avoided the heavier but more fertile soils of the plains and valleys.

The exact place and date of origin of the heavy wheeled plow is still a matter for debate. It may have entered Gaul with the Franks, but if so, it was not widely used until field agriculture acquired more importance than stock raising. Its use required several oxen or other draft animals, and thus contributed to the cooperative nature of cultivation in the manorial system. Unlike the lighter, simpler Roman plow, the wheeled plow was capable of breaking and turning the heavy clay and loam soils of northwestern Europe, which made new resources available to its users.

In the moister climate of northwestern Europe the alternative years of fallow to allow moisture to accumulate were unnecessary. Moreover, the deeper soils could tolerate a steadier pull on their nutrients, especially if the crops planted in them were varied. The first recorded instance of a regular three-course rotation occurs in northern France in the latter part of the eighth century; by the beginning of the eleventh century it was widely practiced throughout northwestern Europe. A typical rotation was a spring crop (oats or barley, sometimes peas or beans), which would be harvested in the summer; an autumn sowing of wheat or rye, the principal bread grains, which would be harvested the following summer; and a year of fallow to help restore fertility to the soil. This basic pattern, however, had many variations.

The three-course rotation had several advantages. The most fundamental was the increased productivity of the soil: for any given quantity of arable land, one-third more could be planted in food crops. It also produced a larger yield per unit of labor and capital; it has been calculated that a plow team sufficient for 160 acres under two-course rotation could work 180 acres under three-course rotation, meaning an increased productivity of 50 percent in terms of the crops actually grown. The three-course rotation, with fall and spring sowing, also spread field work more evenly over the year; it also reduced the risk of famine in the event of crop failure because, if necessary, wheat or rye could be planted in the spring. Finally, with

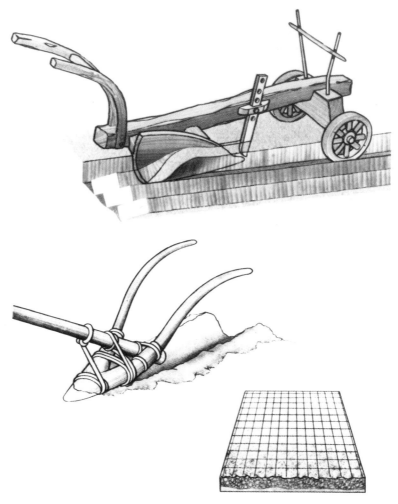

FIGURE 3–2.   The heavy wheeled plow (**top**). This plow, capable of turning the deep loam soils of northern Europe, can be compared with the lighter Mediterranean plow (**bottom**). The latter scratched the surface of the soil in small square plots, whereas the former created ridges and furrows in long strips called "furlongs." (From *Connections,* by James Burke. Victoria and Albert Museum, London.)

more land available for food crops, it was possible to introduce new and more varied plants with favorable effects on nutrition. As a result of its superiority, the three-course rotation spread wherever soil and weather conditions were favorable; by the eleventh century it was in general use throughout northern France, the Low Countries, western Germany, and southern England. In the Mediterranean area, on the other hand, its appearance was exceptional; the classical two-course rotation remained in general use there for field crops until the nineteenth century, although with the growth of urban demand, especially in northern Italy, much land in the vicinity of cities was given over to steady, intensive cultivation of commercial crops, making generous use of urban manure.

Horses were not often used for plowing before the tenth century. Partly this was a matter of cost: horses were more expensive to breed than cattle, consumed more expensive feed, and were in demand by the well-to-do for both warfare and transport. But there was also a more fundamental reason. Before the Middle Ages the harness used for horses was designed in such a way that it cut across the throat and interfered with breathing, thus reducing their effectiveness as draft animals. Sometime before the tenth century the horsecollar, which rested on the horses's shoulders, was introduced in western Europe, probably from Asia. Soon afterward the practice of shoeing horses was also introduced, to protect their hooves, which were more delicate than those of oxen. Thereafter the use of horses as draft animals for plowing and carting spread, but without fully replacing oxen—far from it. There was no question of the physical superiority of the horse; it was both stronger and faster than the ox. On the other hand, it cost more to breed, to feed (horses required oats or a similar grain), and to equip. Contemporary authors calculated that a horse could do about as much work as three or four oxen, but cost three or four times as much to maintain. Its adoption therefore depended on a fine economic calculation and was practical only under certain circumstances. First, a dependable, not too costly supply of oats was needed; that ruled out most areas where the two-course rotation survived because of soil or climate (i.e., most of the Mediterranean basin). In addition, the size of the unit of exploitation had to be sufficiently large to keep the animal fully employed, and sufficiently productive to make him worth his keep. In effect, horse husbandry was confined to northern France, Flanders, parts of Germany, and England, but did not completely replace oxen even in those areas. (Horses were also used in parts of eastern Europe, especially Russia, but in a different system of cultivation and with somewhat different results.) There is thus a close but not a perfect correspondence between the use of horses for plowing and the three-course rotation and wheeled plows. Significantly, those areas were among the most productive agriculturally in the Middle Ages—and still are today.

In addition to these major innovations, medieval agriculture experienced a host of minor innovations and improvements. As a result of new sources of supply and improvements in metallurgy, iron was more plentiful and cheaper in medieval Europe than in the ancient Mediterranean; in addition to its use for knightly armor and weapons, it found increasing use in agricultural implements: not only in the iron cutting edges of wheeled plows, replacing the wooden tips of Mediterranean plows, but also in such simple tools as hoes, pitchforks, and especially axes. Sickles for reaping grain were improved, and the scythe was invented for cutting hay. The harrow, used to break up clods, smooth the surface of the soil, and sometimes cover seeds, had been known in ancient times, but its design was improved with iron parts and its use was far more widespread. The value of animal manure for fertilizing the soil had long been known, but more intensive efforts were made to collect and conserve it. In addition, the practice of marling (adding chalk or lime to the soil) increased the fertility of certain kinds of soils, as did the addition of peat to others. In the thirteenth century, in regions of intensive cultivation, the technique of "green manuring" (plowing under clover, peas, and other nitrogenous plants) was devised to maintain or increase the fertility of the soil. Such techniques, together with the use of vetches, turnips, and clover as fodder crops for intensive grazing and hence heavy manuring, made it possible to introduce a four-course and even more complicated rotations in regions of intensive cultivation.

One can also speak of innovation in terms of the crops grown and livestock raised. Although the science of genetics was far in the future, even simple peasants knew they could grow larger horses, better milk cows, and sheep with longer wool by careful breeding. Over the course of the Middle Ages a number of crops were introduced in Europe, widely diffused, and specially cultivated. Rye, which became the standard bread grain for much of northern and eastern Europe, was one; it was scarcely known, if at all, in ancient times. Much the same can be said of oats, so vital to a horse-powered economy. Peas, beans, and lentils, all previously grown, became more widely diffused and more common with the greater opportunities for cultivation, thus providing more varied and balanced diets. Many garden vegetables and fruits from the Mediterranean and even Africa and Asia were acclimatized in northern Europe. Improved varieties of fruits and nuts were obtained by the technique of grafting, probably an Arab or Moorish invention. From Muslims in Spain and southern Italy, Europeans learned of cotton, sugar cane, citrus fruit, and, most important, rice, which became a staple crop in the Po valley and elsewhere in Italy. Mulberry trees and the culturing of silk worms also came to northern Italy by way of Islamic and Byzantine civilizations. Lacking both olives and wine, northern Europeans learned to grow rape seed for oil and hops for beer. With the growth of textile industries, the demand for woad, madder, saffron, and other natural dyestuffs increased; some small regions specialized completely in these products, importing their foodstuffs from outside.

No single explanation exists for the numerous innovations in both techniques and products. For some innovators, the intention may have been merely to save their own labor or reduce its burden; but the ultimate effect was to make it more efficient. One would scarcely characterize medieval agriculture as individualistic, but in practice it was the individuals who, either alone or in cooperative groups, introduced or adopted innovations that usually benefited from them. This incentive for innovation was the big difference between ancient and medieval agriculture. Similarly, the introduction of new crops or specialization in the production of others reflected both the existence of incentives and the ability of the cultivators to respond to them. Whether produced for the direct consumption of the cultivators, for sale to urban consumers, or as raw materials for growing industries, those commodities are indicative of both rising incomes and more diversified channels of production and distribution, hence of economic development. The most striking evidence of development, however, was the growth of population and its consequences, the rise of cities and the physical expansion of European civilization.

Europe Expands

Numerical accuracy in establishing medieval population is out of the question, but the population of western Europe at about 1000 A.D. has been plausibly estimated at 12 million to 15 million people. (For this purpose western Europe may be regarded as consisting of northern Italy, France, Benelux, the German Federal Republic, Switzerland, the United Kingdom, Ireland, and Denmark.) The population of Christian Europe (other than the Byzantine Empire) at this time—that is, with the addition of Norway, Sweden, most of eastern Europe, and the Christian population of the Iberian peninsula—was probably about 18 million to 20 million. (These figures imply a significantly higher density for western Europe than for the rest of the continent; in fact, it was precisely in the areas of manorial economy, especially northern Italy and northern France, that the population densities were greatest.) By the beginning of the fourteenth century the population of western Europe was probably between 45 million and 50 million, and that of Europe as a whole between 60 million and 70 million. In western Europe the growth can be attributed almost entirely to natural increase; elsewhere, migration from western Europe and the conquest or conversion of non-Christian peoples helped swell the total.

What were the mechanics of such an increase in population? The mathematical condition for a stable total population is an equivalence of crude birth and death rates. If the birth rate rises or the death rate falls, the population grows. Partial evidence from western Europe as well as analogies from other traditional (i.e., predominantly agrarian) societies suggest that crude birth and death rates were in the vicinity of 35 to 40 per thousand per year. (A birth or death rate of 35 means that there were 35 live births or deaths during the year for each 1000 people alive at the midpoint of the year.) Human biologists estimate that the physiological maximum birth rate, under the most favorable conditions, is 50 to 55; but in fact such high rates are rarely encountered. There is no equivalent maximum for the death rate—a rate of 1000 would mean total destruction of the population—but rates of 250 or even 500 might be experienced for very short periods during severe famines or epidemics. If, on the average, the birth rate exceeds the death rate by only three per thousand—for example, a birth rate of 38 or 40 against a death rate of 35 or 37— the ensuing rate of population increase would be 0.3 percent per year, which is sufficient to produce the growth implied by the estimates given earlier.

If we assume that the population of Europe was stable or falling before the tenth century (it certainly fell between the second and seventh), what circumstances would account for a reversal of the conditions that determined it (i.e., a rise in the birth rate or a fall in the death rate)? The most likely explanation is better nourishment as a result of larger, more stable, and more varied food supplies. Death from outright starvation is rare even in today's poorest countries, and no doubt was in medieval Europe as well. But an undernourished population, whether because of insufficient total caloric intake or an unbalanced diet, is more susceptible to disease than a better nourished one. The increase in agricultural productivity as a result of the three-course rotation and other improvements in agricultural technology could easily account for a slight decline in the average death rate, which, if sustained for many years, would bring about a significant rise in population. Furthermore, although we have no firm evidence for it, the average birth rate may have increased slightly as well. Well-nourished parents are more likely to bear healthy children with a greater chance of surviving the rigors of infancy; and favorable economic circumstances may have encouraged earlier marriages, hence a longer child-bearing period.

Other factors may have been favorable to the growth of population, but the evidence is less compelling. Insofar as warfare and pillage were less common and destructive, the security of life would have increased both directly and indirectly, through its effect on production. We know too little about medical practice and sanitary habits to draw any conclusions about their effects, but the manufacture and use of soap grew significantly, at least in the thirteenth century—possibly a minor factor in reducing the death rate. The climate in northern Europe might have ameliorated slightly between the tenth and fourteenth centuries, but if so the influence of this change would have been felt mainly through greater agricultural productivity. In short, it is to the latter that we should ascribe major importance in permitting the growth of population, and improvements in agricultural technology were mainly responsible for this.

How did the increased population distribute itself, and in what activities, productive and otherwise, did it engage? There was, above all, a notable increase in the urban population; we will return to this population and its activities. But only a fraction of the total population, substantially less than half, was absorbed by the growing towns. Much the larger part remained in agriculture, distributing itself in three main ways. First, the average density of existing settlements increased. New land was cleared on the outskirts of fields already under cultivation and, at least in the thirteenth century and especially in the first half of the fourteenth century, the average size of plots was reduced as more villagers had to find a place in the by then saturated settlements.

Second, and more important, formerly wild and unsettled land was cultivated. At the beginning of the tenth century villages in northwestern Europe (and even more so farther north and east) were widely scattered, with large tracts of virgin forest or wasteland in between. A major effort of clearing and reclamation, not unlike that engaged in by European colonists in America in later centuries, was needed to bring these lands into cultivation. A similar effort was undertaken to regain polderlands from the sea in Flanders, Zealand, and Holland. Most of these reclamation efforts were made at the instigation, or at least with the permission, of great lords in whose administrations the lands lay; but to attract settlers for the arduous work of clearing and reclamation, the lords were frequently obliged to renounce the possession of demesne land and labor service from the settlers. The latter thus became rent-paying, but otherwise economically independent, farmers.

The movement to clear forest and reclaim marshes and other wastelands was encouraged and directly assisted by several religious orders, notably the Cistercian brotherhood of monks. Founded in the eleventh century, the Cistercians followed a discipline of extreme asceticism, hard work, and withdrawal from the world. They established their abbeys in the wilderness, and devoted their efforts to making them economically productive, admitting peasants as lay brothers to assist with the work. Under the leadership of Bernard of Clairvaux (St. Bernard), who joined the order in 1112, new chapter houses proliferated throughout France, Germany, and England. By 1152 a total of 328 chapters ranged geographically from the Yorkshire moors to Slavonic territory in eastern Germany.

Finally, European civilization expanded geographically to accommodate its larger numbers. The gradual incorporation of Scandinavia into European civilization and economy is a different matter, because it did not involve a migration of people nor a forced imposition of European institutions. We can also regard the Norman conquest of England as a domestic matter among Europeans, but this was scarcely the case with the reconquest of the Iberian peninsula and Sicily from the Muslims, the *Drang nach Osten* of German settlers in eastern Europe, and least of all the establishment of feudal monarchies in the Near East during the Crusades.

Although the Franks drove the Muslims south of the Pyrenees in the eighth century, and a few miniscule Christian kingdoms held out in the mountainous northern regions, for more than 400 years Islamic states and civilization dominated the greater part of the Iberian peninsula. The Muslim (mainly Moorish) inhabitants were skilled in agriculture, especially horticulture; they revived and extended the Roman irrigation system and made southern Spain one of the most prosperous areas of Europe. The capital, Cordova, was the largest city in Europe west of Constantinople; it was a major intellectual center as well, serving as a bridge for the transmission of knowledge from the ancient world to the emerging civilization of Europe.

The Christian reconquest of the peninsula got underway in earnest in the tenth century, coincident with the growth of European population, and by the thirteenth century nine-tenths of the peninsula was in Christian hands. The reconquest took on a crusading character, and many of the warriors who took part came from north of the Pyrenees. The kingdom of Portugal, for example, was created by Burgundian knights. To support them and settle the wasted territory, the conquerors brought northern peasants with them, encouraged the migration of others, and attempted to transplant the manorial system. The topography and climate of Iberia, so different from that of northern France, was not hospitable to this innovation, however; modifications were introduced, but the end result was a hybrid system that was less productive than either northern manorialism or intensive Moorish agriculture, which the Christian population was unable to maintain.

In the latter part of the eleventh century, when the Christian reconquest of Spain and Portugal was in full swing and Duke William of Normandy successfully asserted his claim to be king of England, other Norman warriors descended on far-off Sicily and undertook its conquest from the Muslims. Before it was conquered by the Muslims Sicily had been a part of the Byzantine Empire; thus its conquest by the Normans brought it, for the first time, into the ambit of the Western economy. For a time after its conquest, with its medley of Greek, Arab, and Norman elements, it was one of the most prosperous areas in Europe. Normans from Sicily also wrested southern Italy, the last remaining Byzantine territory in the West, from Constantinople.

Perhaps the most striking evidence of the economic vitality of medieval Europe was the German expansion into what is now Poland, Czechoslovakia, Hungary, Rumania, and Lithuania. Before the tenth century that area had been sparsely populated, mainly by Slavic tribes employing primitive agricultural techniques along with hunting and gathering. Austria had been a part of Charlemagne's empire, but in the ninth century invading Magyars conquered and pillaged it. In 955 German forces decisively defeated the Magyars, after which the latter settled down in the central Hungarian plain and Austria was resettled by colonists from Bavaria. German missionaries subsequently converted the Hungarians and western Slavs to the Roman church, and the (German) Holy Roman emperors asserted their sovereignty over much of eastern Europe. About the middle of the eleventh century—that is, about a hundred years after the beginning of the demographic upswing in the West—German colonists began spreading eastward across the Elbe River into what became eastern Germany, conquering or displacing the native Wendish (Slavic) population. In the following century, after the devastation wrought by nomadic Mongols, the rulers and the church in Hungary and Poland invited German settlers into their territories, granting them various immunities and allowing them to bring their own legal and economic institutions. Finally, in the thirteenth century the Teutonic Knights were charged with conquering, Christianizing (and incidentally Germanizing) the still pagan lands of Prussia and Lithuania in the eastern Baltic region.

The colonization of this vast region was accomplished in several ways, but much of it involved a rudimentary form of economic planning. Individuals called locators, whose function was not unlike that of a modern real estate developer, made contracts with a great landlord or local ruler to establish a village or group of villages, and perhaps a town. They then toured the more advanced, densely settled parts of Europe, especially western Germany and the Low Countries, to recruit colonists. For settlements in low-lying or marshy areas, such as near the mouths of rivers, colonists from Holland and Flanders who had experience in dyking and draining were preferred. Where forest had to be cleared or wasteland reclaimed, peasants from Westphalia and Saxony predominated. Town-dwelling artisans and traders were also recruited, as the colonization plans envisaged not only purely agricultural settlements but also networks of market towns. The rural settlers brought with them the manorial form of organization and the more advanced agricultural technology that went with it. Rents included both money and in-kind payments to the landlord (usually after the lapse of a stipulated numbers of years, while they made the land productive), but they had more land, fewer burdens, and greater freedom than in the regions from which they came. Locators usually received larger allotments of land than ordinary peasants; sometimes they settled down and became headmen in the villages they established, but frequently they sold their rights and moved on to repeat the process. Religious orders, especially the Cistercians and of course the Teutonic Knights, were also involved in the expansion. The Teutonic Knights founded numerous towns and cities, including Riga, Memel, and Königsberg, and engaged themselves in commercial activities.

The overall economic results of this expansion may be summed up as a diffusion of more advanced technology, a significant increase in population through natural increase as well as immigration, a great extension of the cultivated area (new resources), and an intensification of economic activity. As early as the middle of the thirteenth century grain was being shipped from Brandenburg to the Low Countries and England via the Baltic and North seas; subsequently Poland and East Prussia became major suppliers not only of grain but also of naval stores and other raw materials. Finally, although this consequence goes beyond the purely economic sphere, the German expansion tied eastern Europe more closely to the emerging civilization of the West.

The Crusades, unlike the German push to the East, did not result in a permanent geographic expansion of European civilization; their causation was more complex, involving religious and political motivations to a greater extent than economic. Yet Pope Urban II, in advocating the first crusade in 1095, gave as one of his reasons Europe's "overpopulation," and without the vitality of a growing population and production Europeans would have been unable to mount the considerable military and economic effort the Crusades represented. Significantly, the crusading era ended with the long secular depression of the fourteenth century. Just as a growing economy made it possible for Europeans to undertake the Crusades, however, the latter stimulated the growth of trade and production. Not only was it necessary to finance and supply the crusading armies, but temporary conquests by Christians in the eastern Mediterranean opened up new sources of supply and new markets for Western merchants. It is not true, as was once believed, that the Crusades were responsible for the revival of trade—that had already occurred before the Crusades began; but they were intimately related to its extension and continued growth.

The Revival of Urban Life

The urban population had begun to decline even before the fall of Rome. During the early Middle Ages in northern Europe many urban sites were abandoned altogether; others remained as hollow shells housing a few lay or ecclesiastical administrators and their retainers. They drew their basic supplies from the immediately surrounding countryside, frequently from their own estates. Long-distance trade was confined largely to luxury goods, including slaves, destined for the courts of rich and powerful nobles, both secular and religious; its agents were foreigners, mainly Syrians and Jews, who were granted special protection and passes by their customers.

In Italy, although the cities suffered and shrank during the centuries of invasion and pillage, the urban tradition lingered. Before the eleventh century Italian political, cultural, and economic contacts with the Byzantine Empire (and, after the seventh century, with Islamic civilization) were as strong as or stronger than those with northern Europe. Italian cities were thus in a position to act as intermediaries between the wealthier, more advanced East and the poor and backward West, a position from which they profited both literally and figuratively. Amalfi, Naples, Gaeta, and other port cities in the southern half of the peninsula, which maintained their political affiliation with Constantinople but were sufficiently distant not to be unduly hampered by imperial regulation, were the principal intermediaries between the sixth and ninth centuries. Venice, literally forced to the sea and to maritime trade by the Lombard invasion of the sixth century, which cut it off from its agricultural hinterland, developed rapidly as an entrepôt. Pisa and Genoa were similarly forced to the sea to defend themselves against Muslim raiders in the tenth century; their counterattack was so successful that they soon found themselves in command of the entire western Mediterranean.

Urban growth began first in the port cities, but was not confined to them for long. The Lombard and Tuscan plains formed the natural hinterlands of Venice, Genoa, and Pisa; they were also among the most fertile agricultural regions of Italy, and they, too, clung to the ancient urban tradition of Rome. With the increase of agricultural productivity and the growth of population that it engendered, many peasants migrated to the urban centers, old and new, where they took up new occupations in commerce and industry. Milan was the outstanding example in Lombardy, Florence in Tuscany; but there were many others, smaller but equally bustling (Fig. 3-3). The interaction between town and country was intense. The country provided the surplus population to people the towns, but once there the new urban inhabitants provided larger markets for the produce of the country. Under the pressure of market forces the manorial system, designed for rural self-sufficiency, began to disintegrate. As early as the tenth century tenant labor services were being replaced by money rents; shortly afterward feudal lords began to lease or sell their demesne lands to commercial farmers. The open fields of the manorial system were broken up, enclosed, and subjected to intensive tillage, frequently involving irrigation and heavy manuring. Many of the new agricultural entrepreneurs were urban dwellers who applied to their lands, whether purchased or rented, the same careful calculations of cost and revenue that they had learned in business dealings.

FIGURE 3–3.   City-states of Northern Italy in 1200.

As we have seen, the theorists of the feudal system made no provision for townspeople. Some kings and other great feudal lords tried to treat entire towns as vassals, but the exigencies of urban government, the demands of merchants for freedoms not possessed by other feudal subjects, and above all the pretensions of wealthy men of business did not easily fit into the feudal hierarchy. In the cities of northern Italy the more successful merchants banded together, sometimes with the cooperation of urban-dwelling lesser aristocrats who might also engage in trade, or at least lend money to those who did; they formed voluntary associations to attend to municipal affairs, to protect their common interests, and to settle disputes without recourse to the cumbersome feudal courts. In time these voluntary associations became urban governments, called communes; they bargained with their feudal overlords for charters of freedom, or fought them for the same objective. As early as 1035 Milan won its freedom by force of arms. In Italy, moreover, unlike other parts of Europe, the cities proved strong enough to extend their power over the immediately surrounding countryside, similar to the Greco-Roman city-states of ancient civilization A map of Italy north of the Tiber in the thirteenth century resembles a mosaic whose tiles are the territories of the communes. In 1176 a league of Lombard cities defeated the armies of the Emperor Frederic Barbarossa to confirm their freedom and independence.

Elsewhere in Europe urban development began later and was less intense than in northern Italy. Towns and cities grew—in the Low Countries, in the Rhineland, scattered across northern France, in Provence and Catalonia; the locators of Germany and eastern Europe even carried town plans with them into the wilderness— but with few exceptions they were neither as large nor as concentrated as the cities of northern Italy. Above all, they did not succeed to nearly the same extent in winning autonomy or independence from territorial princes. At the end of the thirteenth century, when Milan had a population of some 200,000, the populations of Venice, Florence, and Genoa each exceeded 100,000, and those of several other Italian cities ranged from 20,000 to 50,000, few cities in northern Europe could reach the latter figure. Paris, which combined the functions of territorial capital and seat of a great court, commercial and industrial town, and university center, may have equaled Milan in population, although some doubt its population exceeded 80,000. As late as 1377 the population of London was no greater than 35,000 to 40,000, and that of Cologne, by far the largest city in Germany, was about the same.

The only region that could compare with northern Italy, in terms of urban development, was the southern Low Countries, especially Flanders and Brabant. Although Ghent, the largest city, had only about 50,000 inhabitants at the beginning of the fourteenth century, the urban population as a whole may have constituted about one-third of the total, approximately the same as in northern Italy. There are other similarities as well. Not only did these two areas have the largest urban populations, but their overall densities were also the greatest in Europe. Their agriculture was the most advanced and intensive, and they contained the most important commercial and industrial centers. The question naturally arises, did men move to cities and turn to commerce and industry because there was no place for them on the land, or did the existence of towns and trade, with their potentially lucrative markets, stimulate the cultivators to greater production and productivity? There can be no definitive answer to this question; undoubtedly, there were reciprocal influences. But the fact that agriculture was always more intensive and productive in the vicinity of towns and cities than in the open countryside suggests an important role for urban demand and markets. It is therefore necessary to consider the development and nature of the market mechanism in greater detail.

Commercial Currents and Techniques

The most prestigious and profitable trade was no doubt that which stimulated the commercial revival between Italy and the Levant. Even before the Italians made it their own, the route had been used by Eastern merchants bringing luxury goods to Western courts. After the Italians took charge luxury goods—spices from as far east as the Moluccas, silk and porcelain from China, brocades from the Byzantine Empire, precious stones, and other goods—still dominated the movement from east to west, but in addition there were such bulky goods as alum from Asia Minor and raw cotton from Syria. In the opposite direction went common cloth of wool and linen, furs from northern Europe, metalwares from central Europe and Lombardy, and glass from Venice. The Venetians had traded with the Byzantine Empire from the very beginnings of their history, but they secured a favored place in the latter part of the eleventh century in return for aid against the Seljuk Turks; as a result, they obtained free access to all ports of the empire without payment of customs duties or other taxes—a privilege not granted even to the empire's own merchants.

Meanwhile Genoa and Pisa, having driven the Muslims from Corsica and Sardinia, descended on their strongholds in North Africa, looted their cities, and extracted specially favorable terms for their own ships and merchants. Subsequently Genoa defeated Pisa for undisputed mastery of the western Mediterranean and challenged Venice for control of the East. During the Crusades the Italian cities, in concert and in rivalry, intensified their penetration of the Levant; they established colonies and special privileged enclaves from Alexandria along the Palestinian and Syrian coasts, in Asia Minor, Greece, the suburbs of Constantinople, and around the shores of the Black Sea from the Crimea to Trebizond. Genoese ships, built on the spot, even sailed the Caspian Sea and Persian Gulf. The fall of the kingdom of Jerusalem and the failure of the Crusades scarcely affected the Italian positions in the East; instead, the Italians made treaties with Arabs and Turks and continued "business as usual."

A special, exotic extension of the Eastern trade that flourished from the mid-thirteenth to the mid-fourteenth century was that with China. During that period the Mongol Empire, the most extensive land empire the world has ever seen, stretched from Hungary and Poland to the Pacific. The Mongol rulers, in spite of their fierce reputations, welcomed Christian missionaries and Western traders. Again, the Italians dominated the trade, with colonies in Peking and other Chinese cities as well as India. Merchants' handbooks described the itineraries—overland through Turkestan, "the great silk route," or Persia, or by sea through the Indian Ocean—in great detail, and gave useful hints as to what merchandise would be in demand.

Marco Polo's account of his adventures was one of the first "best-sellers" in Europe.

At the other end of the Mediterranean trade was more prosaic. It included spices and other luxury products from the East, of course, but more important, at least to the Italians, were their grain supplies from Sicily. That was a regular flow, except in times of war and blockade, necessary for the survival of the grain-poor Italian cities. In addition, other ordinary commodities such as salt, dried fish, wine, oil, cheese, and dried fruits moved from regions of specialized production or temporary surplus to those with chronic or temporary deficits. In spite of the relatively slow communications, alert merchants and active markets saw to it that effective demand did not go unsatisfied for long. Although the great Italian ports dominated this trade also, they shared it, more or less willingly, with Catalan, Spanish, Provençal, Narbonaise, and even Muslim traders (see Fig. 3-4).

The northern seas, though less busy than the Mediterranean, grew steadily in importance in the Middle Ages. In the early Middle Ages the Frisians had been the principal carriers of the slender volume of trade along the North Sea coasts and up the great rivers. As the Baltic became more prominent they were succeeded by the Scandinavians, but in the later Middle Ages the great German trading cities, organized in the Hansa (usually incorrectly called the Hanseatic League), dominated the trade of both the Baltic and the North seas.

The Hansa, which eventually included almost 200 cities and towns, was not formally organized until 1367, in response to the threat of the king of Denmark to restrict their activities; but it had been preceded by many years of informal cooperation among German merchants in foreign cities. In Venice, for example, there was a "German foundation," which provided lodging and board for itinerant German merchants, as well as advice and assistance in marketing their wares. In London the "Steelyard" (*Stalhof*), a district inhabited by resident German merchants, won rights of extraterritoriality and self-government as early as 1281. Similar German colonies existed in Bruges, in Bergen, Norway, in Visby on the island of Gotland, and elsewhere in the Baltic, as well as in the great trading city of Novgorod in Russia. Riga, Memel, and Danzig, among others, were entirely German cities established as enclaves in foreign lands. Their merchants carried the grain, timber, naval stores and other commodities produced by German colonists in the Baltic hinterland to the thriving cities growing up around the North Sea.

As early as the twelfth century regional specialization in production was becoming a marked feature of the medieval economy. The most famous example was the Gascon wine trade, with its headquarters in Bordeaux; but the Flemish woolen industry depended heavily on supplies of raw wool from England, and the Baltic lands became increasingly important as sources of grain to feed the highly urbanized Low Countries. Farther south, Portuguese, French, and English ships brought salt and wine northward, returning with cargoes of dried and salted fish.

Mile for mile, land transport is generally more expensive than transport by water. This was true to an even greater extent before the locomotive steam engine and the internal combustion engine were invented. This accounts for the great importance of seaborne trade prior to the industrial age. In the Middle Ages, however, there was one great exception to this rule—the trade between northern

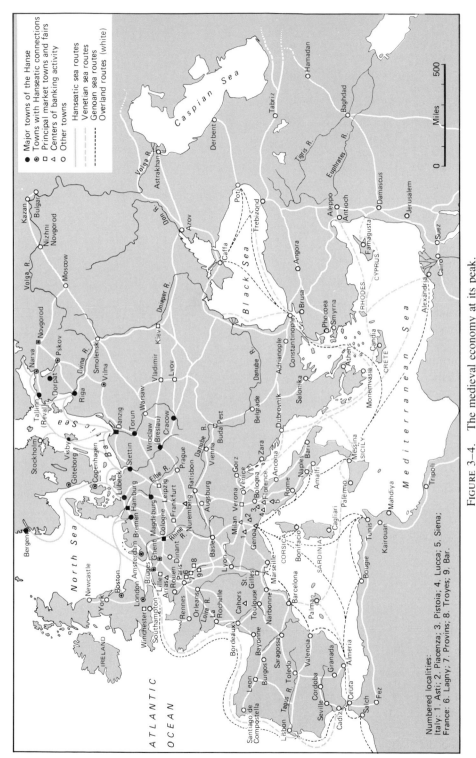

Legend:

- ● Major towns of the Hanse
- ◉ Towns with Hanseatic connections
- □ Principal market towns and fairs
- △ Centers of banking activity
- ○ Other towns

— Hanseatic sea routes
— Venetian sea routes
--- Genoan sea routes
--- Overland routes (white)

FIGURE 3–4. The medieval economy at its peak.

Numbered localities:
Italy: 1. Asti; 2. Piacenza; 3. Pistoia; 4. Lucca; 5. Siena; France: 6. Lagny; 7. Provins; 8. Troyes; 9. Bar.

and southern Europe, especially the trade of northern Italy with Germany and the Low Countries. Before the advances in ship design and navigational techniques of the late thirteenth and fourteenth centuries, which were to have a revolutionary impact in the fifteenth century, the sea route between the Mediterranean and the North Sea was hazardous and not especially profitable. For this reason the great Alpine passes (Brenner, St. Gothard, Simplon, St. Bernard, Mt. Cenis, and others) were more heavily trafficked than the Straits of Gibraltar, in spite of their own obstacles and hazards. Feudal lords, through whose lands the routes passed, put down bandits and improved the roads, for which they charged tolls, although the competition of alternative routes kept them to a reasonable level. Religious brotherhoods organized relay stations and rescue services, of which St. Bernard dogs with their casks of brandy are the most memorable symbol. Professional companies of carters and muleteers provided transport facilities in an atmosphere of lively competition. The most important emporia at the southern end of the route were the cities of the Lombard plain, especially Milan and Verona. There were numerous destinations in the north, from Vienna and Cracow in the east to Lubeck, Hamburg, and Bruges in the extreme north and west; but the majority of the goods changed hands in the great fairs or markets of Leipzig, Frankfurt, and especially the four fair towns of Champagne.

The fairs of Champagne emerged in the twelfth century as the most important meeting place in Europe for merchants from north and south. Under the protection of the counts of Champagne, who provided merchandising facilities and special commercial courts as well as protection on the road for traveling merchants, the fairs rotated almost continuously throughout the year among the four towns of Provins, Troyes, Lagny, and Bar-sur-Aube. Located roughly midway between Europe's two most highly developed economic regions, northern Italy and the Low Countries, they served as meeting ground and place of business for merchants from each; but they also played a role in the trade of northern Germany with southern France and the Iberian peninsula. The commercial practices and techniques that developed in these towns—for example, the "letters of fair" and other credit instruments, and the precedents of their commercial courts—exercised an influence far broader and longer-lasting than the fairs themselves. Even after their decline as commodity trading centers, they continued for many years to serve as financial centers.

In the latter decades of the thirteenth century voyages from the Mediterranean to the North Sea became increasingly frequent; in the second decade of the fourteenth century both Venice and Genoa organized regular annual convoys, the famous Flanders fleets. These seagoing caravans took merchandise from the Mediterranean ports directly to the great permanent market in Bruges (and subsequently to Antwerp), thereby undercutting some of the functions of the Champagne fairs. Although overland trade did not cease entirely (in the fifteenth century Geneva played a role very similar to that of Champagne), a new phase in the economic relations between northern and southern Europe had clearly opened. It involved not only new routes and new means of transport, but also a shift in both the scale of commerce and the mechanisms of business organization. Great trading and financial companies, with headquarters in the major Italian cities and branches throughout Europe, replaced individual traveling merchants as the principal agents of commerce. This development, sometimes called a "commercial revolution," was of primary importance in the next age of European expansion that began in the fifteenth century.

In Carolingian times merchants were usually foreigners—"Syrians" (almost anyone from the Levant) and Jews. With the revival of commerce in the tenth century European merchants became more prominent, but until well into the thirteenth century merchants continued to be itinerant. It was a vigorous life, requiring physical stamina and courage as well as a head for business. By land, merchants frequently traveled in caravans, bearing their own arms or hiring armed guards to ward off bandits. By sea, they were also armed against pirates and had to contend with the possibility of shipwreck as well. It is scarcely surprising that such merchant voyages were called "adventures."

Under the simplest circumstances merchants worked for their own account; their entire capital consisted of the stocks of goods they carried. Very early, however, a form of partnership, the *commenda,* came into use: one merchant, perhaps too old for the rigors of travel, provided the capital for another, who actually undertook the voyage. Profits were divided, usually three-fourths for the sedentary capitalist and one-fourth for the active partner. Such contracts were most common in the sea trade of the Mediterranean, but they were also used in overland travel; usually they were limited to a single (round-trip) venture, but a successful venture was often followed by another contract between the same partners. Sometimes the sedentary merchant would specify the destination and the return cargo, which he might undertake to dispose of in the home port; but it was not uncommon, especially when the "capitalist" was a widow, a foundation or religious establishment, or a trustee acting on behalf of minor children or orphans, for the active partner to make all the key decisions. In Genoa and other Italian cities, as early as the twelfth century, many individuals who were not actually active in trade invested in trade by this means.

As the volume of trade expanded and commercial practices became standardized, a new form of business organization—the *vera società,* or true company—arose to rival, and sometimes supplant, the *commenda.* It had several, sometimes numerous partners, and frequently operated in many cities throughout Europe. The Italians were by far the most prominent in this type of organization; from headquarters in Florence, Siena, Venice, or Milan, they could operate branches in Bruges, London, Paris, Geneva, and several other cities. They frequently engaged in banking along with mercantile operations (or vice versa). The Bardi and Peruzzi companies of Florence were the largest business organizations in the world before the great chartered companies of the seventeenth century; but both were bankrupted in the 1340s as a result of overextensions of credit to Edward III of England and other impecunious sovereigns. In addition to maintaining branches, these great companies had their own ships and wagon and mule trains; some owned or leased metal mines and other mineral deposits.

Smaller merchants who could not afford their own ships devised other means of spreading the risks of long-distance commerce. Shipowners might lease their ships to several merchants in common who traded separately but joined forces to rent the ship. Or a single entrepreneur might lease an entire ship, then retail space in it to other merchants. Various types of sea loans were devised to give nontrading investors an interest in the profits without either making them partners in the enterprise or violating the usury laws. By the end of the thirteenth century maritime insurance was common.

Banking and credit were intimately related to medieval commerce. Primitive deposit banks were set up in Venice and Genoa as early as the twelfth century. Originally intended as mere safe deposits, they soon began to transfer sums from one account to another on oral order and, less frequently, on written order. Although legally forbidden to make loans on fractional reserves, the banks granted overdraft facilities to favored depositors, thereby creating new means of payment. Such banks were found only in the major commercial centers; outside of Italy, these were chiefly Barcelona, Geneva, Bruges, and London. (Lombard Street, the heart of the present financial district of London, got its name from the large number of Italian bankers who kept offices there.) Elsewhere, however, private bankers bought and sold bills of exchange to facilitate long-distance trade. Because of the high risk and expense of shipping coin and bullion, merchants preferred to sell on credit, invest the proceeds in a return cargo, and realize profits only after the latter had been disposed of. Virtually all of the business of the Champagne fairs was conducted by means of credit; at the end of one fair unsettled balances were carried over to the next by means of letters of fair, a kind of bill of exchange. Although bills of exchange developed in connection with commodity trade, they were eventually used purely as financial instruments, with no direct connection to actual commodities.

Another reason for the widespread dependence on credit was the multiplicity and confusion of coinage. Most regions of western Europe used the Carolingian monetary system of pounds, shillings, and pence (in Latin, *libra, solidus, denarius*), but this apparent unity masked a bewildering disunity of actual monies. For one thing, the Genoese lira did not have the same value as the English pound, the French livre, or even the Milanese or Pisan lire. More fundamentally, both the pound and shilling were mere monies of account; no actual coins of those values were struck until very late in the Middle Ages. The most common coins in the eleventh and twelfth centuries were pennies; not only were these inconvenient for large payments, but they were minted by numerous authorities—kings, dukes, counts, even abbeys—with different sizes, weights, and silver contents. Larger silver coins came into use about the beginning of the thirteenth century, but these, too, lacked uniformity of both weight and fineness. Hard-pressed sovereigns with inadequate tax revenues frequently resorted to debasement of the coinage to extend their resources. Under such circumstances the money changers, whose business it was to know the values of the scores of different types of coin, performed an important function in the fairs and commercial cities. From their ranks came many bankers (Fig. 3-5). Not until the second half of the thirteenth century did Europe at last obtain a really stable currency, the famous gold florin first issued in Florence in 1252. (Genoa had minted a similar coin a few months earlier, but it was not as popular; in 1284 Venice began minting similar coins, called ducats or sequins, which were widely used—and imitated—in the eastern Mediterranean.) The florin was ideally suited for mercantile purposes—stable in value, relatively large de-

FIGURE 3–5.   Tuscan banker. A banker with his assistant, seated on his *bancum* (bench) behind his "counter," in which he kept his *conti* (accounts) and over which he counted out money. Many bankers evolved from money changers. (SCALA/Art Resource, New York.)

nomination—but by the time it came along, credit had already become an indispensable part of commercial activity.

Industrial Technology and the Origins of Mechanical Power

Although greatly inferior to agriculture in terms of the numbers involved, manufacturing industry was by no means a negligible sector of the medieval economy. Moreover, it steadily grew in importance with the passing centuries. There may have been some slight regression in technical competence in the early Middle Ages—in architecture and building, for example—but by the year 1000 the average level of technology was at least as high as in ancient times. Thereafter, innovations occurred in steadily increasing numbers so that, from the viewpoint of the history of technology, there is no hiatus between medieval and modern times.

The largest and most ubiquitous industry was no doubt the manufacture of cloth, although the building trades, taken together, may have been a close second. Cloth was made in every country, every province, almost in every household of Europe; but by the eleventh century some areas had definitely begun to specialize in it. Of these the most important was Flanders and the surrounding area in northern France and what is now Belgium. Other centers of importance were northern Italy and

Tuscany (Florence alone employed several thousand cloth workers in the fourteenth century), southern and eastern England, and southern France. Wool was by far the most important raw material and woolen cloth the most important product. Differences in type and quality of cloth produced in different regions account for the widespread trade within Europe. Besides wool, linen was produced in many areas, especially France and eastern Europe. Silk and cotton production were confined to Italy and Muslim Spain.

Although the more skilled workers, such as dyers, fullers, shearers, and even weavers, were organized in guilds, the industry was dominated by merchants (also organized in guilds), who bought the raw materials and sold the final product. The less skilled workers, including spinners (and spinsters), lacked organization and generally worked directly for the merchant. In Flanders and England these merchant-manufacturers sent or "put out" the raw or semifinished materials to the weavers and other artisans, who worked them up in their own homes or shops; but in Italy the work was done in shops or sheds, under the eye of a supervisor. The productivity of labor, compared with that in ancient times, increased severalfold as a result of a trio of interrelated technical innovations: the pedal loom, replacing the simple weaving frame; the spinning wheel, replacing the distaff; and the water-powered fulling mill. Their inventors are unknown, but the devices spread surprisingly fast throughout Europe at the beginning of the twelfth century (Fig. 3-6). Lowered costs of production is no doubt a sufficient reason for their diffusion, but they also reduced the tedium of labor.

Smaller than the textile industries, but strategically more important for economic development, the metallurgical industries and their auxilliaries experienced notable progress in the later Middle Ages. According to a conventional classification, the Iron Age began about 1200 B.C., but throughout classical antiquity objects and implements of iron were rare and expensive, and iron was virtually monopolized for weapons and decorations for the small ruling classes. Even copper and bronze, though somewhat more plentiful, seldom entered the daily lives of common people. In the Middle Ages, on the other hand, the price ratios were reversed, with iron becoming the cheaper metal, and in addition to its continued use for arms and armor it was used in an increasingly wide variety of tools and for other utilitarian purposes. The greater abundance and lower price of iron were partly a result of the greater accessibility of iron ore, and especially fuel (charcoal), in Europe north of the Alps. Improvements in technology, notably the use of water power to actuate bellows and large trip hammers, were also important, however. Toward the beginning of the fourteenth century the first precursors of the modern blast furnace, replacing the so-called Catalan forge, made their appearance. The organization of miners and primary metalworkers in free communities of artisans, in contrast with the slave gangs of Roman times, no doubt facilitated technological change.

Consumer demand should also be taken into account when considering the increased output and pressures for improved technology. When peasants, even serf peasants, and artisans owned their own tools, and their own well-being was in direct proportion to the efficacy of their efforts, it behooved them to buy the best tools and implements they could afford. The use of horseshoes and iron fittings on harnesses, carts, and plows is evidence that the peasants and artisans were aware of this. The

FIGURE 3–6.   Knitting. This picture of the Virgin Mary knitting a garment for her unborn child, taken from a stained glass altarpiece of a church in western Germany, is the first known representation of knitting, a medieval invention. Knitwear was unknown in the ancient Mediterranean world, but was very useful in the colder, damper climate of northern Europe. (Hamburger Kunsthalle.)

ubiquity of the names Smith and Schmidt (or Schmied) in English and German also testifies to the numerous artisans who earned their living by filling their neighbors' demand for metalwares.

Another industry of great practical use that expanded appreciably beyond its classical dimensions was tanning and leather working. It is difficult for a twentieth-century urban dweller, surrounded by synthetic and plastic materials, to appreciate the importance of leather to earlier generations. In addition to its uses in saddles. harnesses, and such, it was used for furniture, clothing, and industrial equipment such as bellows and valves. Similarly, woodworking occupied a much larger place, proportionately, in medieval industry than in earlier or more recent times; its output found literally hundreds of uses, both ornamental and utilitarian.

Far from being tradition-bound and wedded to unchanging routine, as they were formerly depicted in textbooks, medieval men—or some of them—deliberately sought out novelty, both for its own sake and for immediate, practical purposes. It is to medieval tinkers, not classical philosophers, that we owe such useful inventions as eyeglasses and mechanical clocks. The astrolabe and compass came into general use in Europe during the Middle Ages, in connection with the momentous improvements in navigational technique and ship design that help delineate the medieval from the modern age. Similarly, gunpowder and firearms were medieval inventions, although their period of greatest effectiveness came later. Soapmaking, although not a complete novelty, expanded considerably. Papermaking was a new industry whose cultural significance was far greater than its economic weight. And printing from movable type, one of the most important innovations since the dawn of civilization, was also a late medieval invention. But possibly the most characteristic expression of medieval man's deliberate search for new and more efficient means of production can be found in the history of mills and millwork.

Simple horizontal waterwheels, turned by the flow of a current, were used at least as early as the first century B.C. Archeological and documentary evidence for them have been found as far apart as Denmark and China, as well as within the Roman Empire. No one knows where they originated; there are occasional instances of their use for grinding grain during the imperial period, but the Emperor Vespasian (A.D. 69–79) reputedly rejected a design for a water-driven hoist to raise heavy stones for fear of causing unemployment. Labor, whether slave or free, was cheap in the Roman Empire, and builders and entrepreneurs saw no need for labor-saving machinery. Exactly when men changed their notions about the usefulness of such machines is difficult to ascertain, but apparently it was sometime between the sixth and tenth centuries. When William the Conqueror ordered his survey of the resources of England in 1086, his agents counted 5624 watermills in approximately 3000 villages—and England was by no means the most advanced area in Europe, economically or technically. Moreover, most of the mills, there and elsewhere, were far more sophisticated and powerful than the simple horizontal wheel. The majority were vertical, overshot wheels in which the weight of the falling water provides far more force than a gentle current. They had complicated gearing for transmission and modification of the power (Fig. 3-7). By the beginning of the fourteenth century, water power was used not only to grind grain but to grind, crush, and mix other substances, to make paper, full cloth, saw both timber and stone, move bellows and trip hammers for forge and furnace, and wind silk.

Despite their great utility, waterwheels had many limitations. Most important, they required a steady flow or fall of water. Thus, they could not be used in semiarid areas or in low-lying, marshy land. In Venice as early as the middle of the eleventh century a mill wheel actuated by the movement of the tides was in operation. Within the next few centuries many others were erected around the sea coasts of Europe. A still more satisfactory solution, effected in the twelfth century, was the windmill. Given a steady breeze, the windmill could do all the tasks of a watermill, and on the plains of northern Europe, where the winds were more dependable, the streams

FIGURE 3–7.   Waterwheel. This model of a waterwheel is arranged to actuate both a trip hammer and bellows. (From *Connections,* by James Burke. Copyright 1978 by Macmillan London Limited. Reproduced with permission of Little, Brown and Company.)

more sluggish and subject to freezing in winter than farther south, windmills sprouted in profusion. They were especially important in the low-lying provinces of Holland, Zealand, and Flanders where, in addition to other regular uses, they worked pumps in reclaiming the polderlands.

Wind- and watermills required complicated gearing. The millers, millwrights, and various kinds of smiths who built, operated, maintained, and repaired them eventually acquired an expert if empirical knowledge of practical mechanics, which they put to use in a related field, the manufacture of clocks. As early as the twelfth century the demand for water clocks was so strong that there was a specialized guild of clockmakers in Cologne. In the following century the main problems in the design of mechanical (gravity-driven) clocks were solved, and in the fourteenth century every city in Europe of any size and with any civic pride had at least one large clock that not only signaled the hours with ringing bells or chimes, but also staged an entertainment of dancing bears, marching soldiers, or bowing ladies. Between 1348 and 1364 a noted Italian physician and astronomer, Giovanni de' Dondi, built a clock that, in addition to telling the hours, kept track of the movements of the sun, the moon, and the five known planets—two full centuries before the Copernican revolution (Fig. 3-8).

The medieval concern with millwork and clockwork has a significance beyond their immediate economic impact. True, mills saved labor, increased production, and made possible tasks that were previously considered impossible. Clocks made people more aware of the passage of time, and introduced greater regularity and punctuality into human affairs; Genoese business contracts note not only the date but the actual time of signing—a harbinger of the maxim that "time is money." Taken all together, these changes signified a fundamental reorientation of the medieval mentality, a new attitude toward the material world. No longer was the universe seen as inscrutable and man a helpless pawn of nature or of angels and demons. Nature could be understood, and its forces harnessed for our uses. Shortly after Dondi completed his marvelous clock, the French scholar Nicole Oresme (ca. 1325–82), anticipating Kepler, Newton, and other luminaries of the century of genius, compared the universe to a great mechanical clock created and regulated by the supreme clockmaker, God. A century earlier the Oxford scholar-scientist Roger Bacon (ca. 1214–92), who anticipated by four centuries the emphasis of his namesake Francis on experimental method and the utility of science, had prophesied the possibilities of practical science: "machines which will allow us to sail without oarsmen, carts without animals to pull them . . . machines for flying . . . machines which can move in the depths of the seas and rivers. . . ."

FIGURE 3–8. Mechanical clock. This is a modern reconstruction of Dondi's famous clock, originally built in the midfourteenth century. (National Museum of American History, The Smithsonian Institution. Reprinted with permission.)

The Crisis of the Medieval Economy

In 1348 an epidemic of bubonic plague, the infamous Black Death, reached Europe from Asia. Spreading rapidly along the main commercial routes, taking its greatest toll in cities and towns, for two years it ravaged the whole of Europe, from Sicily and Portugal to Norway, from Muscovy to Iceland. In some cities more than half the population succumbed. For Europe as a whole the population was probably reduced by at least one-third. Moreover, the plague became endemic, with new outbursts every ten or fifteen years for the remainder of the century. Adding to the misery engendered by the plague, warfare, both civil and international, reached a new peak of intensity and violence in the fourteenth and fifteenth centuries. In the Hundred Years War (1338–1453) between England and France large areas of western France were devastated by a deliberate policy of pillage and destruction, while in the East the venerable Byzantine Empire finally succumbed to the onslaught of the Ottoman Turks.

The Black Death was the most dramatic episode in the crisis of the medieval economy, but it was by no means the origin or cause of that crisis. By the end of the thirteenth century the demographic increase of the two or three previous centuries had already begun to level off. In the first half of the fourteenth century crop failures and famine became increasingly frequent and severe. Because of these the population may have begun to decrease even before 1348, though this is not proven. The Great Famine of 1315–17 affected the whole of northern Europe, from the Pyrenees to Russia; in Flanders, the most densely populated area, the death rate jumped to ten times its normal figure. The increasing precariousness of the food supply, together with congestion and the inadequacy of sanitary facilities in the towns and cities, rendered the population more susceptible to epidemics, of which the Black Death was the worst.

There is some evidence of climatic deterioration in the fourteenth century. In northern Europe, at least, the winters became longer, colder, and wetter. Grape cultivation disappeared from England; grain would not ripen in Norway. On three occasions the entire Baltic Sea froze over, and in Germany and the Low Countries flooding increased in frequency and severity. As serious as these problems were, they are unlikely to explain entirely the stagnation and decline of the whole economy. A more general explanation is overpopulation for the resources and technology available.

Toward the end of the thirteenth century the extensive forest clearings of earlier centuries came to a halt. In some areas, such as Italy and Spain, there is evidence that deforestation contributed to soil erosion and declining fertility. Farther north, landlords opposed clearings because of their hunting privileges, and peasants needed the remaining forest for firewood and grazing. Numerous disputes occurred, with occasional outbreaks of violence, between lords and peasants over the use of the forests. With no new land available from clearings, pastures, heaths, and meadows were converted to arable. This meant a decrease in livestock, and thus fewer proteins in the diet and less manure for fertilizer. Scarcity of fertilizer had been a persistent problem in the manorial economy, and the diminution of livestock aggravated it; crop yields declined even as more land was brought into cultivation. Efforts to increase productivity, such as the introduction of four-course and other, more complicated crop rotations and the use of green manures, had limited effects in some regions, but the efforts were not made fast enough and their effects were not substantial enough to offset the diminishing returns of overcropped marginal lands.

In the expansive period of the medieval economy, as we have seen, there was a tendency on the part of landlords to commute labor services into money rents and to lease their demesnes to prosperous peasants. As population and urban growth continued, the prices of most agricultural commodities rose while wages fell. Many landlords, either to bolster their own declining revenues or to take advantage of the favorable price-wage ratio, again resorted to demesne farming, sometimes enlarged their demesnes at the expense of pasture and even peasant strips, and attempted to reimpose old labor services. Although the latter efforts met with strong resistance and had only limited success in western Europe, landlords in eastern Europe proved to be stronger. In any case, with the steady fall in wages it was economical for western lords to cultivate their lands with hired labor. Even substantial peasants could do this, thus becoming wealthier; but the great masses of the peasant population found themselves in steadily worsened straits. Partly for this reason, and also because of the increased burden of taxation levied by kings and other territorial rulers, social tensions increased and occasionally burst out in violence and revolt, as in the rising of Flemish peasants and workers against their lords and masters during the Great Famine of 1315–17.

The Black Death greatly intensified the social tensions and conflict. The price-wage scissors abruptly reversed themselves; with the sharp drop in urban population and demand, the price of grain and other foodstuffs dropped precipitately, while wages rose because of the shortage of laborers. The first reaction of the authorities was to impose wage controls; these merely exacerbated the hostility of peasants and workers, who avoided them when possible and revolted when serious efforts were made to enforce them. In the second half of the fourteenth century revolts, revolutions, and civil wars occurred in every part of Europe. Not all were inspired by wage controls, but all were related in one way or another to the sudden change in economic conditions brought on by famine, plague, and war. In 1358 peasants throughout France rose spontaneously against their lords and the government. In England a series of local uprisings preceded a great peasant revolt in 1381 in which a mixture of religious and economic issues almost allowed the revolutionists to triumph. In Italy the violence was generally not greater than that which accompanied the struggles for autonomy by the communes in the eleventh and twelfth centuries; but in 1378 the workers in the woolen industry of Florence temporarily gained control of the city and drove out "the fat people," their masters. Similar revolts of peasants or workers, or both, flared in Germany, Spain and Portugal, Poland, and Russia. Without exception, whatever the extent of their initial success, they were put down with great brutality by the feudal nobility, the governments of the cities, or those of the emerging national monarchies.

Although the revolts seldom achieved their aims, in western Europe the changed economic conditions brought peasants freedom from manorial bondage. Despite the greater political and military strength of the ruling classes, they were unable for long either to enforce claims to labor services or to control wages, since landlords competed with one another to attract peasants to till their land either for wages or for rent. In England, after the turmoil of the late fourteenth century, this resulted in the fifteenth century in what one authority called "the golden age of the English agricultural laborer." Real wages—that is, the ratio of money wages to the prices of consumables—were higher than at any time previously or subsequently until the nineteenth century. Elsewhere in western Europe as well, market forces resulted in the dissolution of the vestigial bonds of serfdom and the rise of wages and living standards for peasants. Low grain prices, resulting from slack urban demand, and the relative abundance of land encouraged stock raising and a shift from grain to root and forage crops. The Great Plague and associated evils of the fourteenth century, dreadful though they were, proved to be a strong cathartic that prepared the way for a period of renewed growth and development beginning in the fifteenth century.

In eastern Europe a different evolutionary course prevailed. The population there had always been less dense than in western Europe, the towns fewer and less populous, and the market forces weaker. After the Great Plague town life virtually withered away, markets declined, and the economy reverted to a subsistence basis. Under these conditions the peasants had no alternative to landlord rule except to flee to unoccupied and uncharted lands, a course that was fraught with perils of its own. As a result the landlords, unchecked by higher authority, forced the peasantry into a position of servitude unknown in western Europe since at least the ninth century.

Towns in western Europe, although severely checked by the plague, survived and eventually recovered. The total volume of production and trade was probably lower at the beginning of the fifteenth century than at the beginning of the four-teenth; but, at various times in the fifteenth century, in different parts of Europe, recovery of population, production, and trade began, and by the beginning of the sixteenth century the totals of all these aggregates were probably greater than at any previous time. Meanwhile a significant realignment of forces had taken place. Guild organizations, reacting to the sharp fall in demand, tightened their regulations so as to control the supply more effectively in cartel terms; they restricted output, en-forced working rules, and restricted new members to the sons or relatives of de-ceased masters. Merchants, seeking to rationalize their operations, invented or adopted double-entry bookkeeping and other methods of control. Fifteenth-century business firms could not rival the Bardi or Peruzzi companies in terms of size, but the largest of them, the Medici bank of Florence, as well as numerous others, adopted a form of organization similar to the modern holding company that reduced the risks of bankruptcy in the event of failure of a branch. Industrialists faced with the rising costs of labor sought new labor-saving methods of production, or mi-grated to the countryside to escape the restrictive rules of the guilds.

Regional shifts in production and trade also occurred as a result of the inten-sified competition. Some cities, such as Florence and Venice, did not shrink from using military force to subdue their rivals and extend their dominion over their neighbors. More subtly, the fair of Geneva gradually replaced in importance those of Champagne in the fourteenth century, then suffered from the competition of Lyons before the end of the fifteenth century. Farther north, Antwerp gradually replaced Bruges as the principal terminus of Italian trade. The German Hansa received a formal organization in 1367, partly as a response to shrinking demand and the attempts of rivals to deprive its merchants of their privileges; for almost a century it dominated the trade of the Baltic and North seas, but before the end of the fifteenth century it was strongly challenged by Dutch and English traders, shippers, and fishing fleets. The Italian cities together maintained their preeminence in trade but lost ground to northern Europe, a prefiguration of further drastic changes in the sixteenth and seventeenth centuries.

4

Non-Western Economies on the Eve of Western Expansion

Europe, especially Western Europe, was the region of the world that, from the sixteenth to the twentieth centuries, experienced the most dynamic growth and change. It was, in large measure, responsible for creating the modern world economy, and its interaction with other world regions determined the mode and timing of their participation in that economy. Before the sixteenth century, however, Western Europe was only one of several more or less isolated regions. This chapter surveys the other regions before their contact with Europeans.

The World of Islam

Islam, the latest of the world's great religions, originated in Arabia in the seventh century A.D. Its founder, the prophet Mohammed, had been a merchant before he became a religious and political leader. By the time of his death in A.D. 632 he had united under his rule virtually all of the Arabian peninsula. Soon after his death his followers exploded with the fury of a desert whirlwind and within a hundred years conquered a large empire stretching from Central Asia across the Middle East and North Africa to Spain. After a few centuries of relative quiescence and the break-up into a number of successor states of the Caliphate, as their empire was known, the Muslims (followers of Islam) expanded again in the twelfth and following centuries (Fig. 4–1), spreading their religion and their customs to Central Asia, India, Ceylon, Indonesia, Anatolia, and sub-Saharan Africa. By this time the Arabs were a small minority among the millions of the faithful, but the Arabic language, in which the holy book, the *Koran*, was written, was the common language for Islamic civilization, although other languages, notably Persian and Turkish, were also used.

The original Arabs were primarily nomadic, although some practiced oasis agriculture and they had a few urban centers, such as Mecca. The lands they conquered were, on the whole, only slightly less arid than Arabia, but they did contain the two great cradles of civilization, the Tigris-Euphrates Valley and the Nile valley. There and elsewhere Muslims practiced irrigation agriculture that, in some areas (e.g., southern Spain), reached high levels of sophistication and productivity. Their conquests also brought them great cities, including Alexandria, Cairo,

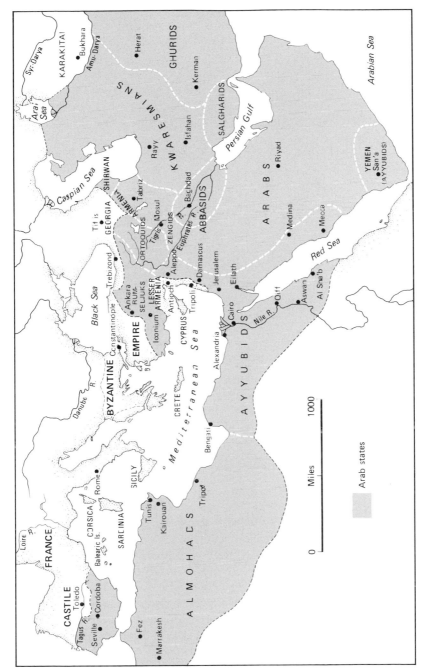

FIGURE 4–1.   The Muslim world, about 1200.

79

and eventually Constantinople, which they renamed Istanbul. In the end Islam developed as a predominantly urban civilization, although many Muslims, Arabs and others, remained nomadic, tending herds of sheep, goats, horses, or camels— rarely cattle, and no pigs at all, as Mohammed had forbidden the consumption of pork.

Although the agricultural potential of their territory was limited, its location conferred great commercial possibilities. The heartland lay between the Persian Gulf and the Mediterranean Sea, and it opened onto the Indian Ocean. It also contained the great caravan routes between the Mediterranean and China. Because Mohammed himself had been a merchant, Islam did not regard mercantile pursuits as inferior activities; on the contrary, merchants were regarded with honor and esteem. Although usury was forbidden, Muslim merchants devised numerous intricate credit instruments, including letters of credit and bills of exchange, to facilitate trade. For hundreds of years the Arabs and their fellow-religionists served as the principal intermediaries in the trade between Europe and Asia. In the process they greatly facilitated the diffusion of technology. Many elements of Chinese technology, including the magnetic compass and the art of making paper, reached Europe by means of the Arabs. They also introduced new crops, such as rice, sugar cane, cotton, citrus fruit, and the watermelon, among other fruits and vegetables.

The Arabs traveled and traded by both land and sea. The Arabian Sea, the northern extension of the Indian Ocean between the Arabian peninsula and the Indian subcontinent, is aptly named, for it was dominated by Arabian merchants and sailors like the legendary Sindbad. Some went as far as China, whose ports contained colonies of Muslim merchants. Muslims also used rivers for transport where possible, and supplemented them, especially in Mesopotamia, with a dense network of canals. Overland the camel, that ''ship of the desert,'' was favored for long-distance carriage, with horses, mules, and donkeys used for shorter trips. Wheeled transport disappeared from the Middle East early in the Christian era, not to reappear until the nineteenth century. Caravans of hundreds, even thousands of camels were not unusual.

One of the principles of Islam was the *jihad*, or holy war against pagans. It accounted in part for the Muslims' remarkable success in making converts, since defeated enemies were given the choice of converting or being killed. Toward Jews and Christians, however, the Muslims had a different policy. Since they were also monotheistic, the Muslims tolerated—and taxed—them (perhaps another reason for success in making converts in those communities). The Jews, in particular, enjoyed great freedom under Islam. Jewish merchants had family members or agents scattered throughout the Islamic world from Spain to Indonesia. Much of our knowledge of medieval Islam, in fact, comes from the Cairo Genizah, a great archive where any piece of paper on which the name of God had been written was deposited—and letters, even business letters between Jewish merchants, usually invoked the blessings of God.

As a result of their conquests in the Greek-speaking Eastern Roman Empire, the Arabs took over much of the learning of classical Greece. During the European Middle Ages they became, along with the Chinese, the world's leaders in scientific and philosophic thought. Many ancient Greek authors are known to us today only through Arabic translations. Modern mathematics is based on the Arabic system of notation, and algebra was an Arab invention. During the intellectual revival of Western Europe in the eleventh and twelfth centuries many Christian scholars went to Cordoba and other Muslim intellectual centers to study classical philosophy and science. At the same time, Christian merchants learned Muslim commercial practices and techniques. Although the pope officially forbad trade with Muslims, Christian merchants—the Venetians in particular—paid little attention to that dictum.

The Ottoman Empire

Among the peoples who accepted Islam as their religion were a number of nomadic Turkish tribes of Central Asia. Lured south and west by the wealth of the Arab Caliphate, they came first as raiders and looters, but eventually they settled down as conquerors. One of them, Tamerlane, known for his ruthless ferocity, conquered Persia (modern Iran) at the end of the fourteenth century. Tamerlane's empire was short-lived, but at the beginning of the sixteenth century, another conqueror, Ismael, founded the Safavid dynasty, which ruled Persia until the eighteenth century.

The most successful of the Turkish conquerors were the Ottomans, who traced their origins to the Sultan Osman (1259–1326). Osman had wrested a small territory in northwestern Anatolia (Asia Minor) from the decrepit Byzantine (Eastern Roman) Empire, which had never fully recovered from its conquest by Western crusaders and the brief tenure of the so-called Latin Empire (1204–1261). The Ottomans gradually extended their hold over all of Anatolia, and in 1354 obtained a toehold in Europe west of Constantinople, which they finally conquered in 1453 (Fig. 4–2). The Ottomans continued to expand in the sixteenth century, annexing the lands in the Near East and Middle East that the Arabs had taken earlier from the Byzantine Empire, as well as North Africa; in Europe they conquered Greece and the Balkans, and in 1683 they reached the gates of Vienna before they were driven back into Hungary.

This vast empire controlled by the Turks did not constitute a unified economy or common market. Although its many provinces had varied climates and resources, the high cost of transport prevented true economic integration. Each region within the empire continued the economic activities it had practiced before conquest, with little regional specialization. Agriculture was the principal occupation of the great majority of the sultan's subjects. The empire endured, unlike most of its predecessors, because the Turks established a regular, relatively equitable system of taxation that provided ample revenue to support the central government's bureaucracy and army. Control and order were maintained by Turkish officials stationed in the provinces and given rents from specified parcels of land, similar in some respects to medieval European feudalism.

In Europe the Turks suffered from a somewhat exaggerated reputation for rapacity and violence. In fact, they behaved rather benignly toward their subjects as long as the tax revenues rolled in and no threats of revolt or rebellion existed. They made few efforts to convert their Christian subjects in Europe to Islam, except in the

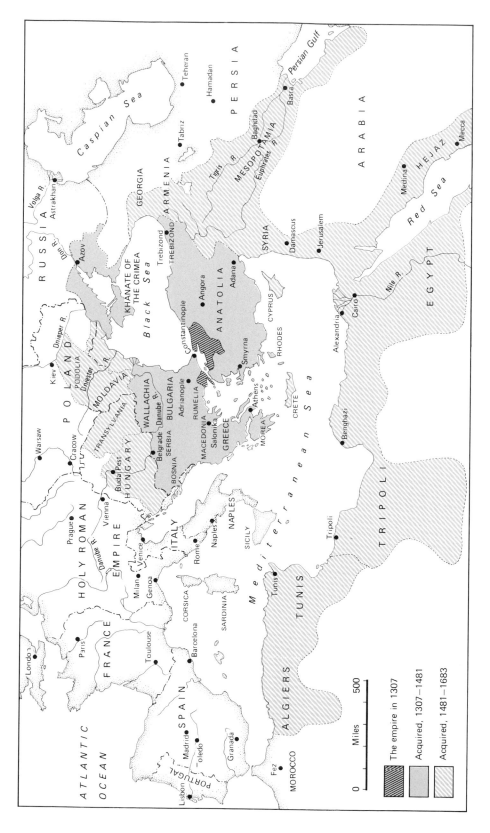

FIGURE 4–2. Growth of the Ottoman Empire, 1307–1683.

special case of the janissaries, elite soldiers who were recruited from Christian households as children and given intensive training under strict military discipline. Jews were also tolerated; when Ferdinand and Isabella expelled the Jews from Spain in 1492 (see below, p. 139) many educated professionals and skilled artisans were happy to accept service with the sultan.

East Asia

The civilization of China, dating from near the beginning of the second millennium B.C., exhibits one of the most self-contained developments of any civilization. Only rarely did foreign—"barbarian"—influences intrude, and when they did they were usually quickly absorbed by and integrated with Chinese traditions. Dynasties rose and fell, sometimes separated by periods of anarchy and "warring states," but the distinctive Chinese civilization, while continuing to develop, did so along lines that seem almost preordained. Confucianism (a philosophy, not a religion) had been fully elaborated as early as the fifth century B.C. Although other philosophies and religions, such as Taoism and Buddhism, also flourished, they did not displace Confucianism as the philosophical basis of Chinese civilization. The bureaucratic tradition of government, carried on by mandarins steeped in the Confucian philosophy, was also established at an early date. In theory, the emperor was all-powerful, and some emperors used their power to the fullest extent; but mostly their wishes were carried out, and frequently shaped, by the mandarins.

The original cradle of Chinese civilization was the middle stretch of the Yellow River valley, where the fertile loess soil deposited by the winds from Central Asia permitted easy cultivation. Its first material basis was millet, a grain native to the region; this was later supplemented by wheat and barley from the Middle East, and still later by rice from Southeast Asia. Chinese agriculture has always been extremely labor intensive, almost "garden-style," making extensive use of irrigation. Draft animals were not introduced until very late. About A.D. 1000, however, a superior variety of rice was introduced that permitted double-cropping (i.e., planting two crops a year on the same land), which greatly increased productivity.

On the basis of this productive agriculture, some urban growth occurred, and a variety of skilled crafts emerged. For example, bronze working was developed to a very high level. The manufacture of silk cloth originated in China at a very early date; the ancient Romans obtained it over the caravan route through central Asia, the Great Silk Road, and China was known to them as Sina or Serica (the land of silk). Porcelain ("chinaware") is also a Chinese invention, as are paper and printing. (The Chinese were already using paper money when Charlemagne minted the first silver pennies. The result, predictable for an economist, was overissue and inflation. The Chinese had already experienced several cycles of inflation and monetary collapse before the West discovered paper money.) The magnetic compass, first used by the Chinese, probably reached the West by way of the Arabs. In general, the Chinese reached a fairly high level of scientific and technical development well in advance of the West.

In spite of its technological and scientific precocity, China did not experience a technological breakthrough into an industrial era. Craft products were destined for the use of the government, the imperial court, and the thin stratum of landowning aristocrats. The peasant masses were much too poor to provide a market for such exotic wares. Even iron, in whose production the Chinese also excelled, was used only for weapons and decorative art, not for tools. Moreover, merchants and commerce had very low status in the Confucian philosophy. Those few merchants who did accumulate some wealth used it to buy land and join the ranks of the aristocracy.

Meanwhile, because of its fertile population as well as its fertile land, the Chinese population grew and spread. From an estimated 50 million around the year A.D 600, the population roughly doubled in the next 600 years. It spread down the Yellow River to the sea, and southward to the Yangtse valley and beyond. Whereas in the seventh century about three-quarters of the population lived in northern China, by the beginning of the thirteenth century more than 60 percent lived in central and southern China. To connect these centers, the government built an elaborate network of roads and especially canals. The Grand Canal, connecting the Yellow and Yangtse rivers, was a stupendous engineering feat. The main purpose of this transportation network was to enable the government to maintain order and collect taxes and tribute, but it also facilitated interregional trade and led to an elementary geographical specialization of labor.

In the thirteenth century a series of events occurred that profoundly affected not only China but virtually the entire Eurasian land mass, including Western Europe. This was the eruption of the Mongols under Genghis Khan from their homeland of Mongolia, north of China (Fig. 4–3). In little more than half a century Genghis and his successors created the largest continuous land empire the world has ever seen, stretching from the Pacific Ocean in the east to Poland and Hungary in the west. In the process they established their countrymen as rulers in Central Asia, China, Russia, and the Middle East. (They overturned the Arab Caliphate in 1258 and left Baghdad in ruins.) Although their name is almost synonymous with rapine and violence, the Mongols did what barbarian conquerors usually do: they settled down and adopted the civilization of their conquered hosts. In Central Asia and the Middle East they converted to Islam and melded with their Turkic allies and the local indigenous populations. In Russia, however, they did not adopt Orthodox Christianity, but maintained their own distinctive life-style until 1480 when the Grand Duke of Moscow, Ivan III, revolted and threw off the "Mongol yoke." In China they followed a middle course; they set themselves up as the Yuan dynasty (1260–1368) in the Chinese fashion and adopted Chinese ways, but attempted to maintain their ethnic distinctiveness, which led to their overthrow after little more than a century.

It was Kublai Khan, Genghis's grandson, whom Marco Polo encountered on his epic journey. By that time the Mongols had given up their warlike ways, and maintained peace and order throughout their domain. In the thirteenth century trade between the Mediterranean and China flourished even more than in the days of the Roman Empire—more in fact than it would again until the nineteenth century. Another Italian trader, a contemporary of Polo, described the Great Silk Road as "perfectly safe by day and night."

The Ming dynasty (1368–1644) reestablished traditional Chinese customs, es-

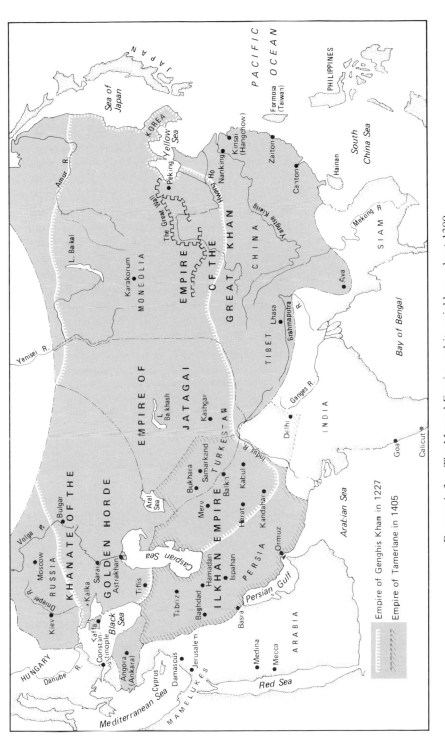

FIGURE 4–3. The Mongol Empire and its neighbors, about 1300.

Empire of Genghis Khan in 1227
Empire of Tamerlane in 1405

pecially Confucianism and the mandarin system. The first half of the Ming era also witnessed considerable economic and demographic growth. During the last years of Mongol rule, and during the revolt against the khans, roads and canals had fallen into disrepair, and the population had declined as a result of floods, drought, and warfare. The government moved energetically to restore the transportation links, and with relative peace the population began to grow again, surpassing 100 million by about 1450. In 1421 the Mings moved their capital from Nanking to Peking (Beijing) in the far north, thus stimulating north-south trade. The cultivation of cotton and the manufacture of cotton cloth were introduced. Regional specialization became more pronounced. Most remarkable of all, the Chinese began to trade overseas. Previously the Chinese had left foreign trade in the hands of foreign merchants, but in the early years of the Ming era Chinese ships and merchants traded with Japan, the Philippines (as they later became known), Southeast Asia, the Malay peninsula, and Indonesia. In the first quarter of the fifteenth century a Chinese admiral, Cheng-ho, led large naval expeditions into the Indian Ocean. The expeditions left colonies of Chinese settlers at ports in Ceylon, India, the Persian Gulf, the Red Sea, and the east coast of Africa. Then suddenly, in 1433, the emperor forbad further voyages, decreed the destruction of ocean-going ships, and prohibited his subjects from traveling abroad. The colonies were left to wither away. One wonders how the course of world history might have differed had the Chinese still been in the Indian Ocean when the Portuguese arrived at the end of the fifteenth century.

Korea and Japan developed in the wake of Chinese civilization, and in large measure in imitation of it. Japan, in particular, was a great imitator of Chinese technology, although, as in more recent times, the acquisition of a foreign technology within the Japanese institutional framework produced novel results. Korea was in political vassalage to China from time to time. Kublai Khan attempted an invasion of Japan from Korea, but his fleet was destroyed by a typhoon, which the Japanese called *kamikaze* ("divine winds"). In the fifteenth and sixteenth centuries Japanese pirates ravaged the coast of China. Early in the seventeenth century, however, after the Tokugawa shogunate had consolidated its rule, the shogun, in imitation of the Ming emperor, prohibited Japanese from traveling abroad (on penalty of death if they returned) and forbad the construction of ocean-going ships.

South Asia

The Indian subcontinent, including modern Pakistan, Bangladesh, and Sri Lanka, is roughly the same size as Europe west of the former Soviet Union (Fig. 4-4). Its population is even more diverse than that of Europe in terms of ethnic origin and language. The terrain and climate are equally varied, from tropical monsoon forest to burning desert and glacial mountains. Throughout its history, from the first civilization on the Indus River in the third millennium B.C. to the present, principalities, kingdoms, and empires have risen and fallen in bewildering array. For the most part this succession of political states has mattered little to common men and women, the peasants whose labor supported the rulers, except that some of

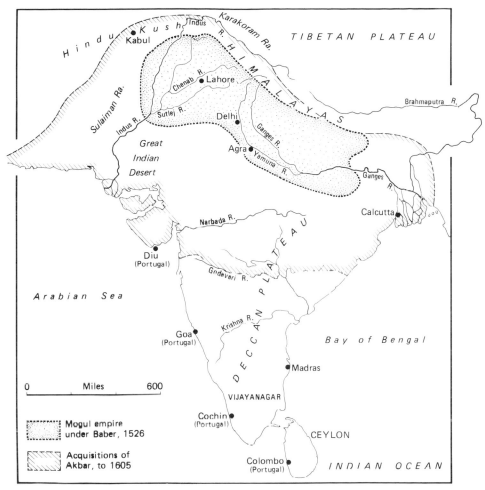

FIGURE 4–4.   India, about 1600.

those rulers were more ruthless and efficient than others in extracting a surplus of tribute and taxation.

The aboriginal population of the subcontinent may have been related to that of Australia. Over the centuries and millennia, however, it was reinforced—or overwhelmed—by waves of migrants and invaders. Most of the newcomers—Bactrian Greeks, Scythians, Parthians, and Mongols, among others—came from the northwest by way of Persia or Afghanistan, but some also came from the northeast from Tibet and Burma. Eventually, with one major exception, the Muslims, these newcomers adopted native ways and local culture, including religion.

Religion had a greater impact on the economy than government, but the complexities of that subject defy succinct explanation. The primitive original religion was Hinduism, which developed with many variants and heterodox sects, including the Jains and Sikhs, who are still active today. Buddhism, whose origin was roughly contemporaneous with that of Confucianism in China, was one such variant; but it had its greatest successes in China, Korea, and Japan, and virtually disappeared from India before modern times. Such was not the case with Islam, which first entered the subcontinent early in the eighth century, then again with renewed dynamism in the thirteenth century and later. Early in the sixteenth century Babar, who claimed descent from Genghis Khan, created the Mughal or Mogul Empire in northern India, which his grandson, Akbar, greatly enlarged (Fig.4–4).

Today the political borders between India and its neighbors do not reflect clear-cut religious divisions, and in former times the admixture of religions was even greater. The enmity between the Muslim kingdoms of the Deccan, in southern India, and the Hindu empire of Vijayanagar facilitated the establishment of bases by the Portuguese at the beginning of the sixteenth century.

One way in which religion impinged on the economy was the caste system of the Hindus. Castes were determined primarily by occupation, but originally there seems to have been an ethnic element as well. In the beginning there were only four *varnas*, or caste orders: the Brahmans, or priestly order; an order of warriors and rulers; one of farmers, artisans, and merchants; and a lowly order of servants. In time, however, the numbers of castes multiplied until there was one (or more) for every occupational category. The hierarchical element in the caste system was very strong, with rigid strictures on social and even physical mingling. Endogamy within a caste was virtually universal. In general, the rule that governed status was the concept of pollution, both literal and figurative: the most polluted occupations had the lowest status, with some being "untouchable" and even "unseeable" (for example, those who washed the clothing of the untouchables had to work at night so they would not be seen). Although the caste system was probably not as rigid as it is sometimes portrayed, it must have been a barrier to both social mobility and the efficient allocation of resources. Another element of the Hindu religion inimical to economic growth was the veneration of cattle—the "sacred cows" that roamed the countryside at will and could not be killed or consumed.

Throughout the ages, and still today, the great majority of the population of the subcontinent lived in villages and engaged primarily in low-productivity, near-subsistence agriculture. Even in relatively recent times, people living in heavily forested areas used a slash-and-burn farming technique, as practiced in northern Europe before the advent of sedentary communities. Elsewhere agricultural techniques, and the crops grown, depended on the characteristics of soil and climate. In the monsoon area the staple crop has been rice, initially obtained from Indochina. In drier lands the staple was wheat or barley, which came from the Middle East, or millet from China or perhaps western Asia. India's authentic native crop was cotton, which is mentioned in the Rig Veda, the Hindu holy book.

Although the majority of the population devoted its time and energy to agriculture, India did not lack skilled craftsmen. Intricate works of art, statuary, and monumental architecture (the Taj Mahal [Fig. 4–5], for example), all of which bear comparison with the best of Greek and Roman art, are the evidence. These craftsmen, however, worked for the rich and powerful; the masses had no purchasing power, and there was no middle class worth mentioning. The little commerce that existed was in the hands of foreigners, mainly Arabs.

FIGURE 4 5. The Taj Mahal. This elegant building, regarded by many as the most beautiful in the world, was constructed on the orders of a seventeenth-century Mughal emperor as a mausoleum for his wife. Thousands of artisans and laborers worked for more than ten years to build it. (Berkson/Art Resource, New York.)

Southeast Asia, from Burma in the northwest to Vietnam in the east and the Malay peninsula in the south, is also known as Indochina because its culture is a blend of Chinese and Indian cultural traditions. It obtained many of its elements of technology and economy from China, but, with the possible exception of Vietnam, the Indian cultural influence was probably greater. Indonesia, as its name implies, was also strongly influenced by India, at first by Hindu and Buddist culture, and later by Islam. Among the agents of diffusion, Buddist monks established monasteries in the wilderness that performed functions not unlike those of the Cistercians in northern Europe, diffusing advanced technology as well as religious culture.

Southeast Asia, including Indonesia, made two major contributions to world civilization: rice, which in time became the staple food not only of China and India but also of large areas of both the Eastern and Western Hemispheres, originated in continental Indochina; and spices—pepper, nutmeg, ginger, cloves, etc.—mostly came from the islands of the Indonesian archipelago, although cinnamon originated in Ceylon.

The recorded history of Southeast Asia is relatively brief, scarcely more than a thousand years old. For earlier eras, historians must rely on archeological evidence, such as the magnificent temple of Angkor Wat in Cambodia, and on inferences from Indian and Chinese records. The bulk of the population lived in the great alluvial valleys of rivers such as the Irrawaddy, the Red, and the Mekong, among others, where irrigated rice cultivation provided subsistence; and on the rich volcanic soils of islands like Java and Bali. Fish from the rivers and seas was another important item in the diet, and figured in local trade in exchange for rice. Pepper and more exotic spices from the Moluccas, the fabled "Spice Islands," had long found markets in India, China, the Middle East, and even Europe. Muslims—Arabs and others—were the principal intermediaries between Indonesia and India; they were also the principal agents in the spread of Islam to Indonesia (all but Bali, which remained true to its Hindu tradition). From India the cargoes were carried by Arabs to Alexandria and other emporia of the eastern Mediterrean where they were sold to Italian merchants, chiefly Venetians, who distributed them in Europe. The desire to circumvent this "monopoly," as other Europeans saw it, was one of the main motives for Portuguese exploration, which led to the discovery of the sea route around Africa.

Africa

The history of North Africa is intimately related to the history of Europe, especially Mediterranean Europe, from the earliest times to the present. Sub-Saharan Africa (black Africa), on the other hand, rarely impinged on European or other world events before the sixteenth or even the nineteenth century. The almost total lack of written records before the arrival of Europeans makes its history problematical. This does not mean, however, that it had no history, or that its history is unimportant. Recent scholarship, using archeological evidence and oral tradition, has discovered a great deal of useful information about the "dark continent."

The recorded history of Africa begins with ancient Egypt, mentioned briefly in Chapter 2. The Phoenicians plied the North African coast, and their colony, Carthage, vied with Rome for control of the Mediterranean. The sudden onslaught of Islam almost transformed that sea into a Muslim lake for a short time in the early Middle Ages. Although separated from Europe by religion as well as by water—the former an impediment to communication and commerce as the latter was a facilitator—North Africa nevertheless continued to play a role in European as well as Islamic and African history. Indeed, it was as a result of Islamic conversions of the sub-Saharan fringe of black Africa that the latter first made contact with the European economy. (Christianity had penetrated Nubia and Abyssinia, or Ethiopia, before the rise of Islam; but thereafter, with the Islamic conquest of Nubia, Abyssinia was effectively cut off from the rest of Christendom.)

The economy of North Africa was similar to that of Mediterranean Europe.

Grain growing predominated where rainfall was adequate (sometimes supplemented by irrigation), and nomadic pastoralism elsewhere. Commerce was lively, but industry was of the household variety. One branch of commerce extended across the Sahara to black Africa. Some trans-Saharan commerce had existed before the Christian era, but it did not become common until camels were introduced (from the Middle East) in the second or third century A.D. Even then the expense of travel restricted commerce to items of high value and low bulk—mainly gold and ivory—and slaves, who were self-propelled. Dates grown in the date palm groves of desert oases were also carried in both directions.

The economy of sub-Saharan Africa is as varied as its climate, topography, and vegetation. Contrary to popular impression, only a portion of the area, mainly in the Congo (or Zaire) basin and the southern coast of West Africa, is covered with tropical rain forests or jungle. Between that and the deserts to the north (Sahara) and south (Kalihari) are vast stretches of savanna—grass and scrub. Inland from the east coast, reaching from Ethiopia in the north to the southern tip of the continent, is a mountainous spine punctuated with great lakes. The great rivers of Africa—the Nile, the Niger, the Zambezi, and others—did not encourage the development of commerce as much as might be expected because of the frequency of falls and rapids.

The population was even more varied than the landscape. Although all the original inhabitants were dark-skinned, or black, there was enormous ethnic, racial, and linguistic variety. Everywhere, however, the tribe was the basic social group above the family. Occasionally larger polities—confederations, kingdoms, and even empires—arose, some like the ancient empire of Ghana lasting for surprising periods of time; but without written records, that necessity of a bureaucratic state, most were quite ephemeral.

The economy was also varied, ranging from the most primitive hunting and gathering to fairly sophisticated field agriculture and stock raising in the savanna and other open spaces. Domesticated plants and animals may have been introduced from Egypt or elsewhere in the Mediterranean as early as the second millennium B.C. Owing to differences in climate and rainfall, the staples of Mediterranean and Middle Eastern agriculture, wheat and barley, did not flourish in sub-Saharan Africa. Because of the prevalence throughout central Africa of the tsetse fly, which carries disease fatal to large domestic animals, the cultivators had no draft animals; thus they relied on hoe culture, using wooden or iron hoes. In the jungle areas they used a slash-and-burn farming technique, moving their fields every few years; they cultivated root crops and bananas (introduced from Southeast Asia, and subsequently diffused to America), and they supplemented their diet with fish from the rivers. Although the level of technology was generally low, that did not prevent the emergence of a specialized caste of ironworkers, for example, or of professional traders.

Trade and commerce were almost ubiquitous, even among hunters and gatherers insofar as they had contact with other social groups. The nomads of the *sahel*, the arid southern fringe of the Sahara, bartered the produce of their flocks—meat, milk, and wool—for the grain, cloth, and metals of the sedentary peoples of the savanna. Salt and fish (dried and salted) were other objects of commerce. In East Africa cowry shells were used as money to obviate the need for barter. Canoes were widely used for porterage on rivers. Elsewhere head porterage sufficed.

The Americas

Scholars generally agree that the native population of the Americas (Amerindians) descended from a Mongoloid (or pre-Mongoloid) people who, at some time in the distant past, crossed a land bridge from Asia to North America over what is now the Bering Strait. There is less agreement on when this occurred; estimates range from a few thousand to more than 30,000 years ago. Moreover, it is unlikely that only a single wave of migration occurred; more likely, the migrations occurred in waves over a period of thousands of years. Ingenious theories have been proposed and even tested to prove that the aboriginal inhabitants might have come by sea across the Atlantic or Pacific; but even if one or a few such voyages had been successful, it is unlikely that the entire pre-Columbian population of the Americas, widespread and speaking many different languages, could have descended from the survivors of those voyages.

Well before the beginning of the Christian era in the Old World, the New World was populated from what are now Canada and Alaska in the north to Patagonia and Tierra del Fuego in the south. The density of population and the level of culture varied considerably, however, from the sparsely populated great plains of North America and the South American Amazonian jungle to the teeming cities of Middle America. Population density varied directly with the productivity of the economy; it was greatest in those areas that practiced settled agriculture and lightest where the inhabitants still lived by hunting and gathering.

The Amerindians had discovered agriculture independently from that of the Old World, but not all of them practiced it. It was most highly developed in Mexico, Central America, and northwestern South America, but it also existed in what is now the southwestern United States and in the eastern woodlands of North America. The staple crop was maize (Indian corn), supplemented by tomatoes, squash, pumpkins, and beans, and, in the Andean highlands, the potato. The Amerindians had no domesticated animals except the dog and, in the Andes, the llama, which could be used as a pack animal but not as a draft animal. The farm technology was thus hoe culture. The Amerindians also had few metals—some alluvial gold used for ornaments, silver, and copper, but no iron. Their tools were made of wood, bone, stone, and especially obsidian, a natural volcanic glass used for cutting and carving. In spite of their apparently primitive technology, they produced some intricate and elaborate works of art as well as monumental architecture.

Markets and trade also existed from an early date. Archeological evidence of long-distance trade dates from the middle of the second millennium B.C. Between the eighth and fourth centuries B.C. the Olmec culture, located along the coast of the Gulf of Mexico, carried on commerce with the central highland area of Mexico. Objects of this commerce included finely carved statuettes and other art objects made from jade and the highly prized obsidian, as well as cacao beans, which were used as a form of money as well as for consumption.

The Mayan civilization of modern Guatemala and Yucatan emerged at about this time or a little later. Its most striking features were large pyramids, not unlike those of Egypt, but surmounted by temples (Fig. 4–6). The Mayans also had a calendar and a form of writing that has only recently been deciphered. Little is known about the organization of the society and the economy, but, as elsewhere, maize was the food staple, and markets were common. Society must have been organized hierarchically to produce its monumental architecture, and the food surplus must have been substantial to free a work force of builders and skilled artisans. Mayan civilization was at its peak between the fourth and ninth centuries of the Christian era. Then the population apparently revolted against their priestly rulers, possibly aided by invaders from the north. The temples, deserted by the faithful, fell into ruins and were overgrown by the surrounding jungle.

Following the Mayas, various other cultures in the highlands of Mexico reached fairly high levels of development. These included the Toltecs, the Chichimecs, and the Mixtecs. About the middle of the fourteenth century the Aztecs, a fierce, warlike tribe whose capital city was Tenochtitlán, the site of modern Mexico City, began to conquer and exploit their neighbors. Since the Aztecs practiced human sacrifice, selecting the victims from among the subject population, it is not surprising that the Spaniards under Cortez found willing allies when they undertook the conquest of Tenochtitlán in 1519.

When the Mayan civilization was at its peak natives along the coast of modern Peru practiced irrigation agriculture using water from the Andes, a technique unknown elsewhere in the Americas. Evidently it was relatively highly productive, because it permitted the growth of dense urban populations who traded among themselves. Some time after about A.D. 1200 the Incas, a highland tribe with their

FIGURE 4–6.   Mayan temple. This imposing edifice is evidence of the wealth and power of the Mayan rulers, as well as the technical skill of their artisans; but the burden of their yoke was heavy, and their subjects eventually revolted and returned to subsistence agriculture. (Ferdinand Anton. Reprinted with permission.)

capital in Cuzco, began a military conquest of the entire highland and coastal region from Ecuador in the north to Chile in the south. Although the Incas had no written language, they were able to keep records and even convey messages over great distances by means of knotted strings. They imposed a highly centralized state bureaucracy over their subjects, including state-owned warehouses for the storage and distribution of grain; but private markets coexisted with the government distribution system.

The Pueblo Indians of the southwestern United States also practiced agriculture and built urban settlements that deserve the designation of towns if not cities. The eastern woodland Indians, who inhabited the area east of the Mississippi River from the St. Lawrence River in the north to the Gulf of Mexico in the south, practiced agriculture along with hunting and fishing, but they lived in villages rather than towns. According to legend, Indians taught the Puritans of New England to fertilize their corn by burying fish with the seeds, which greatly increased the yield.

Elsewhere in the Americas, from the Eskimos of the Arctic Ocean shores to the naked inhabitants of Tierra del Fuego, the vast but thinly populated continents provided a bare subsistence for primitive hunters and gatherers.

5

Europe's Second Logistic

Sometime around the middle of the fifteenth century, after a century of decline and stagnation, Europe's population began to grow once more. Neither the revival nor rates of growth were uniform throughout Europe (as always, there was regional diversity), but by the beginning of the sixteenth century the demographic increase was generalized. It continued unabated throughout the sixteenth century, possibly even accelerating in the latter decades. Early in the seventeenth century, however, this lusty growth encountered the usual checks of famine, plague, and war, especially the Thirty Years War, which decimated the population of central Europe. By the middle of the seventeenth century, with a few exceptions, notably Holland, the population growth had ceased and in some areas actually declined. These termini—roughly the middle of the fifteenth and the middle of the seventeenth centuries—delimit Europe's second logistic. Within them, other important changes, some probably fortuitous and others intimately related to the demographic phenomena, occurred. At the latter date the European, and the world, economies were vastly different from what they had been in the fifteenth century.

The most obvious difference was the greatly expanded geographical horizons. The period of demographic increase corresponded almost exactly with the great age of maritime exploration and discovery that resulted in the establishment of all-water routes between Europe and Asia and, even more momentous for world history, the conquest and settlement of the Western Hemisphere by Europeans. These events, in turn, provided Europe with a greatly expanded supply of resources, both actual and potential, and provoked (together with other causes) significant institutional changes in the European economy, especially with respect to the role of government in the economy.

Another major difference was a pronounced shift in the location of the principal centers of economic activity within Europe. In the fifteenth century the cities of northern Italy retained the leadership in economic affairs they had exercised throughout the Middle Ages. The Portuguese discoveries, however, deprived them of their monopoly of the spice trade. A series of wars involving the invasion and occupation of Italy by foreign armies further disrupted commerce and finance. The decline of Italy was not immediate or drastic, for the Italians had reservoirs of capital, entrepreneurial talent, and highly refined economic institutions to carry them for several generations. In any case, Italy's decline was probably more relative than absolute, because of the great increase in the volume of European commerce.

Nevertheless, by the middle of the seventeenth century Italy had fallen into the backwaters of the European economy, from which it did not fully emerge until the twentieth century.

Spain and Portugal enjoyed a fleeting glory as the leading economic powers of Europe. Lisbon replaced Venice as the great entrepôt of the spice trade, and the Spanish Habsburgs, financed in part by the gold and silver of their American empire, became the most powerful monarchs in Europe. The wealth of the Indies and the Americas was not widely shared within the countries, however; as a result of policies to be described and analyzed in greater detail subsequently, the governments of those countries wasted their resources and stifled the development of vigorous and dynamic economic institutions. Although both nations retained their extensive overseas empires until the nineteenth and twentieth centuries, respectively, they were already in full decline, economically, politically, and militarily, by the middle of the seventeenth century.

Central, eastern, and northern Europe did not participate significantly in the commercial prosperity of the sixteenth century. The German Hansa flourished in the fifteenth century but declined thereafter. Although the main causes of its decline were independent of the great discoveries, the latter probably hastened the decline by strengthening the commercial power of Dutch and English cities. Southern Germany and Switzerland, which had also become commercially prominent in the fifteenth century, retained their prosperity for a time; but since they were no longer on the most important trade routes and had no ports to benefit from the increase in seaborne trade, they slipped backward, relatively speaking, along with the rest of central and eastern Europe. All of central Europe soon plunged into religious and dynastic wars that sapped its energy for economic activity.

The area that gained most from the economic changes associated with the great discoveries was the region bordering on the North Sea and the English Channel: the Low Countries, England, and northern France. Opening on the Atlantic and lying midway between northern and southern Europe, this region prospered greatly in the new era of worldwide oceanic commerce. Throughout the sixteenth century, however, France also engaged in dynastic and religious wars, civil and international, and for the most part its government followed policies unfavorable to business and agriculture. France therefore gained less than the Netherlands and England.

England at the time of the great discoveries was just emerging from the status of a backward, raw materials-producing area into something of a manufacturing country. Its agriculture was also becoming more market-oriented. The Wars of the Roses decimated the ranks of the great nobility but left the urban middle classes and peasants almost untouched. The decline of the great nobility enhanced the importance of the lesser aristocracy, the gentry. The new Tudor dynasty, which came to the throne in 1485, depended heavily on gentry support and granted it favors in return. For example, when Henry VIII revolted from the Roman church and decreed the dissolution of the monasteries, the gentry were the principal beneficiaries, after the crown itself. The action also had the incidental effect of improving the operation of the land market and encouraging the market orientation of agriculture.

Flanders, already the most economically advanced area in northern Europe, recovered slowly from the great depression of the late Middle Ages. Bruges gradually declined as the principal entrepôt for trade with southern Europe, and Antwerp rose to become the most important port and market city in Europe in the first half of the sixteenth century. As a result of dynastic alliances all seventeen provinces of the Low Countries, from Luxembourg and Artois in the south to Friesland and Groningen in the north, fell to the crown of Spain early in the sixteenth century. They were thus in an excellent position to capitalize on the trading opportunities of the Spanish Empire. In 1568, however, the Netherlands revolted against Spanish domination. Spain suppressed the revolt in the southern provinces (modern Belgium), but the seven northern provinces won their independence as the United Netherlands, or Dutch Republic. Economically this episode resulted in a relative decline of the southern provinces, partly because the Spanish government enacted many harsh punitive measures and partly because the Dutch, who controlled the mouths of the Scheldt River, prevented ships from going to Antwerp. Trade shifted to the north, and Amsterdam became the great commercial and financial metropolis of the seventeenth century.

Technological changes in the arts of navigation and shipbuilding were vital to the success of exploration and discovery. The introduction of gunpowder and its application by Europeans to firearms were similarly vital to the success of European conquests overseas. There were concurrent improvements in the arts of metallurgy and some other industrial processes. On the whole, however, the period is not notable for its technological progress. In particular, no major breakthroughs occurred in agricultural technology, such as the introduction of the three-field system and the heavy wheeled plow, although a host of minor improvements were made in crop rotation, new crops, and similar matters.

Population and Levels of Living

In the middle of the fifteenth century the population of Europe as a whole was on the order of 45 or 50 million, that is, about two-thirds of its preplague peak. By the middle of the seventeenth century, authorities agree, the population was in the vicinity of 100 million. In 1600 it must have been at least as large if not larger, in view of the stagnation and possible decline that occurred in the first half of the seventeenth century. What caused this growth and the renewed stagnation and decline?

No single obvious cause for the renewal of population growth presents itself. The incidence of the plague and other epidemic illnesses apparently diminished gradually, possibly as a result of increasing natural immunization or of ecological changes affecting the carriers. The climate may have ameliorated slightly. Higher real wages in the fifteenth century, the consequence of the favorable shift in the population/land ratio as a result of the earlier decline in population, may have encouraged earlier marriages and thus a higher birth rate. In any event, through some combination of reduced death rates and higher birth rates, Europe's population began a sustained increase that continued throughout the sixteenth century, even after the initial favorable conditions had changed.

The growth in population in the sixteenth century, although general, was by no means uniform. Beginning with unequal densitites, and growing at different rates, the populations of the various regions of Europe varied considerably in density at the end of the sixteenth century. Italy, a "mature" economy, and the Netherlands, a dynamic one, had the greatest densities, with 40 or more persons per square kilometer, although some areas, such as Lombardy and the province of Holland had 100 or more. (For purposes of comparison, Italy in recent years had about 190 persons per square kilometer, the Netherlands about 350; the density of western Europe as a whole is about 125 per square kilometer.) France, with approximately 18 million people, had a density of about 34; England and Wales, with 4 or 5 million, had slightly less. Elsewhere the population was spread more thinly: 28 per square kilometer in Germany, 17 in Spain and Portugal, 14 in eastern Europe exclusive of Russia, and only about 1.5 or 2 in Russia and the Scandinavian countries.

As indicated in Chapter 3, these figures clearly show that population density was closely related to the productivity of agriculture. Similar differences are found within countries. For example, Württemberg, one of the most advanced agricultural regions of Germany, had a density of 44 per square kilometer. Southern England was far more densely populated than Wales or the north country, and northern France and the Mediterranean coastal regions of Provence and Languedoc more than the mountainous and infertile Massiv Central. The sparsely populated plateaus of Aragon and Castile contrasted with the teeming valleys and lowlands of Andalusia and Valencia, as did the Appenines and Alpine regions of Italy with the Po valley and the Roman Campagna. Nevertheless, it is possible to speak of over-population of even the mountainous and infertile regions by the latter part of the sixteenth century. Streams of migrants from those regions to the already more densely populated but more prosperous plains and lowlands constitute the evidence. But the plains and lowlands were also overpopulated. In some areas tenures were divided as more and more people sought to make a bare subsistence from the land. In others, the surplus population left the countryside, voluntarily or otherwise. The literature of Elizabethan England carries frequent references to "sturdy beggars" on the highways and in city streets, beggars whose poverty often drove them to crime. For Spain and Portugal, their colonial empires provided an outlet for the excess population—indeed, there were even complaints of labor shortages—and in northern Europe the acquisition of colonies was advocated as a means of dealing with the surplus population. For Europe as a whole, however, overseas migration in the sixteenth and seventeenth centuries was almost negligible; most migrations were domestic, even local.

One consequence of those migrations was that the urban population grew more rapidly than the total. The populations of both Seville and London tripled between 1500 and 1600 (to about 150,000 in both cases), that of Naples doubled (to perhaps 250,000). Paris, already the largest city in Europe with more than 200,000, also increased to about a quarter of a million. Amsterdam grew from about 10,000 at the end of the fifteenth century to more than 100,000 in the early decades of the seventeenth century. (All of these figures are approximate.) Although the percentage rise in the urban population was also general, it was more pronounced in northern Europe than in the Mediterranean lands, which were already more urbanized at the beginning of the period. By the end of the sixteenth century about one-third of the population of Flanders and almost half that of Holland lived in towns and cities.

In some instances an increase in the urban population can be regarded as a favorable indicator of economic development, but this was not necessarily so in the sixteenth century. At that time towns functioned primarily as commercial and administrative rather than industrial centers. Many manufacturing activities, as in the textile and metallurgical industries, took place in the countryside. The handicrafts practiced in the towns were usually organized in guilds, with long apprenticeship requirements and other restrictions on entry. The rural migrants rarely had the skills or aptitudes necessary for urban occupations. In the towns they formed a *lumpenproletariat*, a pool of casual, unskilled labor, frequently unemployed, who supplemented their meager earnings by begging and petty thievery. Their crowded, dirty, and squalid living conditions endangered the whole community by making it more susceptible to epidemic disease.

The plight of both the urban and rural poor was aggravated by a prolonged fall in real wages. Because the population grew more rapidly than agricultural output, the price of foodstuffs, bread grains in particular, rose more rapidly than money wages, a situation that was exacerbated by the phenomenon of the "price revoluton" (see The Price Revolution section, later in this chapter). By the end of the sixteenth century the pressure of population on resources was extreme, and in the first half of the seventeenth century a series of bad harvests, new outbreaks of the bubonic plague and other epidemic diseases, and increased incidence and ferocity of warfare brought the population expansion to a halt. In several areas of Europe, notably Spain, Germany, and Poland, population actually declined during part or all of the seventeenth century.

Exploration and Discovery

There is no reason to suppose that there was any intimate causal relationship between the demographic phenomena in Europe and the maritime discoveries that led to the establishment of direct commerce between Europe and Asia and the conquest and settlement of the New World by Europeans. Population growth was already underway before the significant discoveries occurred, extra-European commerce in the sixteenth and seventeenth centuries was minor in comparison with intra-European commerce, and the importation of foodstuffs (other than spices) was negligible. Nevertheless, the discoveries profoundly affected the course of economic change in Europe.

Notable technological progress in ship design, shipbuilding, and navigational instruments occurred in the later Middle Ages. Three-, four-, and five-masted ships, with combinations of square and lateen sails capable of sailing across the wind, replaced the oared galleys with auxiliary sails of medieval commerce. The hinged sternpost rudder replaced the steering oar. In combination, these changes provided far greater maneuverability and directional control and dispensed with oarsmen. Ships became larger, more manageable, more seaworthy, and had greater cargo capacity, enabling them to make longer voyages. The magnetic compass, probably borrowed from the Chinese by way of the Arabs, significantly reduced the guesswork involved in navigation. Developments in cartography provided greatly improved maps and charts.

The Italians had been leaders in the art of navigation, a leadership that they did not quickly relinquish, as exemplifed by the names Columbus (Colombo), Cabot (Caboto), Vespucci, Verrazano, and others. As early as 1291 a Genoese expedition in oared galleys started down the west coast of Africa in an attempt to reach India by sea, never to be seen again. But the Italians were conservative in ship design, and the lead was soon taken by those who sailed the open sea, especially the Flemish, Dutch, and Portuguese. The Portuguese, in particular, seized the initiative in all aspects of the sailor's art: ship design, navigation, and exploration (Figs. 5-1 and 5-2). The vision and energy of one man, Prince Henry, called the Navigator, were

FIGURE 5–1.   Portuguese discoveries in the fifteenth century.

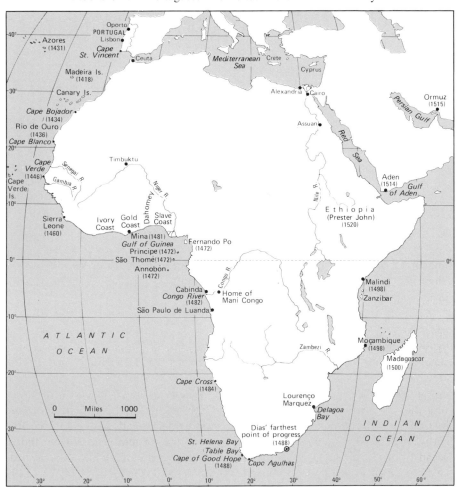

FIGURE 5–2. Portuguese carrack. These large, unwieldy ships, designed especially for the long voyage to India, replaced the smaller, more maneuverable caravels that had done most of the exploration of the African coast during the fifteenth century. (The National Maritime Museum, London.)

chiefly responsible for the great progress in geographical knowledge and discovery made by Europeans in the fifteenth century.

Henry (1393–1460), a younger son of the king of Portugal, devoted himself to encouraging the exploration of the African coast with the ultimate object of reaching the Indian Ocean. At his castle on the promontory of Sagres at the southern tip of Portugal he established a sort of institute for advanced study to which he brought astronomers, geographers, cartographers, and navigators of all nationalities. From 1418 until his death he sent out expeditions almost annually. Carefully and patiently his sailors charted the coast and currents, discovered or rediscovered and colonized the islands of the Atlantic, and established trade relations with the native chiefs of the African coast. Henry did not live to realize his greatest ambition. In fact, at the time of his death his sailors had gone little farther than Cape Verde, but the scientific and exploration work carried out under his patronage laid the foundation for subsequent discoveries.

After Henry's death exploratory activity slackened somewhat for lack of royal patronage and because of the lucrative trade in ivory, gold, and slaves that Portuguese merchants carried on with the native kingdom of Ghana. King John II, who came to the throne in 1481, renewed the explorations at an accelerated pace. Within a few years his navigators pushed almost to the tip of Africa. Realizing that he was on the verge of success, John sent out two expeditions in 1487. Down the coast went Bartholomew Diaz, who rounded the Cape of Good Hope (which he named the Cape of Storms) in 1488; through the Mediterranean and overland to the Red Sea went Pedro de Covilhão, who reconnoitered the western edges of the Indian Ocean from Mozambique in Africa to the Malabar coast of India. The way was paved for the next and greatest voyage, that of Vasco da Gama from 1497 to 1499 around Africa to Calicut in India. As a result of disease, mutiny, storms, and difficulties with both his Hindu hosts and the numerous Arab merchants he encountered, da Gama lost two of his four ships and almost two-thirds of his crew. Nevertheless, the cargo of spices with which he returned sufficed to pay the cost of his voyage many times over.

Seeing such profits, the Portuguese lost no time in capitalizing on their advantage. Within a dozen years they had swept the Arabs off the Indian Ocean and established fortified trading posts from Mozambique and the Persian Gulf to the fabled Spice Islands or Moluccas. In 1513 one of their ships put in at Canton in South China, and by midcentury they had opened trading and diplomatic relations with Japan.

In 1483 or 1484, while the crews of John II were still working their way down the African Coast, a Genoese who had sailed in Portuguese service and married a Portuguese woman asked the Portuguese king to finance a voyage across the Atlantic to reach the East by sailing west. Such a proposal was not entirely novel. General belief held that the earth was a sphere. But was the plan feasible? Christoforo Colombo (Columbus), the Genoese, thought it was, although the weight of opinion was against him. John's advisers had a more nearly correct impression of the size of the globe than did Columbus, who thought that the distance from the Azores to the Spice Islands was little more than the length of the Mediterranean. Although John had authorized privately financed expeditions west of the Azores, he concentrated his own resources on the more likely project of rounding Africa, and rejected Columbus's proposal.

Columbus perservered. He appealed to the Spanish monarchs, Ferdinand and Isabella, who were engaged at the time in a war against the Moorish kingdom of Granada, and had no money to spare for such an unlikely scheme. Columbus tried to interest the realistic and economical King Henry VII of England, as well as the king of France, but in vain. At length, in 1492, Ferdinand and Isabella conquered the Moors, and as a sort of victory celebration Isabella agreed to underwrite an expedition. Columbus set sail on August 3, 1492, and on October 12 sighted the islands later known as the West Indies. Columbus truly thought he had reached the Indies. Though dismayed by the their obvious poverty, he dubbed the inhabitants Indians. After a few weeks of reconnoitering among the islands he returned to Spain to spread the joyful tidings. The following year he returned with seventeen ships, 1500 men, and enough equipment (including cattle and other livestock) to establish a permanent settlement. Altogether Columbus made four voyages to the western seas, and persisted to the end in the belief that he had discovered a direct route to Asia.

Immediately following the return of the first expedition, Ferdinand and Isabella applied to the Pope for a "line of demarcation" to confirm Spanish title to the newly discovered lands. This line, running from pole to pole at a longitude one hundred leagues (about 330 nautical miles) west of the Azores and Cape Verde Islands, divided the non-Christian world into halves for purposes of further exploration, with the western half reserved for the Spanish and the eastern half for the Portuguese. The next year, 1494, in the Treaty of Tordesillas the Portuguese king persuaded the Spanish rulers to set the line about 210 nautical miles further west than the 1493 line. This suggests that the Portuguese may have already known of the existence of the New World, for the new line placed the hump of South America—the beachhead that later became Brazil—in the Portuguese hemisphere. In 1500, on the first major Portuguese trading voyage after da Gama's return, Pedro de Cabral sailed directly for the hump and claimed it for Portugal before proceeding to India.

Meanwhile explorers of other nations followed up the news of Columbus's discovery (Fig. 5-3). In 1497 John Cabot, an Italian sailor who lived in England, secured the backing of Bristol merchants for a voyage on which he discovered Newfoundland and Nova Scotia. The following year he and his son Sebastian led a larger expedition to explore the northern coast of North America, but since they brought back no spices, precious metals, or other marketable commodities, their commercial backers lost interest. Cabot also failed to persuade Henry VII to provide financial support, though the king did give him a modest reward of ten pounds for planting the English flag in the New World. French merchants sent another Italian, Verrazano, to discover a western passage to India in the 1520s. A decade later the Frenchman Jacques Cartier made the first of three voyages that resulted in the discovery and exploration of the St. Lawrence River. Cartier also claimed for France the area later known as Canada; but, failing to find the hoped-for passage to India, the French, like the English, evinced no further immediate interest in the New World except for fishing on the Grand Banks of Newfoundland.

In 1513 the Spaniard Balboa discovered the "South Sea," as he called the Pacific Ocean, beyond the Isthmus of Panama. By the 1520s Spanish and other navigators had explored the entire eastern coast of the two Americas from Labrador to Rio de la Plata. It became increasingly clear not only that Columbus had not discovered the Indies but that there was no easy passage through the center of the new continent. In 1519 Ferdinand Magellan, a Portuguese who had sailed in the Indian Ocean, persuaded the king of Spain to let him lead an expedition of five ships to the Spice Islands by way of the South Sea. Magellan had no thought of circumnavigating the globe, for he expected to find Asia a few days' sailing beyond Panama, within the Spanish orbit as determined by the Tordesillas Treaty. His main problem, as he saw it, was to find a passage through or around South America. This he did, and the stormy, treacherous strait he discovered still bears his name. The "peaceful sea" (Mare Pacificum) into which he emerged, however, yielded not riches but long months of starvation, disease, and eventually death for him and most of his crew. The remnants of his fleet wandered aimlessly in the East Indies for several months. At length one of Magellan's lieutenants, Sebastian del Cano, took the one surviving ship and its skeleton crew through the Indian Ocean and home to Spain after three years, becoming the first men to sail entirely around the earth.

Overseas Expansion and the Feedback to Europe

The first century of European overseas expansion and colonial conquest—that is, the sixteenth century—belonged almost exclusively to Spain and Portugal. The eminence these two nations have achieved in history is a result mainly of their pioneering in the discovery, exploration, and exploitation of the non-European

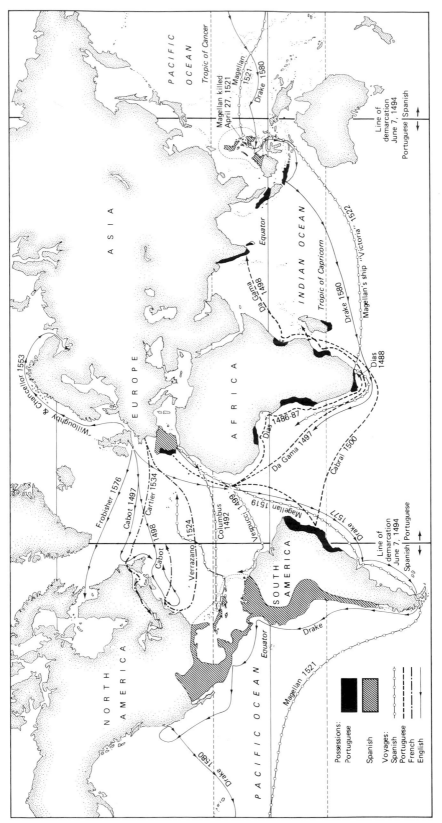

FIGURE 5–3.   World voyages of discovery, fifteenth and sixteenth centuries.

world. Before the sixteenth century they had been outside the mainstream of European civilization; afterward their power and prestige declined rapidly until, by the beginning of the nineteenth century, they had sunk into a state of somnolence approaching suspended animation. In the sixteenth century, however, their dominions were the most extensive and their wealth and power the greatest in the world.

By 1515 the Portuguese had made themselves masters of the Indian Ocean. Vasco da Gama returned to India in 1501 with instructions to halt the Arab trade to the Red Sea and Egypt, by which the Venetians had obtained spices for distribution in Europe. In 1505 Francisco de Almeida went out as first Portuguese viceroy of India. He captured or established several cities and forts on the East African and Indian coasts, and in 1509 completely destroyed a large Muslim fleet in the battle of Diu. In the same year Alfonso de Albuquerque, greatest of the Portuguese viceroys, assumed his duties and completed the subjugation of the Indian Ocean. He captured Ormuz at the entrance to the Persian Gulf and established a fort at Malacca on the narrow strait between the Malay peninsula and Sumatra, a post that controlled the passage to the Celebes and Moluccas islands, from which the most valuable spices came. Finally, in 1515 he captured Ceylon, the key to the mastery of the Indian Ocean. His attempt to capture Aden at the entrance to the Red Sea was repulsed, however, and the Portuguese were unable to maintain an effective monopoly of the spice trade for long. Albuquerque established his capital at Goa on the Malabar coast; Goa and Diu remained Portuguese possessions until 1961. The Portuguese also established trade relations with Siam and Japan. In 1557 they established themselves at Macao on the south coast of China, which they still hold. Because of their small population they did not attempt to conquer or colonize the interior of India, Africa, or the islands, but contented themselves with controlling the sea lanes from strategic forts and trading posts.

Although at first it looked less promising, the Spanish Empire eventually proved to be even more profitable than that of Portugal. Disappointed in their quest for spices and stimulated by a few trinkets plundered from the savages in the islands of the Caribbean, the Spanish quickly turned to a search for gold and silver. Their continued efforts to find a passage to India soon revealed the existence of wealthy civilizations on the mainland of Mexico and northern South America. Between 1519 and 1521 Hernando Cortez effected the conquest of the Aztec Empire in Mexico. Francisco Pizarro conquered the Inca Empire in Peru in the 1530s. By the end of the sixteenth century the Spanish wielded effective power over the entire hemisphere, from Florida and southern California in the north to Chile and the Rio de la Plata in the south (with the exception of Brazil). At first they merely plundered the original inhabitants of their existing movable wealth; when this source was quickly exhausted they introduced European mining methods to the rich silver mines of Mexico and the Andes.

The Spanish, unlike the Portuguese, undertook from the beginning to colonize and settle the areas they conquered. They brought European techniques, equipment, and institutions (including their religion), which they imposed by force on the Indian population. Besides European culture and manufactures, the Spanish introduced natural products previously unknown to the Western Hemisphere, including wheat and other cereal grains (except corn, which traveled in the opposite direction), sugar cane, coffee, most common vegetables and fruits (including citrus fruits), and many other forms of plant life. The pre-Columbian Indians of America had no domesticated animals except dogs and llamas. The Spanish introduced horses, cattle, sheep, donkeys, goats, pigs, and most domesticated fowls.

Some other features of European civilization that were introduced into America, such as firearms, alcohol, and the European diseases of smallpox, measles, and typhus, spread quickly and with lethal effect. The native population may have numbered as many as 25 million at the time of Columbus (some scholarly estimates range much higher), but by the end of the sixteenth century these killers had reduced it to but a few million. To remedy the shortage of labor the Spanish introduced African slaves to the Western Hemisphere as early as 1501. By 1600 a majority of the population of the West Indies were Africans and people of mixed races; slaves were not as important on the mainland, except in Brazil and northern South America.

The transplantation of European culture, together with the modification and occasional extinction of non-Western cultures, were the most dramatic and important aspect of the expansion of Europe. Expansion also produced feedback. European culture itself underwent substantial modifications as a result.

On the economic side expansion resulted in a great increase in the volume and variety of goods traded. In the sixteenth century spices from the East and bullion from the West accounted for an overwhelming proportion of imports from the colonial world. As late as 1594, for example, 95 percent of the value of legal exports from the Spanish colonies in the New World consisted of gold and silver bullion. Nevertheless, other commodities entered the stream of trade, gradually expanded in volume, and by the seventeenth and eighteenth centuries overshadowed the original overseas exports to Europe. Exotic dyestuffs such as indigo and cochineal added color to European fabrics and made them gayer and more salable both in Europe and overseas. Coffee from Africa, cocoa from America, and tea from Asia became staple European beverages. Cotton and sugar, although they were known earlier in Europe, had never been produced or traded on a large scale. When sugar cane was transplanted to America, the production of sugar increased enormously and brought that delicacy within the budget of ordinary Europeans. The introduction of cotton goods from India, at first a luxury reserved for the wealthy, led eventually to the establishment of one of Europe's largest industries, dependent on a raw material imported from America and catering especially to the masses. Chinese porcelain had a similar history. Tobacco, one of America's most celebrated and controversial contributions to civilization, grew rapidly in popularity in Europe in spite of determined efforts by both church and state to stamp it out. In later years tropical fruits and nuts supplemented European diets, and furs, hides, exotic woods, and new fibers constituted important additions to European supplies.

Many foodstuffs previously unknown in Europe, although not imported in large quantities, were introduced and naturalized, eventually becoming important staples of diet. From America came potatoes, tomatoes, string beans, squash, red peppers, pumpkins, and corn (called maize by Europeans), as well as the domesticated turkey, which in spite of its name reached Europe from Mexico. Rice, originally from Asia, became naturalized in both Europe and America.

The Price Revolution

The flow of gold and especially silver from the Spanish colonies greatly increased Europe's supplies of the monetary metals, at least tripling them in the course of the sixteenth century. The Spanish government attempted to forbid the export of bullion, but this proved impossible. In any case, the government was its own worst offender, sending vast quantities to Italy, Germany, and the Netherlands to repay its debts and finance its interminable wars. From those countries, as well as from Spain itself in contraband movements, the precious metals spread throughout Europe. The most immediate and obvious result was a spectacular and prolonged (but irregular) rise in prices. By the end of the sixteenth century prices were, in general, about three or four times higher than they had been at the beginning. Of course, the rise in price varied greatly from region to region and by commodity groups. Prices rose sooner and higher in Andalusia, whose ports were the only legal entrepôts for American gold and silver, than in distant and backward Russia. The price of foodstuffs, especially grain, flour, and bread, rose higher than those of most other commodities. In general, the rise in money wages lagged far behind the rise in commodity prices, resulting in a severe decline in real wages.

The phenomenon of the price revolution has given rise to innumerable, seemingly endless, and mostly unnecessary scholarly discussions concerning its mechanisms, consequences, and even its causes. It has been pointed out that a rise in silver production in central Europe beginning in the latter part of the fifteenth century, and imports of gold from Africa by the Portuguese, added to the money stock and contributed to the price rise. Monetary debasements by impecunious and unscrupulous sovereigns stimulated increases in nominal prices. It has been alleged that the increase in population was a more important factor than increases in the stock of specie in raising prices, an argument that overlooks the distinction between the general (average) price level and relative prices. Consequences attributed to the price revolution range from the impoverishment of the peasantry and nobility to the "rise of capitalism."

In perspective, it appears that many of the consequences attributed to the price revolution are either greatly exaggerated or wrongly attributed. Although the percentage increases in prices over the course of the century are impressive, they pale in comparison with price increases in the second half of the twentieth century on an annual basis. Severe short-term fluctuations—downward as well as upward—probably caused greater havoc than the overall long-run inflation. What is indubitable is that the price revolution, like any inflation, redistributed income and wealth, of both individuals and social groups. Those whose incomes were price-elastic—merchants, manufacturers, landowners who farmed their own land, peasants with secure tenures and producing for the market—benefited at the expense of wage earners and those whose income was either fixed or changed only slowly—pensioners, many rent receivers, and rackrented peasants Although the growth of population did not cause the (absolute) increase in prices, it probably did play a major role in the wage lag, as agriculture and industry proved incapable of absorbing the surplus labor. But the root cause of the decline in real wages was not a monetary problem; rather it was a result of the interrelations between demographic behavior and agricultural productivity.

Agricultural Technology and Productivity

The simple explanation for the cessation of population growth in the seventeenth century is that the population had outgrown its ability to feed itself adequately. Underlying this is a rather more complex explanation: the failure of agricultural technology to advance significantly, with a consequent stagnation or probably even a decline in average agricultural productivity. Few generalizations about European agriculture are wholly valid, however, because of regional diversity; even that in the preceding sentence is subject to qualifications, for the Dutch Netherlands in particular. A few generalizations can nevertheless be advanced with only minor reservations. In the first place, for Europe as a whole and for every major geographical subdivision, agriculture was still the principal economic activity by far, occupying two-thirds or more of the active population in the Dutch Netherlands and up to 90 or 95 percent in eastern and northern Europe. Second, from a human and social point of view, manual labor was by far the most important factor of production. Soil, seeds, and moisture were essential, of course; draft animals and other livestock were almost ubiquitous, if not strictly essential; and fertilizer was highly desirable. But human labor was the most essential input. Plows (in several varieties, according to the type of soil and cultivation), sickles, and flails were the principal instruments of capital equipment, and all required a large complement of manual labor to make them effective.

A final generalization is less certain, and clearly subject to more regional exceptions. For Europe as a whole the average agricultural productivity in the sixteenth century was probably not higher than in the thirteenth century, and it apparently declined somewhat in the seventeenth century. The evidence of the ratios of harvest yield to seed at least suggests this. Unfortunately, we have no good evidence of productivity per unit of land or labor (except for parts of Italy, where production per unit of land may have increased slightly, but probably at the expense of labor productivity). Yield ratios for the principal cereals were no more than 4 or 5 to 1 for Europe as a whole, ranging from 2 or 3 to 1 in parts of eastern Europe to as much as 10 or more to 1 in the most favored areas of the Netherlands and possibly elsewhere. Even these low ratios probably declined somewhat in the seventeenth century in most areas. (Comparable ratios today, using the best practices, are 40 or 50 to 1.) Livestock in general was probably no more than one-third or one-half the weight of modern animals, although they were somewhat larger in the more advanced areas. Milk productivity was comparable.

Yield/seed ratios are not infallible measures of agricultural productivity. The yield per acre of land sown might be increased by a more liberal use of seed, for example, or the productivity per unit of labor might be increased by using less labor with the same amount of seed. It seems unlikely, however, that either increased significantly, and both may have decreased slightly toward the end of the sixteenth or in the first half of the seventeenth century.

Although the direct empirical evidence for a decline in the productivity of both land and labor is tenuous, at best, there are good theoretical reasons for supposing that it occurred. First, instead of using less labor per bushel of seed or per acre of land, more labor was probably applied to the land, because of the increase in population. Although this might result in modest increases in total output, it probably meant a lower average output per man-year (i.e., in the productivity of labor). Second, there is positive evidence that more land was brought under the plow, both by cultivating former wastelands (heaths and marshes, etc.) and by converting pastures to arable farmland. In the case of the wastelands, normally less fertile than those already under the plow, a lower average yield would naturally be expected— that is, a decrease in the productivity of land. In some cases, the yield on converted pastures might be higher temporarily because the animal droppings would increase the fertility of the soil. But the reduction of pastureland brought with it other, less favorable consequences, namely, a reduction in livestock, especially cattle. There is both direct and indirect evidence of a decline in meat consumption in the sixteenth century, with adverse consequences for nutrition and the health of the population. Moreover, the decline in livestock implies a decline in the amount of manure for fertilizer for an already overcropped land. It was a seemingly vicious downward spiral.

To appreciate the full dimensions of the problem, however, it is necessary to consider the several regional variations, not merely for themselves but for their implications for the future.

In the northern and western periphery of Europe—Finland, most of Sweden except the extreme southern tip (Scania), Norway, Scotland, Wales, Cornwall, and much of Ireland—subsistence agriculture predominated. The lands were thinly populated, especially the northernmost parts, which had huge tracts of virgin forests. Primitive slash-and-burn techniques were still applied, although in the more settled regions a slightly less wasteful method, the infield–outfield system, was practiced. Stock raising of a primitive sort was important, especially in the mountainous areas. The principal field crops were rye, barley, and oats (wheat did not thrive in the cold, damp climates with short summers); flax and hemp were grown for their fibers, to be made into rough homespun clothing. Because of the relative abundance of land, tenures were fluid, with most land being held in the name of clans of tribal chiefs or lords. The social organization was hierarchical, but without bondage or ties of servility.

In Europe east of the Elbe and north of the Danube (including European Russia), by contrast, personal bondage or serfdom was the characteristic feature of social relationships at the beginning of the period and increased more or less continuously during it, as powerful lords steadily encroached on the lands and freedom of the few remaining independent peasants, by both legal and illegal means. This was the region of *Gutsherrschaft*, that is, the system of direct exploitation of large estates for the benefit of the territorial lords. The peasants' status, already parlous in the fifteenth century, was steadily reduced in Russia and parts of Poland to one not very different from slavery. They were obliged to give as many as five or six days a week to the lord's service, and in some instances were bought and sold separately from the lands they toiled. Agricultural technology was relatively primitive, employing either the two- or three-field system. The ratio of harvest yield to seed was low even by contemporary standards, averaging no more than 3 to 1. In the lands adjacent to the Baltic Sea, or on navigable rivers leading to it, production for export to the markets of western Europe was a potent stimulus for specialization in grain (mainly

rye) and other marketable crops; elsewhere (i.e., in most of eastern Europe) production was directed primarily at local self-sufficiency.

The Mediterranean area, in spite of a relatively uniform climate and similarities
of soil types, was so diverse it defies generalization. In Italy alone land tenures
ranged from the small but progressive farms of peasant proprietors and independent
tenant farmers in Piedmont and the extreme north to the large estates cultivated by
poverty-stricken sharecroppers and hired laborers in Sicily and the south. In between were a great variety of tenures, with *mezzadria* (sharecropping) being prominent. Italy likewise had the most diversified agriculture of Europe. Cereals, although important, were relatively less so than elsewhere. Rice, which yielded more
than conventional cereals, grew in the lower Po valley and along the Adriatic coast.
Grapes and olives, grown throughout the Mediterranean basin, were especially
important in Italy, which also grew fruits (including citrus fruit in the south),
vegetables, forage crops, and such industrial crops as dye-plants needed in the
textile industry. In spite of its diversification, however, Italian agricultural output
failed to keep pace with the population growth; overcropping and overgrazing took
their toll, with deforestation and soil erosion among the consequences.

Spain presented almost as much variety as Italy, with fertile coastal regions on
the east and south, mountain ranges in the north and elsewhere, and the most
characteristic feature of Spanish geography, the high plateau or *meseta* that spreads
over the central part of the Iberian peninsula. Spanish agriculture received a rich
inheritance from its Muslim predecessors. The Arab and Moorish peoples who had
inhabited Valencia and Andalusia before the Christian reconquest were excellent
horticulturists, and brought the art of irrigation to a high level. Unfortunately, the
Spanish monarchs, fired by religious fanaticism, squandered this inheritance. In the
same year that they conquered the kingdom of Granada and that Columbus discovered America, they decreed the expulsion of all Jews (who were also skilled
agriculturists as well as artisans) from the realm. With the fall of Granada many
Moorish subjects also left, even before they were given the choice of conversion or
flight ten years later. Those who converted, called Moriscos, remained the back-
bone and sinews of the agricultural economy in southern Spain for another century,
before they too were expelled in 1609. The Christians who replaced them were
unable to maintain the intricate irrigation systems and other features of the highly
productive Moorish agriculture. In part it involved incentives as well as knowledge
and ability. Throughout Spain in the sixteenth century the land was gathered into
huge estates owned by the aristocracy and the church, the largest landholder of all.
But these were absentee landlords, who, by means of stewards or other intermediaries, let the land in small parcels to sharecroppers or tenants on short leases, who
lacked both the capital and the incentive to maintain the Moorish system. Many
peasants fell into debt peonage, a status not far from serfdom. Moreover, with the
rise of prices resulting from the inflow of American gold and silver, much land in
both the fertile valleys and the arid *meseta* was diverted to cereal growing. Even so,
grain production did not suffice to feed the population, and Spain depended increasingly on imports of wheat and other grains.

Another major impediment to Spanish agriculture was the rivalry between peasants and sheepowners. Spanish merino wool was in great demand in the Low

FIGURE 5–4. Transhumance routes in Spain.

Countries and other centers of the textile industry. The sheepmen followed the practice of transhumance, that is, the movement of flocks between mountainous summer pastures and lowland winter pastures (Fig. 5-4). Transhumance was not peculiar to Spain. It was practiced in every part of Europe that had mountainous areas unsuited for arable culture, from southern Italy to Norway; it is still used today by the dairy farmers of Switzerland. But the Spanish system was unusual both for the length of its sheepwalks and for its organization. Sheepwalks, protected by royal legislation, covered the whole length of Spain from the Cantabrian Mountains in the north to the valleys of Andalusia and Extremadura in the south. The sheepmasters, organized into a guild or trade association called the Mesta, constituted a powerful lobby at court. Transhumant sheep were easily taxed at strategic toll stations, the wool was valuable, brought in cash revenues (unlike many peasant crops), and was also easily taxed at export. The monarchs, always greedy for tax revenues, granted the Mesta special privileges, such as unlimited grazing on common lands, which were detrimental to agriculture, in return for increased taxes. The privileges of the Mesta, together with other unwise governmental policies, such as the attempt to set maximum prices for wheat during the great inflation known as the price revolution, did nothing to encourage better technical practices in a land tenure system that already discouraged them. The productivity of Spanish agriculture was probably the lowest of western Europe. In the seventeenth century, with population declining, many farms were abandoned altogether.

Elsewhere in western Europe (i.e., France north of the Massiv Central, Germany west of the Elbe, Denmark and Scania, most of England) the system of open fields, an inheritance from the manorial system of the Middle Ages, prevailed. Exceptions must be made for hilly and mountainous areas (e.g., for much of Switzerland) and for large areas of western France in which small enclosed fields (*bocage*) were intermixed with open fields. A special exception must also be made for parts of the Low Countries, which will be described in greater detail. The German term *Grundherrschaft* is sometimes used to describe the system of land tenure. Territorial lords had been transformed into mere landlords; they collected rents in money or in kind, but labor services, already on the wane in the late Middle Ages, were extinguished, although the lords retained special rights and privileges in some areas. Transferability of land ownership became more common, and small peasant proprietors as well as independent tenant farmers increased. It has been estimated that about two-thirds of English peasants had secure tenures—freeholds, copyholds, or life leases. Although some consolidation of properties by large landowners occurred—about 10 percent of the land of England was enclosed in the sixteenth century, mainly for sheep pastures—on balance the peasants gained.

Small holdings and independent tenant farmers were most numerous in the vicinity of cities, where their produce was vital to the supply of the urban population. Elsewhere there were two principal types of tenure, with many variations and gradations. Long-term leases were common in England (some customary leaseholds were even heritable), parts of Germany and northern France. The peasants paid fixed rents either in kind or, more often, in cash, furnished their own livestock, equipment, and seeds, and were independent decision makers, except when constrained by communal custom and decision making in areas of open field agriculture with many strips. The other main type of tenure was sharecropping, called *metayage* in France, where it was especially common south of the Loire River. In that system the landlord furnished all or part of the stock and equipment, shared the risks and decision making (or made the decisions himself), and took a portion of the crop, normally half. (He might also undertake to market the peasant's portion, a situation that lent itself to exploitation and abuse.) A variation of the latter system, called *fermage*, was practiced in north central France and some other parts of Europe. (In fact, the modern English word for farming is derived from *fermage*.) In this system, a substantial *fermier* (farmer) would lease an entire estate, or even several estates, for a fixed cash rental, then sublet the land in smaller parcels to peasants on short-term leases or as sharecroppers. The landlords thus lost all functional connections with agriculture, becoming mere *rentiers* (rent receivers). In the hands of capable *fermiers* this system could produce excellent results in terms of improved techniques and output; but it was also susceptible to rackrenting and exploitation of the peasants.

The most progressive agricultural area in Europe were the Low Countries, especially the northern Netherlands, with its core the province of Holland. At the end of the fifteenth century Dutch and Flemish agriculture was already more productive than the European average, thanks to the opportunity of supplying neighboring cities and workers in the cloth industry. Because of its method of settlement in the Middle Ages the Dutch rural population also possessed greater freedom than that of formerly manorialized regions. In the course of the sixteenth and seventeenth centuries Dutch agriculture underwent a striking transformation that merits its description as the first "modern" agricultural economy. The modernization of agriculture was intimately associated with the equally striking rise of Dutch commercial superiority; without one, the other could not have occurred. The key to the success of the transformation of Dutch agriculture was specialization, a specialization that was made possible in the first instance by the bouyant demand of the prosperous and rapidly growing Dutch cities, but in time enabled Dutch cheeses, for example, to be sold in the markets of Spain and Italy. Instead of trying to produce as much as possible of the goods (nonagricultural as well as agricultural) necessary for their own consumption, as most peasants elsewhere in Europe did, Dutch farmers tried to produce as much as possible for the market, also buying through the market many consumption goods as well as capital and intermediate goods. In some instances farmers marketed their entire output of wheat, purchasing cheaper rye for their own consumption. For the most part, however, Dutch farmers specialized in relatively high-value products, livestock and dairy produce in particular. Raising livestock required growth (or purchase) of large quantities of fodder crops (hay, clover, pulses, turnips, etc.). Specialization in livestock also meant larger quantities of manure for fertilizer; the intensive nature of Dutch agriculture required even more fertilizer, however. So great was the demand for fertilizer that some entrepreneurs found it profitable to specialize in collecting urban night soil and pigeon dung, for example, which they sold by the canal-boatload or cartload—an activity that incidentally kept Dutch cities cleaner and more sanitary than others.

Dutch farmers did not specialize exclusively in dairying and livestock. Horticulture occupied many of them, especially in the immediate vicinity of cities. Some grew barley and hops for the brewing industry, others industrial crops such as flax, hemp, woad, madder, and pastel. Even flowers became subject to specialized commercial exploitation; Dutch bulbs were so highly regarded that speculation in them produced a "tulip mania" in 1637. Nor did Dutch farmers give up cereal cultivation entirely; the urban patriciate was willing to pay a relatively high price for wheaten bread. Nevertheless, thanks to the efficiency of Dutch shipping and the aggressiveness of Dutch merchants, the lower classes (including many specialized farmers) were able to purchase the lesser grains, mainly rye, more cheaply from the Baltic. In the midseventeenth century a large proportion, possibly one-fourth or even more, of Dutch cereal consumption was supplied by imports.

The profitability of Dutch agriculture is attested by the continuing and continuous efforts to create new land by reclaiming it from the sea, by draining lakes and marshes, and by planting peat bogs after the peat had been removed for fuel. This activity had begun in the Middle Ages, but it increased substantially in the sixteenth and seventeenth centuries, and was especially intensive in periods of rising prices for farm products. Nor were only farmers involved. Diking and draining required large capital expenditures; urban merchants and other investors formed companies to reclaim land they then sold or rented to active farmers.

A puzzling question arises. Why were Dutch agricultural techniques not more widely diffused in the sixteenth and seventeenth centuries? Some diffusion did occur. The turnip was introduced into England as early as 1565, as were some fodder crops such as clover; the reclamation of the fenlands of eastern England, begun in the seventeenth century, owed much to the Dutch example, Dutch engineers and technology, and even Dutch capital. Some diffusion also occurred in northern France, adjacent to the southern Netherlands. In more general terms, however, the productivity of nonagricultural occupations was not sufficiently high, and the development of markets not sufficiently extensive, to justify the specialization and intensity of labor and capital that characterized Dutch agriculture.

Industrial Technology and Productivity

In industry as in agriculture, no sharp break occurred between the Middle Ages and the early modern period. Unlike the case of agriculture, however, innovation took place more or less continuously, although at a very slow pace. But a problem arises here: how do we measure innovation and its effects? One obvious way is simply to count the number of inventions or innovations. This is not very satisfactory, however, not only because different innovations have very different effects, but also because of the difficulty of definition. Most of the innovations in the sixteenth and seventeenth centuries (indeed, in any period of history) involved relatively minor improvements in already established techniques. For this reason they frequently go unnoticed by historians. Another possibility is to measure changes in productivity. In 1589 a parson of the Church of England, William Lee, invented a simple machine, the stocking frame for hosiery and other knitwear. Whereas a skilled hand-knitter could achieve a rate of 100 stitches per minute, the stocking frame could average 1000 stitches per minute, and was subsequently improved. Unfortunately, however, few other innovations of the period have left such detailed information, especially the host of minor innovations.

There is another problem. Even when we have a clearly defined and described innovation and can measure its productivity, at least approximately, how do we assess its total economic impact? The greatest invention of the fifteenth century—indeed, one of the greatest of all time—printing with movable type, increased productivity in the book trade enormously, yet its immediate economic impact in terms of value of output or number employed was miniscule. Are we, therefore, to say that its economic significance was negligible? Other innovations of the period, in navigational instruments, firearms and artillery, clock- and watchmaking, were of minor economic significance yet of enormous importance in political and cultural terms—and thus, indirectly, economically as well.

The market orientation of the European economy, greater in industry than in agriculture, encouraged entrepreneurs who could reduce production costs and respond quickly to changes in consumer demand. But there were formidable obstacles to innovation as well. One of the most ubiquitous was the opposition of authorities who feared unemployment as a result of labor-saving innovations and of monopolistic guilds and companies who feared competition. In 1551 the English Parliament passed a law forbidding gig-mills, a device used in the cloth-finishing trade; in this case the market prevailed over the law, as new gig-mills continued to be built. Lee was refused a patent for his stocking frame, and the first ones that he attempted to introduce in Nottinghamshire were destroyed by mobs of hand knitters. Lee himself took refuge in France and established a factory, with the patronage of Henry IV; that failed after the death of his benefactor, but the stocking frame continued to spread. In 1651 the framework knitters of Nottingham applied to Cromwell for a guild charter to exclude unwanted competition! The swivel-loom, a Dutch invention for weaving a dozen or more ribbons simultaneously, was prohibited in England in 1638; but it spread anyway, especially in Manchester and vicinity, where its use created a large number of skilled operatives in advance of the great innovations that revolutionized the cotton industry.

None of the innovations mentioned here involved the use of mechanical power. The deficiencies of power sources and of building materials (mainly wood and stone) were natural obstacles to greater industrial productivity. The sketchbooks of Leonardo da Vinci are concrete evidence of numerous potential innovations that could not be made at the time because of inadequate materials and sources of power. Leonardo was a genius, of course; but there were undoubtedly many other less gifted persons who were frustrated in their attempts to increase the efficiency of human labor by faulty materials and insufficient power. Wind- and watermills had, it is true, already reached a high level of sophistication, as indicated in a previous chapter, but they have obvious limitations. In the seventeenth century, however, water-powered mills for spinning silk (which may have had medieval origins) proliferated in the Po valley and Venezia, and by the end of the century had spread to the Rhone valley in France. The large size and complexity of the machinery required them to be installed in factory-type buildings, making them some of the most important precursors of the modern industrial system.

Not all innovations involved mechanical contrivances. The typical products of the woolen industry in the late Middle Ages were heavy, coarse cloths. In the late fifteenth century Flemish clothmakers introduced a lighter, cheaper fabric called "new draperies" (in French *nouvelle draperie*). Although slow to catch the public fancy at first, their lower prices made them highly competitive in international markets, especially those of southern Europe. After the repression of the revolt in the Spanish Netherlands, and the consequent flight of many Flemish artisans, industries producing the new draperies sprang up in many lands, notably in England, where as early as 1571 there were as many as 4000 Flemish refugees in the city of Norwich alone—most of them weavers. For similar reasons the manufacture of cotton cloth, already produced in Italy in the Middle Ages with raw material from the eastern Mediterranean, gradually spread to Switzerland, south Germany, and Flanders in the sixteenth century. By about 1620 it reached Lancashire.

The textile trades remained, collectively, the largest industrial employer, closely followed by the building trades. This is understandable when one remembers that in a poor, near-subsistence economy such as preindustrial Europe, the basic necessities are food, shelter, and clothing. The cloth industry continued to be widely dispersed, with much production carried on within and for the household and for local markets; but some regions also specialized in production for export. The once great Italian industries suffered from the competition of new and more vigorous rivals, and gradually fell backwards, losing their markets for woolen goods to Dutch, English, and French producers and sharing the market for fine and fancy silks with the French. The Spanish woolen industry expanded briskly in the first half of the sixteenth century but, hampered by excessive taxation and government interference, stagnated and declined thereafter. For the first two-thirds of the century the largest cloth industries, for both woolens and linens, existed in the southern Low Countries, in the provinces of Flanders and Brabant in particular. The Dutch revolt and the brutal repression of the remaining Spanish Netherlands severely damaged both industries, although they recovered somewhat in the seventeenth century because of their privileged position as principal suppliers to the Spanish Empire.

The organization of the textile industries did not change appreciably from the later Middle Ages. The characteristic entrepreneur was the merchant-manufacturer who purchased the raw materials, put them out to spinners, weavers, and other artisans working in their homes, and marketed the final product. Guild organizations, whether of artisans or merchants, apparently did not affect the industry appreciably, at least in England. There the guilds gradually withered away as the woolen industry, in particular, moved to rural areas. In France the royal rulers fostered the guilds as a source of revenue. Whether this adversely affected the fortunes of the industry is a subject that merits further study. In any event, the English industry expanded prodigiously. In the Middle Ages raw wool had been England's principal export. In the sixteenth century exports of unfinished cloth predominated. By 1660 woolen and worsted cloth accounted for two-thirds of the value of all English exports. Moreover, whereas at the beginning of the seventeenth century about three-fourths of English cloth exports were undyed and undressed, by the end of the century virtually all cloth was exported in a finished condition. Well before the rise of modern industry England had already become the largest exporter in Europe's largest industry.

Although the construction industry in general experienced no significant technical changes apart from stylistic changes in monumental architecture, one specialized sector of the industry in one country did undergo a profound transformation, namely, shipbuilding in the Dutch Netherlands. Thanks to the rapid expansion of Dutch commerce, the Dutch merchant fleet experienced a tenfold increase in numbers and an even larger increase in tonnage between the beginning of the sixteenth century and the middle of the seventeenth. At that time it was the largest by far in Europe, three times larger than the English merchant fleet, which was second, and probably larger than all others combined. Considering the relatively short life of wooden sailing ships, this translates into a large demand on the ship-building industry, a demand to which Dutch shipbuilders responded by rationalizing their shipyards and introducing elementary mass production techniques. They used mechanical saws and hoists actuated by windmills, and kept stores of interchangeable parts. Because of their efficiency, they supplied not only their own country's fleet but those of its rivals as well. Since the Netherlands possessed few forests, virtually all of the timber for the shipyards had to be imported, principally from the Baltic area. On the other hand, the large demand for sailcloth and cordage stimulated prosperous subsidiary industries in Holland itself. There were few radical innovations in ship design from the late fifteenth to the nineteenth century, but many small improvements. The size of ships in the Atlantic trades increased from 200 to 600 tons in the course of the sixteenth century. Some warships even reached the unprecedented size of 1500 tons, but the most significant innovation—by the Dutch, of course—was the *fluyt* ship, or flyboat as the English called it, a specialized commercial carrier introduced at the end of the sixteenth century (Fig. 5-5). The equivalent in some respects of the tanker of our times, it was specially designed

FIGURE 5–5. The Dutch fluyt ship. This relatively large, ungainly ship was highly successful as a cargo carrier, replacing the older dual-purpose carrack. Its contemporary, the galleon, replaced the carrack as a warship and dual-purpose vessel. (Netherlands Historical Maritime Museum, Amsterdam. Reprinted with permission.)

for bulky, low-value cargoes such as grain and timber, and operated with smaller crews than conventional ships.

The metallurgical industries, although of relatively minor significance in terms of employment and output, acquired major strategic significance because of the growing importance of firearms and artillery in warfare. In 1450 light firearms played a negligible role, and awkward artillery pieces were confined to siege warfare. By 1600 arquebusses and muskets were the standard weapons of infantry, and large-bore cannons were essential to naval warfare. The metallurgical industries were important also as harbingers of the new age of industrialism. Iron was the most important.

In the Middle Ages wrought iron was obtained from a variety of types of "bloomeries," in which iron ore was heated with charcoal until it became a pasty mass or "bloom," which was then alternately hammered and heated until its impurities were driven out. The process was slow, costly in fuel and ore, and produced its output in small batches. In the fourteenth and fifteenth centuries the height of the furnaces increased progressively, a blast of air provided by water-powered bellows increased the temperature of the charge, and thus evolved the high, or blast, furnace. By the beginning of the sixteenth century the blast furnace was continuously charged at the top with charcoal, ore, and a flux to remove impurities, while from the bottom molten iron was periodically tapped to be cast directly into useful implements (pots, brackets, etc.) or into "pigs" (bars) for further refining. (Cast or pig iron contains a high carbon content—3 percent or more—which makes it very hard but brittle; the pigs, like the blooms, were alternately heated and hammered to remove the carbon, producing wrought iron.) The new method, although indirect, was nevertheless both faster and cheaper, as it made better use of fuel and ore, and could use lower-grade ores. It also required larger amounts of capital, although by far the greater part was tied up in inventories of charcoal and ore rather than in fixed capital as such.

As the blast furnace evolved a number of innovations occurred in ancillary operations. Water-powered bellows, tilt hammers, and stamping mills (for crushing ore) had all been introduced by the middle of the fifteenth century. Later in that century and early in the next, wire-drawing machines and rolling and slitting mills were invented. At the beginning of the sixteenth century the area around Liege and Namur (Wallonia) in the southeastern Low Countries, already an important metallurgical center in the Middle Ages, was the most advanced iron-producing region in Europe and the site of many innovations. Other principal centers were located in Germany, northern Italy, and northern Spain. Total European output was approximately 60,000 tons a year, of which various regions of Germany may have accounted for half. In the next hundred years the blast furnace and its attendant activities spread throughout Europe wherever iron ore, wood for fuel, and water power existed in sufficient quantities and adequate proximity. England was especially precocious. By 1625 its hundred-odd furnaces were producing upwards of 25,000 tons a year. The iron industry was voracious of fuel, however, and in the seventeenth century the high price of charcoal braked the expansion in the established areas of production. As it did so, new and more distant sources of supply came into production in the Swiss and Austrian Alps, in eastern Europe, and especially in Sweden.

Sweden, favored by high-grade iron ore and abundant timber and water power, had a modest iron industry even in the Middle Ages. At the beginning of the sixteenth century exports amounted to about 1000 tons a year. In the seventeenth century Walloon and Dutch entrepreneurs introduced more advanced techniques, and output expanded enormously: exports rose from about 6000 tons in 1620 to more than 30,000 tons at the end of the century, by which time Sweden's iron industry was probably the largest in Europe.

In other metallurgical industries progress was less remarkable, involving primarily an increase in output with conventional techniques and the application of those techniques to new sources of supply. Silver mining in central Europe, well established in the Middle Ages, experienced a boom in the early sixteenth century as a result of the discovery of the mercury amalgamation process for concentrating silver ores; but when that process was transferred (by German mining experts) to the silver mines in the Spanish colonies of Mexico and Peru in the 1560s, the resulting increase in the supply of silver so depressed prices that many European mines were forced to shut down.

Europe was not naturally rich in the precious metals, but the more utilitarian metal ores were relatively abundant. Copper, lead, and zinc were found in many parts of Europe, and had been mined since prehistoric times. Tin was more localized, virtually confined to Cornwall; but it, too, had been an item of commerce long before the Roman conquest of Britain. In the sixteenth and seventeenth centuries, under the pressure of increasing demand, mining techniques were improved, involving deeper shafts, better ventilation, and pumping machinery. German, especially Saxon, miners were the principal innovators, and they carried their skills abroad, to England and Hungary as well as to the New World. In the 1560s the English government granted monopolies in the brass and copper industries to companies that brought in German engineers. Sweden was almost as rich in copper as in iron, and in the seventeenth century, with Dutch capital and technical assistance, was Europe's largest supplier in international markets.

Timber was in great demand, for construction, shipbuilding, metallurgy, and, most important, domestic heating. The shortage of timber throughout the more developed areas of Europe was primarily responsible for integrating Norway and Sweden into the West European economy, both directly and indirectly (i.e., through the demand for metals). The timber shortage was so severe that it involved not only the Baltic area but, in the seventeenth and eighteenth centuries, North America as well. It also led to the search for substitute materials and fuels: brick and stone for building, peat and coal for fuel. Iron and other metals were also substituted for wood, but the increase in demand for them merely intensified the timber shortage. England was among the countries most affected. Certain forests were reserved for the royal navy, but of even greater importance was the growing demand for fuel.

Coal had been mined in Germany and the Low Countries as well as England during the Middle Ages. In spite of its noxious characteristics and frequent laws prohibiting its use, "seacoale" from the banks of the Tyne estuary had become a common household fuel in sixteenth-century London. Gradually it penetrated high fuel-consumption industries such as salt-refining, glass, brick and tile, copper smelting, malting and brewing, and various chemical industries. Attempts were made in the seventeenth century to substitute it for charcoal in iron smelting, but

various impurities (principally sulfur) in raw coal imparted undesirable characteristics to the iron. Even so, the demand for coal from other industries increased
steadily. Output of the English industry grew from about 200,000 tons annually in
the midsixteenth century to 3 million tons at the end of the seventeenth. As the
industry grew, coal from outcroppings along river banks no longer sufficed to
satisfy the demand. Mines had to be sunk; Saxon miners, long-experienced in the
arts of boring, pumping, and mine ventilation, were brought to England to spread
their knowledge.

The overseas discoveries, by providing new raw materials, directly stimulated
new industries; sugar refining and tobacco processing were most important, but
other manufactures ranging from porcelain (in imitation of Chinese ware) to snuff
boxes developed to satisfy newly created tastes. Sugar cane also provided the raw
material for rum distilleries, and in the seventeenth century the affluent Dutch
invented gin, originally intended for medicinal purposes. In addition to such wholly
new industries, a number of older industries in which production had been highly
localized spread to various parts of Europe. In the Middle Ages Italy had been the
principal if not the only producer of such luxury goods as fancy glassware, high-
grade paper, optical instruments, and clocks. The growth of similar industries in
other countries, whose products were frequently of lesser quality but cost less,
accounts in part for the relative decline of Italy. The invention of the printing press
greatly increased the demand for paper (although then, as now, the largest use of
paper was for wrapping and packaging). Before the end of the fifteenth century
more than 200 printing presses had been established and had produced approximately 35,000 separate editions, or about 15 million books. The numbers have
been growing exponentially ever since; in the latter half of the seventeenth century
the catalogs of the Frankfurt book fair, the largest in Europe, listed as many as
40,000 current titles. The Low Countries, especially Antwerp and Amsterdam,
were the most active centers of the industry, but France, Italy, the German Rhineland, and England did not lag far behind.

In spite of this picture of varied, vigorous, and sophisticated industries, one
should bear in mind the still very imperfect degree of specialization in the European
economy, and its extreme dependence on low-productivity agriculture. Many industrial workers, especially in the textile industries, engaged part-time in agriculture,
and most agricultural workers also had secondary occupations as woodworkers,
leather workers, and the like.

Trade, Trade Routes, and Commercial Organization

Of all sectors of the European economy, commerce was undoubtedly the most
dynamic between the fifteenth and the eighteenth centuries. Older textbooks described the sixteenth century as an era of ''commercial revolution.'' As we have
seen, there are earlier candidates for that title, but there is no doubt that a substantial
increase occurred in the volume of long-distance or international trade. Exactly how
much is impossible to state, but the increase in trade probably exceeded by several
times that in population. Extra-European trade contributed to the increase, and also

stimulated some of the increase within Europe; but as was previously noted, trade with Asia and America was but a small fraction of the total. Commerce would certainly have increased even with no discoveries.

It should be remembered that by far the greatest part of commercial exchange, both by volume and by value, was local. Towns and cities received the bulk of their food supplies from their immediate hinterlands and in exchange supplied them with manufactured goods and services. It was mainly small-scale commerce, and varied little either over time or from place to place. More interesting, and more significant for the history of economic development, were the changes that occurred in distant trade.

The principal trade routes and the commodities involved in them, as they existed in the fifteenth century, were sketched in Chapter 3. The most important changes that occurred in the next 200 years, in addition to the opening of the overseas routes, were the shift in the center of gravity of European commerce from the Mediterranean to the northern seas, a slight but perceptible change in the character of the commodities involved in distant trade, and changes in the forms of commercial organization.

The Portuguese invasion of the Indian Ocean was a rude shock to the Venetians and, to a lesser extent, other Italian cities. It is not true, as used to be thought, that the Mediterranean spice trade through Egypt and Arabia ceased abruptly, but the competition of Portuguese spices greatly reduced its profitability. In 1521, in an attempt to regain their monopoly, the Venetians offered to purchase the entire Portuguese import, but were refused. Gradually the initiative in commercial affairs shifted to northern Europe. The Venetians' famed Flanders fleet made its last voyage in 1532, and in the latter part of the century Venetian ambassadors complained of competition from cheaper French and English woolens in the markets of the Near East, which the Italians had regarded as their exclusive domain. The benefits of the Portuguese success did not accrue exclusively to them, however. The first cargo of Portuguese spices appeared on the Antwerp market in 1501, carried from Lisbon not by Portuguese but by Dutch or Flemish merchants. The Spanish and Portuguese, concentrating on the exploitation of their overseas empires, left the business of distributing their imports in Europe, and also of providing most of their exports to the colonies, to other Europeans. Of these the Netherlanders, mainly Dutch and Flemish, were most aggressive.

"The prodigious increase of the Netherlands" (in the words of an envious Englishman) began modestly enough in the fifteenth century, as Dutch fishing fleets in the North Sea began to undercut Hanseatic dominance of the herring trade. (It used to be thought that the herring shoals "migrated" from the Baltic to the North Sea, but more likely the decline of the Hansa, in this as in other trades, occurred because the Dutch were simply more efficient.) The dried and salted fish were first distributed around the shores of the North Sea and up the German rivers, then, in the sixteenth century, to southern Europe and even in the Baltic. Meanwhile the Dutch developed other trades. From Portugal and the Bay of Biscay, they procured salt for the fish and for distribution in northern Europe, occasionally picking up cargoes of wine as well. But the mainstay of Dutch commerce was the Baltic trade, mainly grain and timber, but also naval stores, flax, and hemp. Of the 40,000 ships recorded in the Danish Sound Toll registers as entering or leaving the Baltic between 1497 and 1660, almost 60 percent were Dutch, and the remainder English, Scottish, German, and Scandinavian. Virtually all of the trade between northern Europe and France, Portugal, Spain, and the Mediterranean, and much of the trade between England and the Continent, was in the hands of the Dutch.

The Dutch were equally aggressive in overseas trade. Their war for independence interrupted their trade with Spain, but they continued to trade with the Portuguese Empire through Lisbon. Portugal fell to the crown of Spain in 1580, however, and in 1592 the Spanish authorities closed the port of Lisbon to Dutch shipping. Heavily dependent on maritime commerce, the Dutch immediately began building ships capable of the months-long voyage around Africa to the Indian Ocean. In less than ten years more than fifty ships made the round trip between the Netherlands and the Indies. So successful were these early voyages that, in 1602, the government of the United Provinces, the city of Amsterdam, and several private trading companies formed the Dutch East India Company, which legally monopolized trade between the Indies and the Netherlands.

The Dutch were not the only nation to take advantage of Portugal's weakness. English interlopers had made a voyage as early as 1591, and in 1600 the English East India Company was organized with a monopoly similar to that of the Dutch company. Although the two companies were rivals to some extent, they both regarded the Portuguese as the greater enemy. The Dutch concentrated their attention on the fabulous Spice Islands of Indonesia, and by the middle of the seventeenth century had established their mastery of both the islands and the spice trade more effectively than the Portuguese had ever done. They also took control of the ports of Ceylon. The English, after unsuccessful attempts to obtain a foothold in Indonesia, established fortified trading posts on the mainland of India, which eventually became the "brightest jewel in the British crown." Portugal retained its possessions of Goa, Diu, and Macao, as well as a few ports on the African coasts, but ceased to be a major commercial or naval power in the eastern seas.

The other seapowers also took advantage of Portuguese weakness and Spanish rigidity to invade and create markets in the Western Hemisphere. Early French and English attempts to find a direct route to the East had been disappointing, but in the second half of the sixteenth century new efforts were made to discover a northeast or northwest passage to Asia. The ill-fated voyage in 1553 of Willoughby and Chancellor through Arctic waters into the White Sea failed to find a northeast passage, but it did establish trade relations with the growing Russian Empire and, through it, with the Middle East. At about the same time French, English, and Dutch privateers began carrying on a clandestine trade with Brazil and the Spanish colonies in the New World or, as the occasion presented, raiding Spanish ships and colonial ports. Three brief attempts by the English to found colonies in North America during the reign of Elizabeth I ended in failure, but in the first half of the seventeenth century successful colonies were established in Virginia (1607), New England (1620), and Maryland (1632), as well as on islands seized from the Spanish in the West Indies. In time all of these became important markets for English industries as well as sources of supply for raw materials and consumer goods. In 1608 the French established a permanent settlement at Quebec and claimed the entire Great Lakes area as New France, but the colony did not prosper. In 1660, when the English-speaking colonists in the New World numbered almost 100,000, all Canada contained only 2500 French colonists, fewer than the number of Frenchmen in the few French sugar islands in the West Indies.

In 1624 the Dutch attempted to conquer the Portuguese colonies in Brazil, but after two decades of intermittent strife they were ultimately driven out by the Portuguese colonists themselves, with little help from the mother country. The Dutch retained only Surinam and a few islands in the Caribbean. In the same year the Dutch began their Brazilian conquest, another group of Dutch colonists founded the city of New Amsterdam on the southern tip of Manhattan Island. They laid claim to the entire Hudson valley and surrounding area, founded Fort Orange (Albany), and gave out land under the patroon system of ownership to such families as the Rensselaers and Roosevelts.

Seaborne commerce was by far the most important for international trade, but inland trade, especially river traffic, was not negligible. Local commerce used it extensively, and most commodities even in international trade began their voyages to market overland by cart or on pack animal and downriver by barge. Copper from Hungary, for example, reached the market of Antwerp (later Amsterdam) by land carriage to the Polish rivers, thence by barge to Danzig, where it was transshipped through the Baltic and North seas. Silver from central Europe and the Tyrol followed similar itineraries, whether going to the Baltic, the Mediterranean, or the west. The Rhine, Main, and Neckar rivers were important arteries for the export of the metals and hardware (knives, tools, toys) of South Germany and the Rhineland. The French rivers were equally important.

Metals and some luxury cloths could stand the expense (and wear and tear) of long land journeys. Few other commodities could, unless they were self-propelled, as was the case with cattle. While most of Europe's arable land was increasingly devoted to field crops to feed its growing population, Denmark, Hungary, and Scotland had large open grasslands on which to pasture herds of cattle. Annual cattle drives, foreshadowing those of the American West of the nineteenth century, sent the livestock to fattening pens and markets in the cities of northern Germany and the Low Countries, to South Germany and northern Italy, and to England.

The character of the commodities involved in distant trade changed somewhat in the sixteenth and seventeenth centuries. In the early Middle Ages these had been mainly luxury goods for the well-to-do. Later, with the growth of towns, more mundane articles entered the lists. By the sixteenth century a large proportion of the volume of goods moving in international trade consisted of such staples as grain, timber, fish, wine, salt, metals, textile raw materials, and cloth. At the end of the seventeenth century half of English imports, by volume, consisted of timber; more than half of the exports, also by volume, were coal, although cloth exports were far more valuable. The trade in bulky staples was made possible principally by the improvements in ship design and construction, which lowered costs of transport. A reduction in the risks of maritime travel, both natural and manmade, by better navigational techniques, and by the action of navies in putting down pirates also contributed in the same direction.

In the intercontinental trade the situation corresponded more closely to the older model, although even here shifts occurred in the seventeenth and especially the eighteenth centuries. The commerce in pepper, a luxury at the beginning of the sixteenth century, gradually took on the character of a staple trade. As the importance of the precious metals declined in the seventeenth century, and other countries acquired colonies in the Western Hemisphere, sugar, tobacco, hides, and even timber became increasingly prominent among Europe's imports. Europe's exports to the colonies consisted of manufactured goods, for the most part; these were not bulky, but the remaining space available was filled partly by emigrants. The situation in the eastern trade was quite different. From the beginnings of direct European contact, Europeans had difficulty in finding merchandise to exchange for the spices and other desirable wares. For this reason much of Europe's "trade" was, in effect, plunder. Where plunder was not possible or feasible, Asians accepted firearms and munitions, but mostly they demanded gold and silver, which they hoarded or converted into jewelry. On balance, Asia was a sink for European monetary metals. Not until the conquest of India by England in the eighteenth century was the balance reversed.

One very special branch of commerce dealt in human beings: the slave trade. Although the Spanish colonies were among the largest purchasers of slaves, the Spanish themselves did not engage in the trade to any great extent but granted it by contract, or *asiento,* to the traders of other nations. The trade was dominated at first by the Portuguese, then in turn by the Dutch, the French, and the English. Usually the trade was triangular in nature. A European ship carrying firearms, knives, other metalwares, beads and similar cheap trinkets, gaily colored cloth, and liquor would sail for the West African coast, where it exchanged its cargo with local African chieftains for slaves, either war captives or the chief's own people. When the slave trader had loaded as many chained and manacled Africans as his ship would carry, he sailed for the West Indies or the mainland of North or South America. There he exchanged his human cargo for one of sugar, tobacco, or other products of the Western Hemisphere, with which he returned to Europe. Although the death rate from disease and other causes for slaves in transit was dreadfully high (frequently 50 percent and sometimes more), the profits of the slave trade were extraordinary. European governments took no effective steps to prohibit it until the nineteenth century.

The organization of trade varied from country to country and according to the nature of the commerce. Intra-European trade inherited the sophisticated and complex organization developed by the Italian merchants in the later Middle Ages. In the fifteenth century colonies of Italian merchants could be found in every important commercial center: Geneva, Lyons, Barcelona, Seville, London, Bruges, and especially Antwerp, which in the first half of the sixteenth century became the world's greatest entrepôt. Native merchants, and those from other countries as well, learned Italian business techniques such as double-entry bookkeeping and the uses of credit—so well, indeed, that by the first half of the sixteenth century the Italians could no longer assert their predominance. The greatest business dynasty of the sixteenth century was the Fugger family, with headquarters in Augsburg in South Germany.

The first Fugger known to history was a weaver. Some of his descendants became putters-out (merchant-manufacturers) in the woolen industry, eventually

FIGURE 5–6. Jacob Fugger II, "The Rich." Fugger is shown here in his office with his chief clerk, Mathias Schwartz. The large volumes behind him are labeled with the names of the cities with which he did business: Venice, Cracow, Milan, Innsbruck, Nuremburg, Lisbon, etc. (Braunschweigisches Landesmuseum fur Geschichte und Volkstum. Reprinted with permission.)

getting into wholesale trade in silk and spices with a warehouse in Venice. By the end of the fifteenth century they were actively engaged in financing the Holy Roman emperors, as a result of which they obtained control over the output of the Tyrolean silver and copper mines and the copper mines of Hungary. Under Jacob Fugger II (1459–1525) the family firm operated branches in several German cities and in Hungary, Poland, Italy, Spain, Lisbon, London, and Antwerp (Fig. 5-6). From Lisbon and Antwerp they largely controlled the distribution of spices in central Europe, for which they exchanged the silver needed to purchase the spices in India. They also accepted deposits, dealt extensively in bills of exchange, and were heavily involved in financing the monarchs of Spain and Portugal—a business that eventually led to their downfall.

The Fuggers were preeminent in the sixteenth century—Jacob II was called "a prince among merchants"—but many others were only slightly less prominent, in Italy and the Low Countries as well as Germany. Even Spain had a few notable merchant dynasties. The form of organization they favored was the partnership, usually formalized by written contracts specifying the rights and obligations of each partner. By means of correspondence among widely separated partners or agents they kept abreast of developments, political as well as economic, in all parts of

Europe and beyond. It was said that Queen Elizabeth's government was the best informed in Europe because of her financial agent in Antwerp, the merchant Sir Thomas Gresham. Merchant newsletters were the forerunners of the great news-gathering agencies, or "wire services," of today.

Commercial organization in England, a peripheral country in the fifteenth century, reflected an earlier form than the more highly developed economies on the Continent; but it made rapid progress and by the late seventeenth century was one of the most advanced. In the Middle Ages the trade in raw wool, by far the most important export, was handled by the Merchants of the Staple, a regulated company that functioned something like a guild. There was no joint stock; each merchant traded for his own account (and for his partners, if any), but they had a common headquarters and warehouse (the staple) and obeyed a common set of rules. The wool trade was still important, though declining, in the fifteenth and sixteenth centuries; the staple, where the wool was taxed and sold to foreign merchants, was located in Calais, an English possession until 1558. Replacing the Staplers in importance, the Merchant Adventurers, another regulated company, handled the trade in woolen cloth. (Some merchants were members of both companies.) They established their staple in Antwerp, contributing more than a little to the growth of that market, and received certain privileges in return. In 1564 the company received a royal charter conferring on it a legal monopoly of cloth exports to the Low Countries and Germany, the most important markets.

In the latter half of the sixteenth century the English set up a number of other companies with monopolistic trading charters: the Muscovy Company (1555), an outgrowth of the Willoughby-Chancellor expedition; the Spanish Company (1577); the Eastland (Baltic) Company (1579); the Levant (Turkey) Company (1583); the first of several African companies in 1585; the East India Company (1600); and a French company (1611). The establishment of special companies for trade with France, Spain, and the Baltic, in particular, indicates one (or both) of two things: the small amount of direct trade between England and those countries before the existence of the companies (and possibly afterwards as well), and the extent to which such trade as did exist was in the hands of Dutch or other merchants. It is significant that the Dutch did not see the need for such monopolistic concerns, except for the Dutch East India Company (1602).

Some of these companies adopted the regulated form, but others became joint-stock companies; that is, they pooled the capital contributions of the members and placed them under a common management. This was done in the distant trades, in which the risks and capital required to outfit a single voyage exceeded the amounts that one or a few individuals were willing to assume or furnish. The Muscovy and Levant companies were first formed on a joint-stock basis, but as trade relations developed and became more stable they became regulated companies. The Muscovy Company, trading through the port of Archangel, handled most of western Europe's commerce with northern Russia until the tsar withdrew its privileges in favor of the Dutch in 1649. The East India Company also adopted the joint-stock form. At first each annual voyage was a separate venture, which might have different groups of stockholders from year to year. In time, as it became necessary to establish permanent installations in India and to provide continuous supervision of its affairs, the company adopted a permanent form of organization in which a stockholder could withdraw only by selling his shares to another investor. The Dutch East India Company adopted the permanent form as early as 1612.

The existence of a single great entrepôt in northwestern Europe—first Bruges, then Antwerp, then Amsterdam, each larger and more imposing than the former—is doubly significant. First, their mere existence, in contrast to the periodic fairs of the Middle Ages, is evidence of the growth in the size of markets and of market-oriented production. But the fact that there was only one at a time, and that as one rose another declined, indicates the limits to that development. True, there were other emporia of some significance—London, Hamburg and other Hanseatic cities, Copenhagen, Rouen, and others—but none had the full range of commercial and financial services of the one great metropolis. The reasons are related to the limited extent of markets and the existence of external economies in commercial and especially financial transactions. When the total volume of commercial or financial turnover is relatively small, it is cheaper to concentrate them in a single location.

The organization of the entrepôt was already highly sophisticated at the beginning of the fifteenth century in Bruges, and became more so as it migrated to Antwerp and Amsterdam. The first requirement is a burse, or marketplace (Fig. 5-7). (The modern word *burse* and its equivalents in various languages—*bourse, börse, borsa, bolsa*—meaning an organized or regulated market for trade in either commodities or financial instruments, derives from the merchants' meeting hall in Bruges, which was identifiable by a sign showing three money bags, or purses.) As a rule the goods displayed were not actually exchanged on the spot; they were merely samples that could be inspected for quality. After orders were placed the goods would be shipped from warehouses. The use of credit was widespread, with most payments being made with financial instruments such as the bill of exchange, or by assignment in banks, instead of with hard cash. The banks were mostly private affairs, including many merchant firms such as the Fuggers, who carried on a banking business on the side, until the famous Amsterdamsche Wisselbank, or Bank of Amsterdam, was founded in 1609. This was a public bank in that it was founded under the auspices of the city itself. It was also an exchange bank, rather than a bank of issue and discount. Funds could be deposited here and transferred from one account to another on the books; but the bank did not issue banknotes or make loans to merchants by discounting commercial paper. Its principal function, which it performed well, was to provide the city and all the Dutch and foreign merchants who flocked there with a stable, reliable means of payment.

The regime of the colonial trades differed markedly from intra-European trade. The spice trade of the Portuguese Empire was a crown monopoly; the Portuguese navy doubled as a merchant fleet, and all spices had to be sold through the *Casa da India* (India House) in Lisbon. Portuguese seamen were allowed to bring other commodities as personal possessions, which they could subsequently sell—a practice that led to some dangerously overloaded ships on the return voyage—but, strictly speaking, no commerce existed between Portugal and the East except that organized and controlled by the state.

The situation was different beyond the Cape of Good Hope, however. There Portuguese merchants took part in the "country trade" (between ports on the Indian

FIGURE 5–7.   The Amsterdam Exchange. This painting by Emmanuel de Witte shows the interior court of the Amsterdam Exchange, or burse. (By De Witte-loan: Willem van der Vorn Foundation, Museum Boymans-van Beuningen, Rotterdam.)

Ocean, in Indonesia, and even in China and Japan) in competition with Muslim, Hindu, and Chinese merchants. For a time, as a result of a ban on direct trade with Japan imposed by the Chinese emperor, they had a virtual monopoly on trade between China and Japan. In the spice trade Goa was the eastern terminus as Lisbon was the western. The spices, of which pepper was quantitatively most important, were purchased in markets throughout the Indian Ocean and in the Spice Islands and brought to Goa to be loaded on homebound ships under the eyes of royal officials. Since Portugal produced few goods of interest to eastern markets, the outward cargoes consisted mainly of gold and silver bullion, along with some firearms and munitions. Overall, although the spice trade was lucrative for the government, it did little to develop or strengthen Portugal's own economy.

The trade between Spain and its colonies was similar. Technically, trade with the colonies was a monopoly of the Crown of Castile. For practical purposes the government turned it over to the *Casa de Contratación* (House of Trade), a guildlike organization in Seville that operated under the watchful eyes of government inspectors. All shipping between Spain and the colonies went out in convoys that, as eventually organized, departed from Seville in two contingents in the spring and late summer, wintered in the colonies, and returned as one fleet the following spring. The official reason for the convoy system was to protect the bullion supply from privateers and, in time of war, enemies; but it was also a convenient but ineffective means of attempting to prevent contraband trade. How much contraband there actually was is impossible to say, but it must have been substantial in view of the pitifully small amount of legal exports. Although there were fluctuations, the average number of ships in the convoys each year in the latter part of the sixteenth century was only about eighty, a small fraction of the number engaged in the Baltic trade, for example. At that time the European population in the New World amounted to well over 100,000. Even though they were, to a large extent, self-sufficient in terms of food supply, they still demanded European wines and olive oil, not to mention manufactured goods such as cloth, firearms, tools, and other hardware. It has been estimated that approximately half of all official bullion imports into Seville was required to purchase return cargoes, with an additional 10 percent or so absorbed by shipping and other commercial services. The crown, for its part, demanded the *quinto real* (royal fifth) of all bullion imports but, with other taxes, actually claimed about 40 percent of the total. As in the case of Portugal, Spain's fabulous empire did little to further the development of Spain's own economy and, as a result of short-sighted government policies, actually retarded it. We now turn to a consideration of those policies.

6

Economic Nationalism and Imperialism

The economic policies of nation-states in the period of Europe's second logistic had a dual purpose: to build up economic power to strengthen the state, and to use the power of the state to promote economic growth and enrich the nation. In the words of Sir Josiah Child, British merchant and politician of the late seventeenth century, "profit and power ought jointly to be considered." Above all, however, the states sought to obtain revenue, and frequently their need for revenue led them to enact policies that were detrimental to truly productive activities.

In pursuit of their objectives policymakers had to deal with the conflicting desires of both their own subjects and rival nation-states. In medieval times municipalities and other local government units had possessed extensive powers of economic control and regulation. They levied tolls or tariffs on goods entering and leaving their jurisdictions. Local guilds of merchants and artisans fixed wages and prices, and otherwise regulated working conditions. The policies of economic nationalism represented a transfer of these functions from the local to the national level, where the central government attempted to unify the state economically as well as politically.

At the same time they were seeking to impose economic and political unity on their subjects, the rulers of Europe were aggressively competing with one another for extension of territory and control of overseas possessions and trade. They did so partly to make their countries more nearly self-sufficient in time of war, but the very attempt to gain more territory or trade at the expense of others often led to war. Thus, economic nationalism aggravated the antagonisms engendered by religious differences and dynastic rivalry among the rulers of Europe.

Mercantilism: A Misnomer

Adam Smith, a Scottish philosopher of the Enlightenment and the founder of the modern science of economics, characterized the economic policies of his day (and of earlier centuries) under a single rubric, the *mercantile system*. In his view they were perverse, because they interfered with the "natural liberty" of individuals and resulted in what modern economists call a malallocation of resources. Although he condemned the policies as unwise and unjust, he attempted to systematize them—hence the term *mercantile system*—partly, at least, to highlight their absurdities. Drawing chiefly on British examples, he declared that the policies were devised by merchants and foisted on rulers and statesmen who were ignorant of economic affairs. Just as merchants are enriched to the degree their income exceeds their expenditures, nations, they argued (in Smith's construction), would enrich themselves to the extent that they sold more to foreigners than they purchased abroad, taking the difference, or the "balance of trade," in gold and silver. Hence, they favored policies that would stimulate exports and penalize imports (both of which were favorable to their own private interests), to create a "favorable balance of trade" for the nation as a whole.

For more than a century after Smith published his epochal *Inquiry into the Nature and Causes of the Wealth of Nations* in 1776, the term *mercantile system* had a pejorative connotation. In the latter part of the nineteenth century, however, a number of German historians and economists, notably Gustav von Schmoller, radically reversed that notion. For them, nationalists and patriots living in the wake of the unification of Germany under Prussian hegemony, *merkantilismus* (mercantilism) was above all a policy of state-making (*Staatsbildung*) carried out by wise and benevolent rulers, of whom Frederick the Great was the principal exemplar. In Schmoller's words, mercantilism "in its innermost kernel is nothing but state-making—not state-making in a narrow sense but state-making and national-economy-making at the same time."[1]

Subsequent scholars attempted to harmonize and rationalize these two basically divergent, even antagonistic, ideas. Thus, one can find in textbooks such definitions of mercantilism as the "theory" or the "system" of economic policy characteristic of early modern Europe or, more cautiously, as a "loosely knit body of ideas and practices which prevailed in western European countries and their overseas dependencies from around 1500 to perhaps 1800."[2] In view of these popular misconceptions and oversimplifications, it can scarcely be overemphasized that precious little "system" underlay economic policy, other than the need for revenue by financially hard-pressed governments, or that the theoretical underpinnings of economic policy were notoriously weak if not altogether absent; certainly there was no general consensus on either theory or policy.

There were, to be sure, some common themes or elements of economic policy, resulting from the similarity of needs and circumstances of the policy-making authorities, that is, the effective rulers or ruling classes. These are sketched out later. But at least as significant were the differences occasioned by different circumstances, and especially by the differing natures and compositions of the ruling classes. These are briefly touched on here, and elaborated more fully in the sections that follow.

Despite similarities, each nation had distinctive economic policies derived from peculiarities of local and national traditions, geographic circumstances, and most

---

[1] Gustav von Schmoller, *The Mercantile System and Its Historical Significance* (New York and London, 1896), p. 69.

[2] Edmund Whittaker, *Schools and Streams of Economic Thought* (Chicago, 1960), p. 31.

important, the character of the state itself. Advocates of economic nationalism all claimed that their policies were designed to benefit the state. But what was the state? It varied in character from the absolute monarchies of Louis XIV and most other continental powers to the burgher republics of the Dutch, the Swiss, and the Hanseatic cities. In no case did all or even a majority of the inhabitants participate in the process of government. Since the nationalism of the early nation-states rested on a class, not a mass, basis, the key to national differences in economic policy should be sought in the differing composition and interests of the ruling classes.

In France and other absolute monarchies the wishes of the sovereign were paramount. Although few absolute monarchs had understanding or appreciation of economic matters, they were accustomed to having their orders obeyed. The day-to-day administration of affairs was carried out by ministers and lesser officials who were hardly more familiar with problems of industrial technology and commercial enterprise, and reflected the values and attitudes of their master. Elaborate regulations for the conduct of industry and trade added to the cost and frustration of doing business, and encouraged evasion. On large issues absolute monarchs often sacrificed both the economic welfare of their subjects and the economic foundations of their own power through ignorance or indifference. Thus, in spite of its great empire, the government of Spain continually overspent its income, hamstrung its merchants, and steadily declined in power. Even France under Louis XIV, the most populous and powerful nation in Europe, could not easily support the continued drain of its wealth for the prosecution of Louis' territorial ambitions and the maintenance of his court. When he died France hovered on the brink of national bankruptcy.

The United Netherlands, governed by and for the wealthy merchants who controlled the principal cities, followed a more informed economic policy. Living principally by trade, they could not afford the restrictive, protectionist policies of their larger neighbors. They established free trade at home, welcoming to their ports and markets the merchants of all nations. On the other hand, in the Dutch Empire the monopoly of Dutch traders was absolute.

England lay somewhere near the center of the spectrum. The landed aristocracy intermarried with wealthy merchant families and mercantile-connected lawyers and officials, and great merchants had long taken a prominent part in government and politics. After the revolution of 1688–89 their representatives in parliament held ultimate power in the state. The laws and regulations they made concerning the economy reflected a balance of interests, benefiting the landed and agricultural interests of the nation while encouraging domestic manufactures and assisting shipping and trading interests.

The Common Elements

In the Middle Ages most feudal lords, especially sovereigns, owned "war chests" that were literally that: huge armored chests in which they accumulated coins and bullion to finance both anticipated and unexpected hostilities. By the sixteenth century the methods of government finance were somewhat more sophisticated, but the preoccupation with plentiful stocks of gold and silver persisted. This gave rise to a crude form of economic policy known as ''bullionism''—the attempt to accumulate as much gold and silver within a country as possible, and to prohibit their export by fiat, with the death penalty for violators. Spain's futile attempts to husband its New World treasure was the most conspicuous example of this policy, but most nation-states had similar legislation.

Since few European countries had mines producing gold and silver (and those that did, mainly in central Europe, were forced to shut down by the inundation of Spanish treasure in the midsixteenth century), the acquisition of colonies that possessed them was a major goal of exploration and colonization. Once again, the Spanish bonanza was the model to be emulated. The colonies of France, England, and Holland produced little gold or silver, however, so the only way for these countries to obtain supplies of the precious metals (apart from conquest and piracy, to which they also resorted) was through trade.

It was in this connection, as Adam Smith pointed out, that merchants were able to influence the councils of state, and it was they who devised the argument for a favorable balance of trade. Ideally, according to this theory, a country should only sell and should purchase nothing abroad. Practically, however, this was manifestly impossible, and the question arose: What should be exported and what imported? Because of the high incidence of poor harvests and periodic famines, governments sought plentiful domestic supplies of grain and other foodstuffs, and generally prohibited their export. At the same time they encouraged manufactures not only to have something to sell abroad but also to further self-sufficiency by broadening the range of their own production.

To encourage domestic production, foreign manufactures were excluded or forced to pay high protective tariffs, although the tariffs were also a source of revenue. Domestic manufactures were also encouraged by grants of monopoly and by subsidies (*bounties* in English terminology) for exports. If raw materials were not available domestically, they might be imported without import taxes, in contradiction of the general policy of discouraging imports. Sumptuary laws (laws governing consumption) attempted to restrict the consumption of foreign merchandise and to promote that of domestic products.

Large merchant navies were valued because they earned money from foreigners by providing them with shipping services and encouraged domestic exports by providing cheap transport—at least in theory (Fig. 6-1). Moreover, when the chief difference between a merchant ship and a warship was the number of guns it carried, a large merchant fleet could be converted to a navy in case of war. Most nations had ''navigation laws,'' which attempted to restrict the carriage of imports and exports to native ships, and in other ways promoted the merchant marine. Governments also encouraged fisheries as a means of training seamen and stimulating the shipbuilding industry, as well as of making the nation more self-sufficient in food supply and furnishing a commodity for export. The extensive herring fisheries of the Dutch were a prime example. Underlying the emphasis on merchant marines was the notion that a fixed and definite volume of international trade existed. According to Colbert, the principal minister of Louis XIV, all the trade of Europe was carried by 20,000 ships, more than three-fourths of which belonged to the

FIGURE 6–1.   The quay at Amsterdam. In the seventeenth century the Dutch merchant fleet was the envy of Europe, and Amsterdam was its principal port. This contemporary painting by Jacob van Ruisdael shows the busy port in action. (The small ships in the foreground are lighters, which ferried merchandise to and from larger ships anchored in the outer harbor.) (Copyright, The Frick Collection, New York.)

Dutch. Colbert reasoned that France could increase its share only by decreasing that of the Dutch, an objective he was prepared to make war to achieve.

Theorists of all nations stressed the importance of colonial possessions as an element of national wealth and power. Even if colonies did not have gold and silver mines, they might produce goods not available in the mother country that could be used at home or sold abroad. The spices of the Indies, the sugar and rum of Brazil and the West Indies, and the tobacco of Virginia served such purposes.

These were some of the notions concerning economic policy current in the sixteenth and seventeenth centuries. Usually they were not this clearly and simply spelled out, and they never commanded universal adherence, much less constituted a ''theory'' or ''system'' to guide the actions of rulers. In actual practice the legislation and other interventions of governments in the economic sphere consisted of a series of expedients, usually lacking an economic rationale and frequently producing unintended, deleterious results, as the following survey shows.

Spain and Spanish America

In the sixteenth century Spain was the envy and the scourge of the crowned heads of Europe. As a result of dynastic marriage alliances, its King Charles I (1516–56) inherited not only the kingdom of Spain (actually, the separate kingdoms of Aragon and Castile), but also the Habsburg dominions in central Europe, the Low Countries, and Franche-Comté. In addition, the kingdom of Aragon brought with it Sardinia, Sicily, and all of Italy south of Rome, and that of Castile contributed the newly discovered, still-to-be-conquered empire in America. In 1519 Charles became Holy Roman Emperor as Charles V.

This formidable political empire appeared to rest on substantial economic bases as well. Although Spain's agricultural resources were not the best, it inherited the elaborate Moorish system of horticulture in Valencia and Andalusia, and the wool of its Merino sheep was prized throughout Europe. It also had some flourishing industries, notably cloth and iron. Charles's possessions in the Netherlands boasted both the most advanced agriculture and some of the most prosperous industries in Europe. The Habsburg domains in central Europe contained, in addition to agricultural resources, important mineral deposits including iron, lead, copper, tin, and silver. Most spectacularly, gold and silver from its New World empire began to flow to Spain in large quantities in the 1530s and steadily increased to their peak levels in the last decade of the century, before they gradually subsided in the seventeenth century.

In spite of these favorable circumstances, the Spanish economy failed to progress—indeed, from about midcentury it regressed—and the Spanish people paid the price in the form of lowered standards of living, increased incidence of famine and plague, and ultimately, in the seventeenth century, depopulation. Although many factors have been adduced to account for the "decline of Spain," the exorbitant ambitions of its sovereigns and the short-sightedness and perversity of their economic policies must bear a large share of the responsibility.

Charles V deemed it his mission to reunify Christian Europe (Fig. 6-2). To this end he fought the Turks in the Mediterranean and Hungary, struggled against the rebellious Protestant princes of Germany, and feuded with the Valois kings of France, who had territorial ambitions in Italy and the Netherlands and felt threatened by the surrounding Habsburg dominions. Unable to gain permanent success on any of these fronts, he abdicated the throne of Spain in 1556, a tired and broken man. He had hoped to pass on his possessions intact to his son Philip, but his brother Ferdinand succeeded in wresting the Habsburg lands in central Europe and the title of Holy Roman Emperor after Charles's death in 1558. Philip II (1556–98) continued most of his father's crusades and even added England to Spain's list of enemies, with disastrous consequences when the "invincible" armada of 1588 was decisively routed. Scarcely a year passed that Spanish troops were not involved in warfare in some part of Europe, in addition to their role in conquering and governing America. Moreover, in addition to their bellicose tendencies, the Spanish monarchs demonstrated a penchant for monumental architecture and lavish court ceremonies.

FIGURE 6–2.   Empire of Charles V.

To finance their wars and conspicuous consumption, Charles and Philip relied, in the first instance, on taxation. In spite of their relative poverty, the Spanish people in the sixteenth century were the most heavily taxed of any in Europe. Moreover, the incidence of taxation was extremely uneven. Already at the end of the fifteenth century 97 percent of the land of Spain was owned by about 2 or 3 percent of the families (including the church), and this disparity increased during the sixteenth century. The great landowners, almost all of "noble" blood (*grandés, títulos, hidalgos,* and *caballeros*) in addition to the royal family itself, were exempt from direct taxation; thus, the burden fell principally on those least able to pay—artisans, tradesmen, and especially the peasants.

The crown obtained an unexpected source of revenue with the discovery of gold and silver in its American empire. Imports before 1530 were scarcely significant, but thereafter they rose steadily from about a million ducats per year in the 1540s to more than 8 million in the 1590s. (The figures refer to legal imports only, subject to taxation; illegal imports may have been almost as much.) As noted, the government acquired about 40 percent of the legal imports. Even so, in the last years of Philip's reign his share of the precious metals accounted for no more than about 20 to 25 percent of his total revenue.

To make matters worse, total revenue rarely equaled the vast government expenditures. This forced the monarchs to resort to yet a third source of finance, borrowing. (They also had recourse to other expedients, such as selling patents of nobility to wealthy merchants, but this sacrificed long-run tax revenue for one-time gains.) Borrowing was not a novelty for Spanish, or other, monarchs. Ferdinand and Isabella had borrowed to finance their successful war against Grenada, for example; and, according to popular legend, Isabella pawned her jewels to help finance Columbus's voyage. But under Charles and Philip deficit finance became a regular practice, like an addiction to a habit-forming drug. In fact, Charles, early in his reign, had borrowed huge sums from the Fuggers and other German and Italian bankers to bribe the electors who named him Holy Roman Emperor. The interest on those debts, and others that he incurred, mounted continuously. The lenders, who came to include Flemish and Spanish as well as German and Italian bankers, and even some well-to-do merchants and nobles, obtained contracts that specified particular tax revenues or portions of the next shipment of American silver as security for their loans. As early as 1544 two-thirds of the regular annual revenue was pledged for debt payment, and in 1552 the government suspended all interest payments. In 1557 the burden had become so heavy that the government repudiated a substantial portion of its debts, an event frequently referred to as "national bankruptcy." But governments, unlike business firms, are not liquidated when bankrupt. Instead, short-term debts were reorganized as long-term obligations, both the principal and the rate of interest were reduced, and the cycle began anew, but always under more onerous conditions for the borrower. On eight occasions (in 1557, 1575, 1596, 1607, 1627, 1647, 1653, and 1680) the Spanish Habsburgs declared royal bankruptcy. Each time resulted in a financial panic, the actual bankruptcy and liquidation of many bankers and other investors, and disruption of ordinary commercial and financial transactions.

Financial mismanagement was not the only way the government hobbled the economy, although many of its interventions were occasioned by its fiscal needs. Royal favoritism in behalf of the Mesta, the sheepowners guild, has already been mentioned (see p. 111). This favoritism culminated in a decree of 1501 that reserved in perpetuity for sheep pasturage all land on which sheep had ever grazed, regardless of the wishes of the owners. By such measures, the government sacrificed the interests of the cultivators, and ultimately those of the consumers, for the sake of increased taxes from the privileged sheepowners.

In a similar measure, Ferdinand and Isabella in 1494 created the *Consulado* of Burgos, a merchant guild, and conferred on it a monopoly of the export trade in raw wool. Burgos, although a flourishing market town, was more than one hundred miles from the nearest port. All wool destined for export, from whatever part of Spain, had first to be transported to Burgos and thence, by mule trains, to Bilbao on the Viscayan coast for shipment to northern Europe. The merchants of Burgos thus obtained a collective monopoly on Spain's most valuable export commodity, at the expense of domestic producers as well as northern consumers. The *Consulado* of Burgos also served as a model for the *Casa de Contratación,* set up in Seville less than a decade later to control the trade with America. Throughout their reign Ferdinand and Isabella favored the extension of guild control, and thus of monopoly, to increase tax revenues. Their successors, no less financially straitened, did nothing to lessen that control.

The absence of any systematic long-range economic policy is vividly illustrated in the histories of two of Spain's most important economic activities, cereal production and cloth manufacturing. Cereal production, although hindered by the privileges accorded to the Mesta, prospered during the first third of the sixteenth century as a result of both the increase in population and the mild rise in prices occasioned by the initial influx of American treasure. As the price increase accelerated, the government responded to consumer complaints by imposing maximum prices on bread grains in 1539. Since costs continued to increase, the result was that arable land was devoted to other purposes than growing grain, and the grain shortage became worse. To counteract the shortage the government admitted foreign grain, previously prohibited or subject to high tariffs, duty free; but this discouraged cereal growers even more. Much land went out of production altogether, and Spain became a regular importer of bread grains.

The situation in the cloth industry was much the same. At the beginning of the sixteenth century Spain exported fine cloth as well as raw wool. The expansion of domestic demand and, especially, that of the colonies in America raised costs as well as prices. The supply could not keep up with increasing demand. In 1548 foreign cloth was admitted duty free, and in 1552 the export (except to the colonies) of domestic cloth was prohibited. The immediate result was a severe depression in the cloth industry. The export prohibition was rescinded in 1555, but by that time the loss of foreign markets and the inflationary cost increases had deprived Spain of its competitive advantage. Spain remained a net importer of cloth until the nineteenth century.

Conceivably, with a truly enlightened economic policy, Charles V could have created lasting prosperity for his vast empire by converting it into a free trade area or a customs union. There is no evidence, however, that such a thought ever crossed his mind. In the first place, each region, principality, and kingdom within the empire was conscious of its own traditions and privileges, and probably would have resisted such a move. More important, from the point of view of the policymaker, the monarch was too dependent on customs revenues to abolish the internal tariffs and tolls on commerce among the various components of the empire. Even after the union of the crowns of Castile and Aragon the citizens of one were treated as foreigners in the other; each maintained its own tariff barriers against the other, and even their separate coinage systems. Other Habsburg possessions were in no better position. The merchants and industrialists of the Low Countries owed their substantial penetration of Spanish markets to their superior competitiveness rather than to any special privileges.

Even in their religious policies the Spanish monarchs contrived to damage the well-being of their subjects and weaken the economic bases of their own power. Early in their reign Ferdinand and Isabella obtained permission from the papacy to establish a Holy Office, a branch of the infamous Inquisition, over which they exercised direct

ואמרו עבדי המלך אל עני עבדי יהוה בני יהוה קם לכם לפני נשלח את האנשים

ויביאו את יושם

FIGURE 6–3. Expulsion of the Jews. This illustration shows the King of Castile's advisers urging him to expel the Jews from his kingdom, around 1300. Ferdinand and Isabella actually did so in 1492, with deleterious consequences for their country. (Weidenfeld & Nicolson Archives.)

royal authority. The initial targets of the Spanish Inquisition were backsliders among the *conversos*, Jews who had converted to Catholicism, in fact or only nominally, even though practicing Jews were still officially tolerated. Many Jews and *conversos* were among the wealthiest and most cultivated of Spanish commoners; their numbers contained many merchants, financiers, physicians, skilled artisans, and other economically successful persons. Some wealthy *conversos* intermarried with the nobility; even Ferdinand had some Jewish blood. The climate of fear created by the Inquisition led many *conversos* and Jews alike to emigrate, taking with them their wealth as well as their talents (Fig. 6-3). Then, in 1492, shortly after the successful conquest of Granada, the Catholic kings decreed that all Jews must either convert to

Catholicism or leave the country. Estimates of the numbers that left range from 120,000 to 150,000, but the damage to the economy was far greater, proportionally, than the ratio of the refugees to the total population.

The monarchs followed a similar policy with respect to their other religious minority, the Muslim Moors. At the capitulation of the Moorish kingdom of Granada the Catholic kings had decreed a policy of religious toleration toward the Moors (contrary to their almost simultaneous persecution of the Jews); but within less than a decade they began to persecute the Moors as well. In 1502 they decreed the conversion or expulsion of all Moors. Since the majority of Moors were humble agricultural laborers, they had no resources with which to emigrate and became nominal Christians, the Moriscos. For more than a century they remained, barely tolerated, many still faithful to their original religion; they did much useful labor, especially in the rich agricultural provinces of Valencia and Andalusia. In 1609 another Spanish government, seeking to camouflage the news of another military defeat abroad, ordered the expulsion of all Moriscos. Not all were actually deported, but many were, and the government thereby deprived itself of another badly needed economic resource.

Spain's policies toward its American empire were as short-sighted and self-defeating as its domestic policies. As soon as something of the nature and extent of the New World discoveries began to be realized, the government imposed a policy of monopoly and strict control. In 1501 foreigners (including Catalans and Aragonese) were prohibited from settling in or trading with the new colonies. In 1503 the *Casa de Contratación* was created in Seville with a monopoly of trade. All merchant ships had to sail with the armed convoys, as previously described. These convoys were very expensive and inefficient, although they did succeed in one of their principal objectives—protection of the bullion shipments. Not until 1628 was a bullion fleet intercepted, by the Dutch; the English did it again, in 1656 and 1657, each time provoking a major financial crisis.

The monopolistic and restrictive policies proved so unworkable that the government soon had to back down. In 1524 it allowed foreign merchants to trade with, but not settle in, America. This provided such a bonanza for Italian and German merchants that, in 1538, the government rescinded the policy and restored the monopoly to Castilians. But many of the Castilian firms that participated in the trade through the *Casa de Contratación* were actually mere fronts for foreign, especially Genoese, financiers. From 1529 to 1573 ships from ten other Castilian ports were allowed to trade with America, but they were obliged to register their cargoes in Seville, and to land their homeward-bound shipments there; because of the increased costs, this permission accomplished little. Instead, the policies of monopoly and restriction encouraged evasion and smuggling, by both Spanish and other shippers. In 1680, as a result of the silting of the Guadalquivir River, which prevented large sailing ships from reaching Seville, the monopoly of American trade was moved to Cadiz; but by that time the bullion shipments were a mere trickle, and the glory days were over.

Policy within the empire was no more enlightened. Intracolonial trade was discouraged, although some did take place, especially between Mexico and Peru. Vineyards and olive orchards were officially prohibited for the benefit of domestic producers and exporters. Although a few industries were permitted, such as the silk industry of New Spain (Mexico), the general policy was to reserve the market for manufactured goods in the colonies for the producers in the metropolis; but since Spain's own industries were in more or less continual decline, the net effect was to stimulate demand for the products of Spain's European rivals.

The essential absurdity of Spain's colonial economic policies is highlighted by its treatment of its sole Pacific possession, the Philippine Islands. Although within the Portuguese orbit, as determined by the papal line of demarcation, the Philippines became a Spanish possession by virtue of Magellan's discovery. Filipinos and other Asians carried on trade among themselves and with neighboring Asian areas, including China; but the only trade permitted by the Spanish authorities with Europe was indirect, through Mexico and Spain itself. Each year a single ship (in principle, although there were interlopers), the Manila galleon, set forth from Acapulco laden mainly with silver from Peru and Mexico, destined eventually for China and other Asian recipients. The round trip required two years; the ship wintered in Manila, where it loaded spices, Chinese silks and porcelains, and other luxury products of the East. Goods not sold in the Mexican and Peruvian markets were shipped overland to Vera Cruz, where they were picked up by the *flota* for the trip to Spain. Unsurprisingly, few commodities could bear the high cost of such an itinerary.

Portugal

One of the most remarkable feats of Europe's age of expansion was the achievement by Portugal, a small, relatively poor country, of dominion over a vast seaborne empire in Asia, Africa, and America. At the beginning of the sixteenth century Portugal's population amounted to scarcely more than a million inhabitants. Outside of the few, small cities, the economy was predominantly of the subsistence variety. Along the sea coast fishing and salt-panning were the most important non-agricultural occupations. Foreign trade was of minor but growing significance. Exports were almost entirely primary products: salt, fish, wine, olive oil, fruit, cork, and hides. Imports consisted of wheat (in spite of its small population and agrarian orientation, the country was not self-sufficient in bread grains) and such manufactured products as cloth and metalwares.

How did such a small, backward country acquire mastery of its huge empire so quickly? The question cannot be answered simply or briefly. Many factors were involved, not all of them subject to precise measurement. One was good fortune: at the time Portugal made its breakthrough into the Indian Ocean the polities in that area were unusually weak and divided, for reasons independent of developments in Europe. Another, less accidental but nevertheless fortuitous, factor was the accumulated knowledge and experience of the Portuguese in ship design, navigational techniques, and all related arts—a continuing legacy of the work and dedication of Prince Henry. Yet another factor is more speculative, but important nonetheless: the zeal, courage, and rapacity of the men who ventured across the seas in the service of their God and king, and in search of riches.

In the first flush of their Asian discoveries and success, the Portuguese paid little attention to their African and American possessions. The spice trade and its auxiliaries promised quick and lavish returns to king and commoner alike, whereas the development of the sultry, savage tropics of Brazil and Africa would clearly be expensive and uncertain long-term ventures. For the sixteenth century as a whole an estimated annual average of about 2400 persons, most of them young able-bodied males, sought their fortunes overseas, mainly in the East. In the 1530s, however, the Portuguese crown became alarmed by the activities of French free-booters along the Brazilian coast, and undertook to secure Portuguese settlers for the mainland. The king made land grants to private individuals, not unlike the grants of the English crown to Lord Baltimore and William Penn in the seventeenth century, and hoped in this way to secure settlers at little expense to himself. The early colonies did not flourish, however; the local Indian population, sparse, primitive, and frequently hostile, provided neither markets for Portuguese produce nor reliable labor for the Brazilian economy. Not until the 1570s, with the transplantation of sugar cane from the Madeiras and Sao Tomé Island, and the techniques for its cultivation with African slave labor, did Brazil become an integral part of the imperial economy. Soon afterward, in 1580, Portugal fell to the crown of Spain, and although Philip II promised to preserve and protect the Portuguese imperial system, it suffered from the depredations of the Dutch and others in both East and West. Portuguese plans for developing and exploiting an African empire were repeatedly postponed until the twentieth century.

The Portuguese crown's legal monopoly of the spice trade provoked mocking references to the "Grocer King" and the "Pepper Potentate," but the reality behind those terms was quite different from what one might suspect. In the first place, Portugal never secured effective control of the sources of supply of spices. True, in the first years of its explosive entrance into the Indian Ocean it did severely disrupt the traditional overland carriage of spices to the eastern Mediterranean, thereby temporarily depriving the Venetians of their lucrative distribution trade; but the traditional routes were eventually reestablished, and by the end of the sixteenth century they supported a larger volume of commerce than ever before—larger even than that of the Portuguese fleets. For this there were two principal reasons. First, the Portuguese were simply spread too thin. Even at the peak of their maritime strength in the 1530s they possessed only about 300 ocean-going vessels, and some of those were employed on the Brazilian and African routes. It proved impossible to police the greater part of two oceans with so few men and ships. Second, the crown was obliged to rely either on royal officials for enforcement of its monopoly or on contractors who leased, or "farmed," a portion of the monopoly. In both cases it suffered from inefficiency and fraud. The royal officials, although endowed with extensive powers, were not well paid, and frequently supplemented their meager salaries by taking bribes from smugglers or engaging in illicit trading themselves. The crown contractors, of course, had ample incentive to violate their contracts when possible.

The spice trade was the most famous, but it was only one of many branches of commerce that the Portuguese kings tried to monopolize for fiscal reasons. Even before the opening of the Cape route the Portuguese crown monopolized trade with Africa, whose most valuable exports were gold, slaves, and ivory. With the discovery of the Americas the demand for slaves increased enormously, and the Portuguese kings were the first to benefit; the actual slave traders were private contractors who operated under license from the crown, paying it a share of the profits. In the eighteenth century the discovery of gold and diamonds in Brazil presented the crown with a new Eldorado. As before, it tried to monopolize the commerce and prohibit the export of gold from Portugal, but without success. English warships, which had special status in Portuguese waters because of treaty arrangements, were common vehicles of the contraband trade.

The crown's attempts at monopoly did not stop with the exotic products of India and Africa, but extended to such domestically produced staples as salt and soap and, among the most profitable, the tobacco of Brazil. And what the crown could not monopolize it tried to tax. This effort was notable with Brazil's principal export, sugar; but all commodities involved in both international and intraimperial trade were heavily taxed. At the beginning of the eighteenth century almost 40 percent of the value of goods legally shipped from Lisbon to Brazil represented customs duties and other taxes.

The motive of both monopoly and taxation was, of course, to gain revenue for the crown. But, given the inefficiency and venality of the royal agents, evasion was relatively easy and widespread. Moreover, the higher the rate of taxation, the greater was the incentive to evade. It was a vicious circle as far as the crown was concerned. As a result the Portuguese kings were forced to borrow, as their Spanish counterparts had. For the most part they borrowed for short terms at high interest rates against future deliveries of pepper or other highly salable commodities. The lenders were most often foreigners—Italians and Flemings—or the king's own subjects, the "new Christians."

"New Christians" was the term euphemistically applied to Portuguese citizens of Jewish ancestry. Some of them actually converted to Christianity, but many secretly maintained their old belief and customs or at least they were widely suspected of so doing. King Manuel had ordered the forcible conversion of Jews in 1497 in imitation and at the instance of the Spanish monarchs, but for several decades no repressive measures were taken to enforce the edict. Indeed, "new" and "old" Christians, Jews and gentiles, continued to live together in harmony and even to intermarry to such an extent that by the end of the sixteenth century it was estimated that as much as one-third of the Portuguese population had some Jewish blood. Eventually, however, Portugal obtained its own branch of the Inquisition, and its zeal in preserving and promoting the one true faith rivaled that of its Spanish counterpart. Citizens were encouraged to inform on one another; the informer's identity was not revealed, and the burden of proof was on the accused. Even such innocent acts as changing into a clean shirt or blouse on a Saturday could be used as "evidence" of proscribed beliefs. As a result of the practices of the Inquisition an atmosphere of mutual suspicion and distrust plagued Portuguese life for centuries, and Portugal lost much wealth and many skilled workers and professional people to more tolerant countries, the Dutch Netherlands in particular.

Central, Eastern, and Northern Europe

The whole of central Europe, from northern Italy to the Baltic, was nominally united in the Holy Roman Empire. In fact, the territory was organized into hundreds of independent or quasi-independent principalities, both lay and ecclesiastical, ranging in size from the estate of a single imperial knight to the Habsburg crownlands of Austria, Bohemia, and Hungary. After the Protestant Reformation, in particular, during which many secular and even some ecclesiastical lords adopted the new religion to gain control of church property, the authority of the emperor was sharply curtailed. Even within their own territories the Habsburgs, who were also hereditary emperors, had difficulty enforcing their authority over regional aristocracies and municipal bodies. The struggle between local particularism and the centralizing tendencies of the more powerful monarchs and princes occupies a large part of the history of early modern Europe, especially in central and eastern Europe; and in that struggle economic factors sometimes played a crucial role.

In Germany the advocates of economic nationalism propounded a series of principles or maxims that almost deserve to be called a system, or at least a quasi-system. Writers in this tradition are usually referred to as cameralists, from the Latin word *camera*, which in the German usage of that time meant the treasure chest or treasury of the territorial state. Most of those writers were active or former civil servants—that is, servants of the territorial princes who were striving for both political and economic autonomy. Some notion of the tenor of the policies they advocated can be gained from the title of one of the most influential of their books, *Oesterreich über Alles wann es nur will* ("Austria over all if it only will"), by Philipp W. van Hornigk (1684). In their concern for strengthening the territorial state they advocated measures that would, in addition to filling the state's coffers, reduce its dependence on other states and make it more self-sufficient in time of war: restrictions on foreign trade, promotion of domestic manufactures, reclamation of wastelands, provision of employment for the "idle poor" (which in some instances amounted to forced labor), and so on. In the eighteenth century special professorial chairs in *Staatswissenschaft* (science of the state) were established in several German universities to train future civil servants. For the most part the German states were too small and lacked the necessary resources to become truly self-sufficient; there were, however, a few examples of policies that succeeded in enhancing the power and authority of the territorial rulers, although at the expense of the welfare of their subjects.

The most spectacular instance of a successful policy of centralization is to be found in the rise of Hohenzollern Prussia. It was this success that led some historians and economists to reverse the prevailing condemnation of the policies of economic nationalism (see preceding, p. 131). The Hohenzollern dynasty became rulers of the Electorate of Brandenburg, centered on the city of Berlin, in the fifteenth century. The Hohenzollerns gradually expanded their possessions by means of inheritance, notably by the acquisition of East Prussia in 1618. The Thirty Years War caused great devastation but, beginning with the accession of Frederick William, the "Great Elector," in 1640, a succession of able rulers built Brandenburg-Prussia into one of the largest and most powerful nations in Europe, the precursor of modern Germany. The means they used included some of the standard instruments of so-called mercantilist policy, such as protective tariffs, grants of monopoly and subsidies to industry, and inducements to foreign entrepreneurs and skilled workers to settle in their underpopulated territories (notably French

Huguenots after the revocation of the Edict of Nantes in 1685); but more important for the success of their endeavor was their careful management of the state's own resources. Through centralization of their administration, requirements of strict accountability on the part of the corps of professional civil servants that they built up, punctilious collection of taxes, and frugality in expenditure, they created an efficient state mechanism that was quite exceptional in the Europe of its day. Their one notable extravagance was their army, which at times accounted for more than half the state budget. A later Prussian general remarked that Prussia was "not a country with an army, but an army with a country that served as its headquarters and food supplier." But, frugal and cautious as they were, the Hohenzollern kings rarely committed their army to battle, and then only for limited engagements. In 1740, for example, young Frederick II ("the Great") launched an unprovoked surprise invasion of Habsburg Silesia, thereby igniting the War of the Austrian Succession; but as soon as the Austrians acquiesced to his occupation of the mineral-rich province, he withdrew from the war, leaving his allies to fight on.

The Prussian kings used their army to their advantage not only militarily and politically, but also economically. Because of its awesome reputation, they were able to obtain subsidies from their allies, thereby avoiding the necessity for borrowing, a process that cursed the reigns of most other absolute monarchs. They also made good use of their crown domains, which included, besides agricultural estates, coal mines, iron foundries, and other productive enterprises; through good management and careful accounting, these domains were made to produce up to 50 percent of total state revenues. As efficient and powerful as the state was, however, the country's economy was only moderately prosperous by the standards of the day. The overwhelming majority of the productive population was still engaged in low-productivity agriculture, and Prussia was far from the great industrial power that Germany became at the end of the nineteenth century.

At the opposite extreme from the rise of Prussia was the disappearance of the kingdom of Poland. Prior to 1772 Poland was the third largest state in Europe in area and fourth largest in population; but in that year its more powerful neighbors, Russia, Prussia, and Austria, began the process of partition that by 1795 eliminated Poland from the political map. As with the rise of Prussia, the decline and fall of Poland was caused more by military and political factors, such as the weak elective kingship and the *liberum veto*, by which any single member of the *sejm* (parliament) could nullify the actions of the entire session, than by purely economic ones; but the poverty and backwardness of the economy was a contributing factor. About three-quarters of the population were legal serfs, bound to the land and with no rights other than those accorded them by their masters. The Polish nobility was fairly numerous, amounting to about 8 percent of the total population; but the great majority was also poor and virtually landless. Most of the land, the prime source of wealth in the country, was controlled by fewer than two dozen families. In the sixteenth and seventeenth centuries Poland exported large quantities of grain to the West, principally through Danzig to the Amsterdam market; but as agricultural production in the West increased in the eighteenth century, demand for Polish grain declined, and the country reverted to subsistence agriculture.

Although the absence of an effective central authority made a coherent eco-

FIGURE 6–4.   Courland and its neighbors.

nomic policy impossible for Poland, some of its constituent parts did have one. The Duchy of Courland is an example (Fig. 6-4). Under the energetic Duke James (or Jacob) in the midseventeenth century (1638–1682), Courland, which occupied a part of the area of present-day Latvia, became the very model of a mercantilist state. James attempted to promote industry by means of protective tariffs and subsidies, built a merchant fleet and a navy, and even purchased the island of Tobago in the West Indies and a small colony at the mouth of the Gambia River in West Africa. This fledgling development was cut short, unfortunately, by the Swedish–Polish war of 1655–60, during which James was captured and his capital pillaged. Like Sisyphus, he returned to his task after the war, but Courland failed to develop a dynamic economy, and disappeared from the map along with Poland in 1795. Courland's experience illustrates the limited effectiveness of deliberate state policy in the early modern era.

The limitations on the ability of the state to shape the economy were highlighted even more by the history of Russia, the largest and one of the most powerful states in Europe. In the sixteenth and seventeenth centuries Russia developed, politically and economically, largely in isolation from the West. Virtually landlocked, it had very little long-distance trade, although after 1553 a trickle went in and out via the far northern port of Archangel, open for only three months each year. The vast majority of the population engaged in subsistence agriculture, in which the institution of serfdom loomed large and actually increased in severity over the centuries. Meanwhile, in spite of numerous revolts, civil wars, and palace coups, the authority of the tsar grew stronger. In 1696, when Peter I ("the Great") became sole ruler, his power within the Russian state was unchallenged.

Peter set out deliberately to modernize—that is, to westernize—his country, including its economy. In addition to such petty measures as obliging his courtiers to wear Western-style clothing and to shave their beards, he traveled extensively in the West, observing industrial processes as well as military fortifications and procedures. He gave subsidies and privileges to Western artisans and entrepreneurs to settle in Russia and practice their crafts and commerce. He built the city of St. Petersburg, his "window on the West," on land recently conquered from Sweden at the head of the Gulf of Finland, an arm of the Baltic Sea. This gave him a more convenient port than Archangel, and he set out to build a navy. Underlying all of Peter's policies and reforms was his desire to expand his influence and territory and to make Russia a great military power. (The country was at war, usually offensive, for all but two years of Peter's long reign.) To this end he instituted a new and, he hoped, more efficient system of taxation and reformed his central administration whose function was, as he said, "to collect money, as much as possible, for money is the artery of war." When domestic industries could not meet his demand for military goods he set up state-owned arsenals, shipyards, foundries, mines, and cloth factories, staffed in part by Western technicians who were supposed to train a native labor force; but since the native labor supply consisted mainly of illiterate serfs, who were bound to their occupations whether they liked it or not, the effort met with little success. Only in the copper and iron industries of the Ural Mountains, where ore, timber, and water power were plentiful and cheap, did viable enterprises emerge from their hothouse atmosphere. After Peter's death most of the enterprises he had established withered away, his navy fell into ruin, and even his system of taxation, extremely regressive in that the burden fell mainly on the peasants, yielded inadequate revenue to support the army and the ponderous bureaucracy. One of his successors, Catherine (also "the Great"), was responsible for two innovations in state finance that had deleterious effects on the economy: foreign borrowing and enormous issues of fiduciary (paper) currency. Meanwhile the truly productive forces in the economy, the peasants, toiled away with their traditional techniques, gaining a bare subsistence for themselves after the exactions of their masters and the state.

In the sixteenth and seventeenth centuries Sweden played a role as a great political and military power that is surprising in view of its small population. Its success resulted partly from its abundance of natural resources, especially copper and iron, both essential for military power, and partly from the administrative efficiency of its government. The Swedish monarchs early achieved a degree of absolute power within their kingdom that was unrivaled elsewhere in Europe, even in such absolutist states as France and Spain. Moreover, they used their power wisely on the whole—with the exception of their rash military ventures, which ultimately led to their defeat and retrenchment—at least in the economic sphere. They abolished the internal tolls and tariffs that hindered commerce in other countries, standardized weights and measures, instituted a uniform system of taxation, and undertook other measures that favored the growth of commerce and industry. Not all policies were equally favorable—for example, the restriction of foreign trade to Stockholm and a few other port cities—but on the whole they gave free reign to both native and immigrant entrepreneurs (especially Dutch and Walloon, who brought special skills and knowledge as well as capital) to develop Sweden's resources. In the eighteenth century, after the decline of its political power, Sweden became the leading supplier of iron on the European market.

Italy has been excluded from this survey of the policies of economic nationalism because, for most of the early modern period, it was the victim of great power rivalries. Repeatedly invaded, occupied, and dominated by the military forces of France, Spain, and Austria, its city-states and small principalities had little opportunity to initiate or execute independent policies. One exception, however, the Venetian Republic, managed to retain both political independence and a modicum of economic prosperity until it was overrun by the French in 1797. At the end of the fifteenth century Venice was at the height of its commercial supremacy, with extensive possessions in the Aegean and Adriatic as well as on the mainland of Italy (Fig. 6-5). The advance of the Ottoman Turks, the discovery of the sea route to the Indian Ocean, and the gradual shift of Europe's economic center of gravity from the Mediterranean to the North Sea all combined to force Venice to the defensive. The Venetians reacted to changed circumstances by reallocating their capital and other resources. In the sixteenth century they developed an important woolen industry to supplement their already famous luxury products, such as glassware, paper, and printing. When the woolen industry encountered stiff competition from the Dutch, the French, and the English in the seventeenth century, many Venetian families invested in agricultural improvement on the mainland. The government, an oligarchy composed of representatives of the most important families, attempted to stave off commercial and industrial decay, but with little success. The average value of Venetian commerce and industry steadily declined. At the end of the seventeenth century production of woolen cloth was less than 12 percent of what it had been at the beginning of the century. Venice stagnated while Europe expanded.

Colbertism in France

The archetypical example of economic nationalism was the France of Louis XIV. Louis provided the symbol—and the power—but responsibility for policy making and implementation devolved on his principal minister for more than twenty years (1661–83), Jean-Baptiste Colbert. Colbert's influence was such that the French coined the term *colbertisme*, more or less synonomous with mercantilism as that word is used in other languages. Colbert attempted to systematize and rationalize the apparatus of state controls over the economy that he inherited from his predecessors, but he never fully succeeded, even to his own satisfaction. The main reason for this failure was his inability to extract enough revenue from the economy to

FIGURE 6–5. The port of Venice. The once major entrepôt of Venice steadily receded in the seventeenth and eighteenth centuries, but did not lose its commercial functions entirely. This early eighteenth-century view of the port is by its native artistic genius, Canaletto. (Reproduced by courtesy of the Trustees, The National Gallery, London.)

149

finance Louis' wars and extravagant court. That, in turn, resulted partly from France's haphazard system of taxation—if it could be called a system at all—which Colbert was never able to reform.

In principle, under the medieval theory of kingship, the king was supposed to be supported by the produce of his own domains, although his subjects, acting through representative assemblies, could grant him "extraordinary" tax revenues in time of emergency, such as war. In fact, by the end of the Hundred Years War several such "extraordinary" taxes had become permanent parts of the royal revenues. Moreover, by the end of the fifteenth century the king had won the power to raise tax rates and to impose new taxes by decree, without the consent of any representative assembly. By the end of the sixteenth century, as a result of increased taxes, the price inflation, and the real growth of the economy, royal tax revenue had increased sevenfold in the course of the century and tenfold since the end of the Hundred Years War in 1453. Even this fiscal bonanza did not suffice to cover the expenses of the Italian campaigns, the long series of wars between the Valois kings of France and the Habsburgs that spanned the first three-fifths of the sixteenth century, and the civil and religious wars that followed. Thus, the kings were obliged to resort to other expedients to raise funds, such as borrowing and the sale of offices.

French kings had borrowed in the Middle Ages, especially during the Hundred Years War, but not until the reign of Francis I (1515–47) did a royal debt become a permanent feature of the fiscal system. Thereafter the debt rose steadily except on those occasions when the crown arbitrarily suspended interest payments and wrote down the value of the principal. The effect of such partial bankruptcies was to make it even more difficult for the monarchy to borrow; but borrow it did, at ever more onerous interest rates. In addition to borrowing, the crown raised revenue through the sale of offices (judicial, fiscal, and administrative). The sale of offices was not unknown in other lands, but in France it became standard practice. Some authorities aver that it produced as much as one-third of royal revenues; that is probably an exaggeration, but it is safe to say that in many years it produced as much as 10 or 15 percent of those revenues. The practice succeeded in its immediate purpose, but in the long run its effect was wholly deleterious. It created a host of new offices that were without function or whose functions were inimical to the masses (in some instances two or more individuals were appointed to the same office) with increased charges to the government and, ultimately, the taxpayers; it placed in those offices men without competence or even interest in discharging their duties, thus encouraging inefficiency and corruption; it allowed wealthy commoners access to the *noblesse de la robe*, diverting their wealth from productive enterprise to the uses of the state and exempting them from further taxation.

In spite of the multiplication of offices and officials, the crown was obliged to rely on private enterprise to collect the bulk of its taxes through the institution of tax farmers. These individuals, usually wealthy financiers, contracted with the state to pay a lump sum of money in return for the privilege of collecting certain specified taxes, such as *aides* (excise taxes applied to a wide variety of commodities), the hated *gabelle* (originally an excise tax on salt, which became a fixed impost regardless of the amount of salt purchased or consumed), and especially the numerous tariffs and tolls that were collected on merchandise in transit, within the country as well as on the frontiers. Colbert wished to reform the system, especially by abolishing the internal tariffs and tolls, but the crown's need for revenue was too great, and he could not. In the latter part of the eighteenth century, under the influence of the Enlightenment and the Physiocrats, some of Colbert's successors, notably the economist Jacques Turgot, actually attempted to reform the system and create internal free trade; but the opposition of the vested interests, including officials, tax farmers, and the aristocracy, forced him out of office. In the end it was the failure of the fiscal system to produce sufficient revenue that led to the assembly of the Estates-General of 1789, the beginning of the end of the Old Regime.

In addition to their attempts to reform and increase the proceeds of the tax system, Colbert, his predecessors, and his successors tried to increase the efficiency and productivity of the French economy in much the same way that a drill sergeant tries to enhance the performance of his soldiers. They issued numerous orders and decrees with respect to the technical characteristics of manufactured items and the conduct of merchants. They fostered the multiplication of guilds with the avowed intention of improving quality controls, even when their real objective was to obtain more revenue. They subsidized *manufactures royales,* both to supply their royal masters with luxury goods and to establish new industries. To secure a "favorable" balance of trade, they created a system of prohibitions and high protective tariffs.

The French kings began their attempts to centralize their power over the country, and with it their control of the economy, in the aftermath of the Hundred Years War. Louis XI (1461–83) prohibited French merchants from attending the fairs of Geneva and at the same time granted special privileges to those of Lyons, which may have contributed to the growth of the latter. He also extended royal control over municipal guilds, but that was mainly to increase the royal revenues. One result of the Italian wars was to stimulate aristocratic demand for the luxury consumer goods that the king and his officers had encountered there. Francis I and his successors recruited Italian artisans and established them in privileged *manufactures royales* for the production of silk, tapestries, porcelain, fancy glassware, and such; these had a significant cultural and artistic importance in succeeding centuries, but, except for the silk industry, their immediate economic impact was negligible. The religious civil wars that ensued from 1562 to 1598 occasioned much damage and destruction, and rendered coherent, consistent economic policies impossible.

The man who, even more than Colbert, should be regarded as the founder of the French tradition of *étatisme* (statism) in economic affairs was the Duke de Sully, principal minister of Henry IV (1589–1610). Sully is generally regarded as an energetic, efficient administrator who both increased revenues and held down expenses, but his ambiguous legacy is best symbolized by two measures (usually attributed to Henry) taken in 1598, soon after Henry had consolidated his powers as king. On the one hand, in the Edict of Nantes Henry granted limited toleration to Protestants (Sully was one of the principal advisers who persuaded Henry to convert to Catholicism to strengthen his hold on the throne, but Sully himself remained a Protestant). On the other, he arbitrarily, by decree, wrote down both the principal and the interest rates on outstanding royal debts—in effect, a royal declaration of partial bankruptcy. Although a strong advocate of royal absolutism, as a shrewd financier Sully opposed the subsidies involved in the creation of *manufactures royales,* but Henry created them anyway; of the forty-eight in existence at his death in 1610, forty had been established since 1603. More characteristic of Sully's achievements was his success in raising the yield of the royal monopolies on the production of saltpeter, gunpowder, ammunition and especially salt. These monopolies had existed on the statute books for many decades, but enforcement was lax; Sully enforced them rigorously, with the result that the yield of the *gabelle,* for instance, nearly doubled during his tenure.

Richelieu and Mazarin, Sully's successors as principal minister under Louis XIII and during the minority of Louis XIV, lacked both interest and ability in financial and economic affairs. With their principal objective (after the maintenance of their own positions) the aggrandizement of France in the international arena, they allowed state finances to slip back into the deplorable conditions that prevailed before Sully. Colbert's first task, therefore, was to restore some semblance of order to the shattered state finances, which he did in characteristic fashion by abrogating approximately one-third of the royal indebtedness. Colbert's historical renown, however, derives from his ambitious but largely unsuccessful attempts to regulate and direct the economy. Colbert was not a great innovator; historical precedents existed for virtually all of his policies. What distinguished his regime, in addition to his comparatively long tenure as the trusted lieutenant of Louis XIV, was the vigor of his efforts and the fact that he wrote copiously about them.

One of Colbert's principal objectives was to make France economically self-sufficient. To this end he promulgated in 1664 a comprehensive system of protective tariffs; when this failed to improve the trade balance he resorted in 1667 to still higher, virtually prohibitive tariffs. The Dutch, who carried a large proportion of French commerce, retaliated with discriminatory measures of their own. Such measures of commercial warfare contributed to the outbreak of actual war in 1672, but the war ended in a stalemate, and in the peace treaty that followed France was obliged to restore the tariff of 1664.

Colbert's measures of industrial regulation were less directly related to the goal of self-sufficiency, but not entirely foreign to it. He issued detailed instructions covering every step in the manufacture of literally hundreds of products. In itself, the practice was not new, but Colbert also established corps of inspectors and judges to enforce the regulations, which added considerably to costs of production. The regulations were resisted and evaded by producers and consumers alike but, to the extent that they were successfully enforced, they also hindered technological progress. Colbert's Ordinance of Trade (1673), which codified commercial law, was far more beneficial to the economy.

As a part of his grand design Colbert also sought to create an overseas empire. The French had already, in the first half of the seventeenth century, established outposts in Canada, the West Indies, and India but, preoccupied with European power politics, had failed to give them much support. Colbert went to the opposite extreme, smothering the colonies with a mass of detailed, paternalistic regulations. He also created monopolistic joint-stock companies to conduct trade with both the East and West Indies (as well as similar companies for trade with the Baltic and Russia, the Levant, and Africa); but, unlike their Dutch and English models, which resulted from private initiative with the cooperation of the governments, the French

companies were in effect government proxies in which private individuals, including members of the royal family and nobility, had been induced or coerced into
investing. Within a few years they were all moribund.

Colbert, although a staunch Catholic, supported the limited toleration of the
Huguenots granted by the Edict of Nantes. After his death his weak successor
acquiesced in Louis' determination to stamp out the Protestant heresy, which culminated in the revocation of the edict in 1685 and the subsequent flight of many
Huguenots to more tolerant climes. That, along with the continuation of Colbert's
stifling paternalism and Louis' disastrous wars, threw France into a serious economic crisis from which it did not emerge until after the War of the Spanish
Succession.

The Prodigious Increase of the Netherlands

Dutch economic policies differed significantly from those of the nation-states pre
viously considered. For this there are two principal reasons. First, the structure of
government of the Dutch Republic was quite unlike that of the absolute monarchies
of continental Europe. Second, the Dutch economy depended on international commerce to a much greater degree than that of any of the Netherlands' larger
neighbors.

The Union of Utrecht of 1579, the agreement among the seven northern
provinces that eventually became the United Netherlands or Dutch Republic, was
more in the nature of a defensive alliance against Spain than the constitution of a
nation-state. The States-General, the legislative body of the Republic, concerned
itself exclusively with foreign policy, leaving domestic matters in the hands of the
provincial States and town councils. Moreover, all decisions had to be reached by a
unanimous vote in which each province had one voice; failure to agree required that
the delegates return to their provincial States for consultation and instruction. The
provincial States, for their part, were dominated by the chief towns. The towns were
governed by self-perpetuating town councils of from twenty to forty members, who
were the effective rulers (the burgher-oligarchs) of the Dutch Republic. Originally
the members of this oligarchy had been selected from among the wealthier merchants of the towns (at least in the maritime provinces of Holland and Zealand; in
the less urbanized provinces of the east and north the provincial nobility and well-
to-do farmers played more prominent roles). There was a general tendency, pronounced by the middle of the seventeenth century, for the members of this ruling
group, known as "regents," to be drawn from a *rentier* class of landowners and
bondholders, rather than active merchants. Nevertheless, the regents were usually
descended from merchant families, intermarried with them, and were conscious of
and responsive to their needs and desires. (See Fig. 6-6.)

The Dutch established their mercantile preeminence by the beginning of the
seventeenth century, and it continued to grow until at least the middle of the
century. The bases of Dutch commercial superiority were the so-called "mother
trades," those that connected the Dutch ports with others of the North Sea, the
Baltic, the Bay of Biscay, and the Mediterranean. Within that region Dutch ship-

FIGURE 6–6.   Dutch merchant. Daniel Bernard, a prosperous Dutch merchant of the mid-seventeenth century, is the subject of this portrait by Bartholomeus van der Helst. (By Van der Helst: Museum Boymans-van Beuningen, Rotterdam.)

ping accounted for as much as three-quarters of the total. From the Baltic they brought grain, timber, and naval stores to be distributed throughout western and southern Europe in exchange for wine and salt from Portugal and the Bay of Biscay, for their own manufactured goods, mainly textiles, and for herring. The herring fishery occupied a unique place in the Dutch economy, with as much as one-quarter of the population depending on it either directly or indirectly. Dried, smoked, and salted herring was in great demand in a Europe perennially short of fresh meat. As early as the fifteenth century the Dutch had perfected a method of curing the fish at sea, which allowed their fishing fleets to remain abroad for several weeks instead of returning to port every night. Fishing in the North Sea off the coasts of Scotland and England, they soon undercut the Hanseatic and Scandinavian fisheries in the Baltic and distributed their catch up the German rivers, in France, England, the Mediterranean, and the Baltic itself.

The Dutch specialized in carrying the goods of others, along with their herring exports, but they also exported some other products of their own. Dutch agriculture, although it occupied a far smaller proportion of the labor force than elsewhere, was the most productive in Europe and specialized in high-value produce such as butter, cheese, and industrial crops. The Netherlands lacked natural resources such as coal and mineral ores, but it imported raw materials and semifinished goods, such as rough woolen cloth from England, and exported them in finished form. The ship-building industry, developed to a high level of technical perfection, depended on timber from the Baltic; but it supplied not only the Dutch fishing, merchant, and naval fleets, but those of other countries as well. Similarly, the sailcloth and cordage industries obtained flax and hemp from abroad.

The northern Netherlands, especially Holland and Zealand, benefited in great measure from free immigration from other parts of Europe. In the immediate after-math of the Dutch Revolt large numbers of Flemings, Brabanters, and Walloons, of whom a disproportionate number were merchants and skilled artisans, flooded into the northern cities. The ease with which Amsterdam achieved its rank as the principal entrepôt of Europe was partly a result of the influx of merchants and financiers from fallen Antwerp, who brought both their capitalist knowhow and their liquid capital. In subsequent years the Netherlands continued to gain both financial and human capital by the inflow of religious refugees from the southern Netherlands, Jews from Spain and Portugal, and, after 1685, Huguenots from France. These migrations both symbolized and contributed to a policy of religious toleration in the Netherlands, unique in its time. Although Calvinist fanatics occasionally sought to impose a new religious orthodoxy, the merchant oligarchs succeeded in maintaining religous as well as economic freedom, for Catholics and Jews as well as Protestants.

The Dutch concern for freedom was real, particularly with respect to freedom of the seas. As a small maritime nation surrounded by vastly more populous, powerful neighbors, the Netherlands (led, as usual, by the province of Holland and the city of Amsterdam) resisted the pretensions of Spain to control the western Atlantic and the Pacific oceans, of Portugal to the South Atlantic and Indian oceans, and of Britain to the "British Seas" (including the English Channel). The Dutch jurist Hugo de Groot (Grotius) wrote his famous treatise, *Mare Liberum* ("Freedom of the Seas"), destined to become one of the foundations of international law, as a brief in the negotiations leading to a truce with Spain in 1609. In the frequent, more or less continual wars of the seventeenth century, the Dutch insisted on their rights as neutrals to carry merchandise to all combatants and were prepared to make war themselves to protect those rights. (For that matter, individual Dutch merchants were not above trading with the enemy, a practice that was tacitly accepted by the government.)

The Dutch commitment to freedom in matters of commercial and industrial policy was slightly more equivocal. Generally speaking, the cities (who were the effective units) followed free trade policies. No tariffs encumbered exports or imports of raw materials or semifinished goods, which were to be processed and reexported; tariffs and taxes on consumer goods were for revenue, not protection of domestic industries. The trade in precious metals, in particular, was entirely free, in striking contrast to the policies of other nations. Amsterdam, with its bank, bourse, and favorable balance of payments, quickly became the world emporium for gold and silver; it has been estimated that between one-fourth and one-half of the annual imports of silver from the Spanish Empire eventually wound up in Amsterdam, even during the Dutch War for Independence.

Freedom was also the rule in industry. Although guilds existed, they were neither as widespread nor as powerful as in other countries; most major industries operated entirely outside the guild system. More restrictive were the regulations imposed by the larger towns and cities on their surrounding districts, which prevented the growth of rural industries. The major exception to the absence of regulation in Dutch trade and industry was the government-sanctioned "College of the Fishery," which regulated the herring fishery. The ships of only five cities were permitted to take part in the "Great Fishery" (as opposed to the local fresh-herring fisheries for domestic consumption). The College licensed vessels to control quantity, and also imposed strict quality controls to maintain the reputation of Dutch herring. These restrictive policies paid handsomely as long as the Dutch maintained their near-monopoly in the European market, but as other nations gradually adopted Dutch technology the policies contributed to the stagnation and eventual decline of the herring trade, which was symptomatic (and in part a cause) of the decline of the Dutch economy as a whole.

The most striking departure of the Dutch from their general rule of freedom was with respect to their colonial empire. As the English ambassador to the Netherlands truly stated in 1663, "It is *mare liberum* in the British Seas but *mare clausum* on ye coast of Africa and in ye East Indies." In contrast to Spain and Portugal, in which trade with the overseas empire was regarded as a royal monopoly, the States-General of the Netherlands turned over not only the control of trade but also the powers of government to privately owned joint-stock companies, the East India Company for the Indian Ocean and Indonesia, and the West India Company for the west coast of Africa and North and South America. Although chartered initially as purely commercial ventures, the companies soon discovered that to succeed in that capacity, in competition with Portuguese, Spanish, English, and French rivals, to say nothing of the aspirations and desires of the peoples with whom they wished to trade, they needed to establish territorial control. To the extent that they were successful in this, they became "states within a state"; monopoly of trade, with respect to both their own nationals and competition with other nations, naturally followed.

"Parliamentary Colbertism" in Britain

Economic policies in England (and, after the union of the Scottish and English parliaments in 1707, in Great Britain) differed from those of both the Netherlands and the continental absolute monarchies. Moreover, whereas the general character of economic policies in other European nations remained more or less constant from the beginning of the sixteenth to the end of the eighteenth century, those of England and Britain underwent a gradual evolution corresponding to the evolution of constitutional government. Henry VIII (1509–47) was as much an absolute monarch in England as any of his fellow sovereigns were in their countries. But whereas royal absolutism increased in most continental countries in the sixteenth and seventeenth centuries, a contrary development occurred in England, resulting in the establishment of a constitutional monarchy under parliamentary control after 1688.

Another contrast between England and the Continent illuminates the nature and consequences of economic policy. In Spain and France, for example, the fiscal demands of the crown made it impossible for the government to pursue consistently a rational policy of economic development. In England the fiscal demands of the crown brought it into repeated conflicts with Parliament until the latter finally triumphed. Unlike representative assemblies on the Continent, the English Parliament had never given up its prerogative to approve new taxes. Although economic and financial questions were not the only or even the most important causes of the English Civil War, Charles I's attempt in the 1630s to govern without Parliament and to collect taxes without parliamentary approval was a major factor leading to the outbreak of armed insurrection. Similarly, after the Stuart dynasty was restored in 1660, the profligacy of Charles II and James II and their financial chicanery (e.g., the "Stop of the Exchequer" of 1672, in which the government diverted revenues intended for repayment of royal debts to the prosecution of an unpopular war with the Dutch) exacerbated religious and constitutional issues. After the installation of William and Mary in 1689 as constitutional monarchs Parliament took direct control of the government's finances and in 1693 formally instituted a "national" debt distinct from the personal debts of the sovereign.

The so-called Glorious Revolution of 1688–89 constitutes a major turning point not only in political and constitutional history, but in economic history as well. In the matter of public finance alone, the 1690s saw, in addition to the establishment of a funded debt, the creation of the Bank of England, a recoinage of the nation's money, and the emergence of an organized market for public as well as private securities. The success of the new financial system was not immediate; it was wracked in the early years by a number of crises, culminating in the famous South Sea Bubble of 1720. By the middle decades of the eighteenth century, however, when Britain was engaged in a series of both European and colonial wars with France, its government could borrow money at only a fraction of the cost of its rival. Moreover, the ease, cheapness, and stability of credit for public finance reacted favorably on private capital markets, making funds available for investment in agriculture, commerce, and industry.

An earlier historian referred to English economic policy making between the Glorious Revolution and the American Revolution as "Parliamentary Colbertism." Like the term *mercantilism*, it is both inaccurate and misleading. It is inaccurate because it ignores the substantial role of Parliament in policy making *before* 1688; and it is misleading in suggesting that Parliament ever aspired to, much less obtained, the degree of intervention in the economy that Colbert did. It does, nevertheless, have the merit of indicating that, in England, economic policy making was not the prerogative of an absolute monarch (and his minions), but responded to the varied and sometimes conflicting interests of those groups—titled aristocrats, landed gentry, wealthy merchants, professionals, courtiers, and others—who were effectively represented in Parliament.

It is out of the question in a brief survey to recount the myriad ways in which

Parliament influenced, or attempted to influence, the economy, such as the laws that required corpses to be buried in woolen shrouds (at the behest of the woolen industry; what better way to stimulate demand for its products than to bury them?), or those to stimulate the fishing industry that decreed more "fish days" (meatless) for Protestant than for Catholic England. Instead, we shall consider a few characteristic major pieces of legislation, including one generally regarded as successful in achieving its objectives, and others that either had negligible impacts or provoked adverse consequences.

The Statute of Artificers of 1563 has often been pointed out as an archetypical piece of mercantilist legislation, carefully thought out, providing a long-range plan for the economy as a whole. In fact, it was nothing of the kind; it was a reaction to a temporary situation, "a hotchpotch of compromise between the efforts of the Queen's councillors and numerous amendments originating in the House of Commons."[3] (In that sense it might be regarded as "typical" of mercantilist legislation.) Its main concern was with social stability. Its principal provisions required all able-bodied persons to engage in productive labor, with the first priority being agriculture, the second the cloth industry, and then certain other crafts and industries deemed to be of national importance. It established a norm of seven years' apprenticeship for all arts and crafts, including farming, and specified the social ranks from which apprentices should be selected. Combined with subsequent legislation regulating wages and poor relief, it would have, if effectively enforced, almost totally halted occupational and social mobility, and thus economic development. But "effective enforcement" was the key to almost all English (and other) economic legislation. In the case of the Statute of Artificers, and most similar English legislation, enforcement was left to the justices of the peace, part-time unpaid royal officials who had their own interests to look out for. Except in the rare instances when these interests coincided with those of the government, enforcement was lax at best, and nonexistent as a rule.

Less typical, perhaps, but more revealing as an example of the aims and consequences of the policies of economic nationalism is the case of the infamous Cokayne Project. In the Middle Ages England's major export was raw wool. In the course of the fifteenth and sixteenth centuries exports of rough, unfinished cloth, the monopoly of the Merchant Adventurers, surpassed those of raw wool. The principal market for this cloth was the Low Countries, where it was finished, dyed, and reexported throughout Europe. In 1614 Sir William Cokayne, a merchant, alderman of the City of London, and confidant of (or money-lender to) King James I persuaded the latter to revoke the monopoly of the Merchant Adventurers, forbid the export of undyed cloth, and grant a monopoly of the export of finished cloth to a new company of which Cokayne, of course, was the leading member. The rationale was that the finishing processes were the most lucrative branches of cloth manufacture; by reserving them for England the project would increase domestic employment and income, enhance the royal revenue from the export tax, and strike a blow at the Dutch. The Dutch retaliated, however, by forbidding the importation of dyed cloth from England; moreover, the finishing and dyeing trades required highly skilled artisans, who were in short supply in England. Cloth exports declined,

---

[3] Donald C. Coleman, *The Economy of England, 1450–1750* (Oxford, 1977), p. 181.

unemployment spread in the woolen industry, and a general depression ensued. In 1617 the government restored the monopoly of the Merchant Adventurers, but the commercial crisis continued, exacerbated by the renewed outbreak of war on the Continent. In 1624, under pressure from the House of Commons, the government opened the cloth trade to all.

The most famous and effective of all the policies of Parliamentary Colbertism were the Navigation Acts. Even Adam Smith granted them his grudging admiration, but on the grounds of national defense, for he explicitly noted that they decreased national income. Navigation laws, the general purpose of which was to reserve a country's international trade for its own merchant marine, were not unique to England or, in England, to the seventeenth century. Almost all countries had them; the first such law was passed in England in 1381, and was repeated frequently thereafter. Generally speaking, however, such laws were ineffective for two reasons: they lacked adequate enforcement mechanisms and, more fundamentally, the merchant marines they were intended to benefit lacked capacity and competitive ability. In 1651, however, the Long Parliament of the Commonwealth government passed a law that was intended not only to protect the English merchant marine but also to deprive the Dutch of their near-monopoly of both shipping and fishing in English waters. The Dutch were sufficiently concerned that they declared war the following year. Although the Navigation Act was not the only reason for the declaration, its repeal was one of the objectives for which the Dutch argued, unsuccessfully, in the negotiations that ended the stalemated war. In 1660, after the restoration of Charles II, Parliament renewed and strengthened the act. As amended from time to time thereafter, the Navigation Law not only sought to protect the English merchant marine and fishing fleet, but also became the cornerstone of England's colonial system.

Under the terms of the law all goods imported into Great Britain had to be carried in either British ships or ships of the country from which the goods originated. (British ships were defined as those of which the owners, master, and three-fourths of the crew were British subjects. The law also tried to protect the shipbuilding industry by requiring that the ships be built in Britain; but that provision proved difficult to enforce, and for many years Dutch shipbuilders supplied a considerable proportion of the British merchant fleet.) Moreover, even British ships were required to bring goods directly from the country of origin, rather than from an intermediate port; in this way the law sought to weaken Amsterdam's position as an entrepôt as well as to cut into the Dutch carrying trade. The coasting trade (from one British port to another) was reserved entirely for British ships, as was the importation of fish. Trade with the British colonies (in North America, the West Indies, India) also had to be carried in British bottoms. (Colonial ships were regarded as British if they met the previously noted stipulations.) In addition, all colonial imports of manufactured goods from foreign countries (e.g., metalwares from Germany) had to be landed first in Great Britain; in effect, this reserved the colonial market for British merchants and manufacturers. Likewise, the staple colonial exports, such as tobacco, sugar, cotton, dyestuffs, and eventually many other commodities, had to be shipped through Great Britain, rather than directly to foreign ports.

The Navigation Laws were not easily enforced, especially in the colonies; more than one New England fortune grew on the profits of illicit trade. Although the laws were intended to injure the Dutch as much as they were to benefit the English, the Dutch maintained their maritime and commercial supremacy until well into the eighteenth century; even then their decline was relative rather than absolute, and a result more of other causes (especially warfare) than of English competition. Nevertheless, the Navigation Acts probably did promote the growth of the English merchant marine and maritime trade, as they were intended to do (but, as Adam Smith pointed out, at a cost to British consumers). They could not have done so, however—as earlier, similar legislation had not—if English merchants and shippers had not already become involved in aggressive pursuit of foreign markets, and were thus willing and able to take advantage of the privileges the laws conferred.

The Navigation Acts had yet another, unintended, effect: the loss of a large part—and the economically most progressive and prosperous part—of the "old" British Empire (Fig. 6-7). Although they were not the sole or even the most important cause of the American Revolution, they were at the heart of the "old colonial system" and for most Americans they symbolized the disadvantages, real and imagined, of colonial dependence. From their parlous beginnings in the early seventeenth century, England's North American colonies had grown prodigiously. In numbers alone the record is impressive: from only a few thousand in 1630, they exceeded a quarter of a million by the beginning of the eighteenth century and 2 million on the eve of the revolution. The sinister counterpart of this achievement must, however, be deplored: the displacement and eventual extinction of most native Americans and the enslavement of thousands of black Africans.

The growth of income and wealth was even more impressive than the growth of population as, after the suffering and disasters of the early years, they specialized along lines of comparative advantage and traded extensively with one another, with the mother country, and, illegally, with the Spanish Empire and parts of continental Europe. Virginia and the Chesapeake Bay area specialized in tobacco, South Carolina in rice and indigo, the middle colonies in foodstuffs, some of which they sold to the southern colonies and New England. The latter had a more diversified economy, with commerce and the carrying trade being especially important. Although the Navigation Acts governed colonial commerce, enforcement was not especially effective until after the Seven Years War (1763); even then it was not terribly burdensome, just enough to give those who sought political independence for other reasons a rallying cry.

One more example of British legislation must suffice. In the latter part of the seventeenth century the East India Company began importing inexpensive, lightweight, brightly printed cloths from India, called calicoes, which quickly became popular. The woolen industry in 1701 persuaded Parliament to pass the first Calico Act, prohibiting the importation of printed cotton cloth. A new industry quickly sprang up, the printing of imported plain cotton goods. The woolen industry again became alarmed, and in 1721 Parliament obligingly passed a second Calico Act, which forbade the display or consumption of printed cotton goods. This, in turn, stimulated a domestic cotton textile industry based on imported raw cotton, which eventually became the cradle of the so-called Industrial Revolution. By the end of the century the manufacture of cotton had displaced that of woolen goods to become Britain's leading industry.

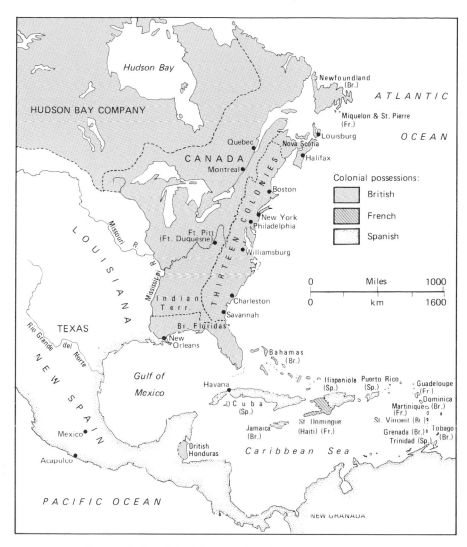

FIGURE 6–7. Colonial possessions in North America, 1763.
*Florida was a British colony from 1763 to 1780.

To summarize, in Britain the growth of Parliamentary power at the expense of the monarchy brought with it better order in the public finances, a more rational system of taxation than was found elsewhere in Europe, and a smaller state bureaucracy. The ideal was still that of a "regulated" economy, as on the Continent, but the means of regulation were quite different. Parliamentary control was most effective in economic relations with the outside world (facilitated by Britain's island nature) and Parliament followed a policy of strict economic nationalism. Domestically, although Parliament wished to control the economy, it generally lacked the ability. As a result British entrepreneurs enjoyed a degree of freedom and opportunity that was virtually unique in the world.

7

The Dawn of Modern Industry

By the beginning of the eighteenth century several regions of Europe, mainly in western Europe, had acquired sizable concentrations of rural industry, mostly but not exclusively in the textile trades. In the early 1970s a new term was invented to describe the process of expansion and occasional transformation of these industries: proto-industrialization. The term was first employed with reference to the linen industry of Flanders. That was a rural, cottage-based industry organized by entrepreneurs in Ghent and other market towns, who exported its output, linen cloth, to distant markets, especially those of the Spanish Empire. The workers, family units of husband, wife, and children, usually cultivated small plots of ground as well, although they also bought additional supplies in markets. The term has subsequently been refined and extended in both space and time to other, similar industries. In some instances—for example, the Lancashire cotton industry—it has been seen as the prelude to a fully developed factory system. In others, however, such as the Irish and even the Flemish linen industries, no such transition occurred.

The essential features of a proto-industrial economy are dispersed, usually rural workers organized by urban entrepreneurs (merchant-manufacturers) who supply the workers with raw materials and dispose of their output in distant markets. The workers must also purchase at least a part of their means of subsistence. Perceptive readers will note that this definition seems to apply to those industries described in Chapters 3 and 5 as cottage industry, domestic industry, and the putting-out system. Indeed, critics of the term proto-industrialization regard it as superfluous. If there is a significant difference, it is in the emphasis placed on distant markets; most traditional cottage or domestic industry catered only to local markets.

Proto-industrialization and the related terms refer primarily to consumer goods industries, especially textiles. Well before the advent of the factory system in the cotton industry, however, other large-scale, highly capitalized industries existed, producing capital or intermediate goods, and sometimes even consumer goods. The French *manufactures royales* have already been mentioned (p. 151); they were usually located in large factorylike structures where skilled artisans worked under the supervision of a foreman or entrepreneur, but without mechanical power. Similar "proto-factories" were built by noble landowner–entrepreneurs in the Austrian Empire (Bohemia and Moravia) and elsewhere. Large landowners also acted as entrepreneurs in the coal industry, mining the deposits located on their estates. The Duke of Bridgewater, who owned a mine at Worsley, hired the self-taught engineer

James Brindley to build a canal from his mine to Manchester in 1759–61. Ironworks, usually located in rural areas near timber (for charcoal) and iron ore, sometimes employed hundreds, even thousands of workers. Lead, copper, and glassworks also frequently had large-scale organizations, as did shipyards. The state-owned Arsenal of Venice, dating from the Middle Ages, was one of the earliest large-scale industrial enterprises in history. The intricate organization of Dutch shipyards has already been described (p. 117). The English government built the Woolwich Arsenal, near London, and private entrepreneurs also maintained sizable installations at several locations.

Impressive though these achievements were, they were overshadowed in the eighteenth century by the rise of new forms of industrial enterprise.

Characteristics of Modern Industry

One of the most obvious differences between preindustrial and modern industrial societies is the greatly diminished relative role of agriculture in the latter. The counterpart of its diminished importance, however, is the greatly increased productivity of modern agriculture, which enables it to feed a large nonagricultural population. A related difference is the high proportion of the modern labor force engaged in the tertiary, or service, sector (especially professional as opposed to domestic services); the proportion is now 50 percent or more, in contrast to the 30 or 40 percent engaged in manufacturing and related industries. Yet this is a relatively recent development, particularly notable in the second half of the twentieth century. During the period of industrialization proper, extending roughly from the beginning of the eighteenth century (in Great Britain) to the first half of the twentieth century, the characteristic feature of the structural transformation of the economy was the rise of the secondary sector (mining, manfacturing, and construction), observable in the proportion of both the labor force employed and the output.

The transformation was first noted in England, then Scotland, and Great Britain has been truly described as "the first industrial nation." A more picturesque but less useful term, "the industrial revolution," has been applied to the last few decades of the eighteenth century and the first few in the nineteenth; as will become evident, the term is both inaccurate and misleading. More important, its use deflects attention from contemporaneous but different types of development in continental Europe. If Great Britain had never existed, or had sunk beneath the ocean in a gigantic tidal wave, Europe (and America) would have industrialized, albeit along somewhat different lines. Nevertheless, this chapter is devoted to the beginning of the process of industrialization in eighteenth-century Britain (Fig. 7-1).

In the course of this transformation, which is more accurately, if prosaically, designated as the "rise of modern industry," certain characteristics gradually emerged that clearly distinguish "modern" from "premodern" industry. These are (1) the extensive use of mechanically powered machinery; (2) the introduction of new, inanimate sources of power (or energy), especially fossil fuels; and (3) the widespread use of materials that do not normally occur in nature. A related feature is the larger scale of enterprise in most industries.

FIGURE 7–1.   English industry in 1700. (Compare Figure 7–11, p. 185.)

The most significant improvements in technology involved the use of machinery and mechanical power to perform tasks that had been done far more slowly and laboriously by human or animal power, or that had not been done at all. To be sure, elementary machines like the wheel, the pulley, and the lever had been used from antiquity, and for centuries humankind had used a fraction of the inanimate powers of nature to propel sailing ships and actuate windmills and waterwheels for rudimentary industrial processes. During the eighteenth century a notable increase in the use of water power occurred in such industries as grain milling, textiles, and metallurgy; and in recent times we have witnessed the proliferation of a wide variety of prime movers, from small electric motors operated on household current to huge nuclear reactors. But the most important developments in the application of energy in the early stages of industrialization involved the substitution of coal for wood and charcoal as fuel, and the introduction of the steam engine for use in mining, manufacturing, and transportation. Similarly, although metallic ores had been converted into metals for centuries, the use of coal and coke in the smelting process greatly reduced the cost of metals and multiplied their uses, whereas the application of chemical science created a host of new "artificial" or synthetic materials.

The Industrial Revolution: A Misnomer

Probably no term from the economic historian's lexicon has been more widely accepted by the public than "industrial revolution." This is unfortunate, because the term itself has no scientific standing and conveys a grossly misleading impression of the nature of economic change. Nevertheless, for more than a century it has been used to denote that period in British history that witnessed the application of mechanically powered machinery in the textile industries, the introduction of James Watt's steam engine, and the "triumph" of the factory system of production. By analogy, the term has also been applied to the onset of industrialization in other countries, although without general agreement on dates.

The expression *révolution industrielle* was first used in the 1820s by French writers who, wishing to emphasize the importance of the mechanization of the French cotton industry then taking place in Normandy and the Nord, compared it with the great political revolution of 1789. Contrary to widespread belief, Karl Marx did not use the term in its conventional sense. It acquired currency only after the publication in 1884 of Arnold Toynbee's *Lectures on the Industrial Revolution in England*. Toynbee was a social reformer, not a scholar; his major interest lay in remedying what he believed to be the moral degradation of the British working classes.[1] Invited to lecture at Oxford, he devoted his lectures to the interrelations of economic events and economic policy, especially to the emergence of laissez-faire policies, which he regarded as a disaster for workers. Toynbee did not prepare the lectures for publication; rather, they were assembled from notes by some of his student disciples after his premature death in 1883. Nevertheless, the book sold readily, and the expression immediately caught the public's fancy.

Early descriptions of the phenomenon emphasized the "great inventions" and the dramatic nature of the changes. As an 1896 textbook put it, "The change . . . was sudden and violent. The great inventions were all made in a comparatively short space of time. . . . In a little more than twenty years all the great inventions of Watt, Arkwright, and Boulton had been completed, steam had been applied to the new looms, and the modern factory system had begun," a description that A. P. Usher dryly characterized as exhibiting "all the higher forms of historical inac-

---

[1] "Our object is . . . to improve the great mass of the population"; Arnold Toynbee, *Lectures on the Industrial Revolution in England: Popular Addresses, Notes and Other Fragments* (1884; reprinted New York, 1969), p. 150. Toynbee was the uncle of the slightly more famous Arnold J. Toynbee, author of *A Study of History* (10 vols., 1934–54).

curacy."[2] Early interpretations also emphasized what were assumed to be the deleterious consequences of the new mode of production. Although increases in productivity as a result of the use of mechanical power and machinery were acknowledged, most accounts stressed the use of child labor, the displacement of traditional skills by machinery, and the unwholesome conditions of the new factory towns. For most of its history, for most people, the term "industrial revolution" has had a pejorative connotation.

Serious scholars recognized the inadequacies of the term and protested its use, but without avail. As early as 1919 Usher prophesied, "The term has captured the imagination, and despite misleading connotations it will doubtless hold its place in the literature," then added, "but interpretation becomes more and more necessary."[3] In 1924 George Unwin wrote, "when, on looking back, we find that the revolution has been going on for two centuries and had been in preparation for two centuries before that . . . we may begin to doubt whether the term. . . , though useful enough when it was first adopted, has not by this time served its turn."[4]

The dates implicit in Toynbee's *Lectures,* 1760 to 1820, were arbitrarily determined by the reign of George III, on which Toynbee had been invited to lecture. Some scholars, aware that the rapidity of change had been exaggerated in the conventional treatments, argued for a longer period for the "revolution," such as 1750 to 1850 (equally arbitrary), and even for no terminal date at all. On the other hand, John U. Nef, who described the idea of an industrial revolution as "essentially false," nevertheless found that an "unprecedented acceleration of industrial progress began, not in 1750 or 1760, but in the 1780s."[5] Nef's discovery was taken up by Walt W. Rostow and given an even greater aura of precision when he assigned the dates 1783–1802 for England's "take-off."[6] (The phrase "take-off" into self-sustained growth, coined and popularized by Rostow, is essentially a pseudoscientific substitute for "industrial revolution," and is equally misleading.)

Despite these efforts both to lengthen and to shorten the span of the "revolution," the conventional dating received the imprimatur of no less an authority than T. S. Ashton, the most famous economic historian of eighteenth-century England. This is doubly ironic, because Ashton, unlike most of his predecessors, viewed the outcome of the period as an "achievement" rather than a "catastrophe," and because he had no special fondness for the term. (The dates, and possibly the title of his book, were selected by the publisher, for whom it was one in a chronological series.) Ashton himself wrote: "The changes were not merely 'industrial,' but also social and intellectual. The word 'revolution' implies a suddenness of change that is

[2] A. P. Usher, *An Introduction to the Industrial History of England* (Boston and New York, 1920), p. 249.

[3] Ibid.

[4] *Studies in Economic History: The Collected Papers of George Unwin,* R. H. Tawney, ed. (London, 1927), p. 15.

[5] John U. Nef, *Western Civilization since the Renaissance: Peace, War, Industry and the Arts* (New York, 1963), pp. 276, 290. Nef first proposed 1785 as the initial date in "The Industrial Revolution Reconsidered," *Journal of Economic History,* 3 (May 1943): 1–31.

[6] W. W. Rostow, *The Stages of Economic Growth: A Non-Communist Manifesto* (Cambridge, 1960), p. 38.

not, in fact, characteristic of economic processes. The system of human relationships that is sometimes called capitalism had its origins long before 1760, and attained its full development long after 1830: there is a danger of overlooking the essential fact of continuity."[7]

Prerequisites and Concomitants of Industrialization

As Ashton wrote, the changes were not merely industrial, but social and intellectual as well. Indeed, they were also commercial, financial, agricultural, and even political. In this "seamless web" of historical change it is difficult to assign priorities or weights, especially when methods and units of measurement are unreliable or nonexistent; but there is reason to believe that the intellectual changes were the most fundamental, in the sense that they permitted or encouraged the others.

Already in the Middle Ages some individuals had begun to contemplate the practical possibilities of harnessing the forces of nature (see p. 75). The later scientific achievements associated with Copernicus, Galileo, Descartes, and Newton (to mention only a few) reinforced such ideas. In England the influence of Francis Bacon, one of whose aphorisms was "knowledge is power," led to the founding in 1660 of the Royal Society "for Improving Natural Knowledge." Some scholars regard the application of science to industry as *the* distinguishing characteristic of modern industry. Despite its attractiveness, this view has its difficulties. In the eighteenth-century dawn of modern industry the body of scientific knowledge was too slender and weak to be applied directly to industrial processes, whatever the intentions of its advocates. In fact, it was not until the second half of the nineteenth century, with the flowering of chemical and electrical sciences, that scientific *theories* provided the foundations for new processes and new industries. It is indisputable, however, that as early as the late seventeenth century the *methods* of science—in particular, observation and experiment—were being applied (not always successfully) for utilitarian purposes. Nor were such efforts limited to men of scientific training. Indeed, one of the most remarkable features of technical advance in the eighteenth and early nineteenth centuries was the large proportion of major innovations made by ingenious tinkerers, self-taught mechanics and engineers (the word *engineer* acquired its modern meaning in the eighteenth century), and other autodidacts. In many instances the term *experimental method* may be too formal and exact to describe the process; *trial-and-error* may be more apposite. But a willingness to experiment and to innovate penetrated all strata of society, including even the agricultural population, traditionally the most conservative and suspicious of innovation.

Just as England was the first nation to industrialize on a large scale, it was one of the first to increase its agricultural productivity. By the end of the seventeenth century England was already in advance of most of continental Europe in agricultural productivity, with only about 60 percent of its workers involved primarily in food production. Although the actual number of workers in agriculture continued

[7] T. S. Ashton, *The Industrial Revolution, 1760–1830* (Oxford, 1948), p. 2.

to grow until the middle of the nineteenth century, the proportion declined steadily to about 36 percent at the beginning of the nineteenth century, to about 22 percent in the midnineteenth century (when the absolute number was at its maximum), and to less than 10 percent at the beginning of the twentieth century.

The means by which England increased its agricultural productivity owed much to trial-and-error experimentation with new crops and new crop rotations. Turnips, clover, and other fodder crops were introduced from the Netherlands in the sixteenth century (see p. 114), and became widely diffused in the seventeenth. Probably the most important agricultural innovation before scientific agriculture was introduced in the nineteenth century was the development of so-called convertible husbandry, involving the alternation of field crops with temporary pastures (frequently sown with the new fodder crops) in place of permanent arable land and pastures. This had the double advantage of restoring the fertility of the soil through improved rotations, including leguminous crops, and of carrying a larger number of livestock, thus producing more manure for fertilizer as well as more meat, dairy produce, and wool. Many landowners and farmers also experimented with selective breeding of livestock.

An important condition for both the improved rotations and selective breeding was enclosure and consolidation of the fields (Fig. 7-2). Under the traditional open field system it was difficult, if not impossible, to obtain agreement among the many participants on the introduction of new crops or rotations; and with livestock grazing in common herds it was equally difficult to manage selective breeding. Notwithstanding strong incentives for enclosure, it had many opponents, predominantly cottagers and squatters who had no open field holdings of their own, but only customary rights to graze a beast or two on the common pasture. The most famous enclosures were those carried out by acts of Parliament between 1760 and the end of the Napoleonic Wars, for it was these that gave rise to a literature of protest (Fig. 7-3). Enclosure by private agreement, however, had been going on almost continuously from the late Middle Ages; it was especially active in the late seventeenth and the first six decades of the eighteenth centuries. By that time more than half of England's arable land had been enclosed.

The new agricultural landscape that emerged to replace the nucleated villages surrounded by their open fields consisted of compact, consolidated, and enclosed (walled, fenced, or hedged) farms, mainly in the range of 100 to 300 acres. Concomitant with the processes of enclosure and technological improvement, a gradual tendency toward larger farms emerged. By 1851 about one-third of the cultivated acreage was in farms larger than 300 acres; farms smaller than 100 acres accounted for only 22 percent of the land. Even so, the occupiers of the small farms outnumbered those of the others almost 2 to 1. The reason for this is that the small farmers were owner-occupiers who farmed with the help of family labor; the larger farmers were capitalistic tenant farmers who rented their land for a cash payment and hired landless agricultural laborers. It used to be thought that the enclosures "depopulated" the countryside, but in fact the new techniques of cultivation associated with them actually increased the demand for labor. Not until the second half of the nineteenth century, with the introduction of such farm machinery as threshers,

FIGURE 7-2. Enclosures. The open fields surrounding the village of Ilmington in the English Midlands were enclosed in 1778. In this aerial photograph from the 1950s the ridges and furrows of the medieval open fields are still clearly visible. Compare with the map of the village of Shilbottle in Figure 3-1. (From *Medieval England: An Aerial Survey*, by M. W. Beresford and J. K. S. St. Joseph. Copyright 1969 by Cambridge University Press. Reprinted with permission.)

harvesters, and steam plows, did the absolute size of the agricultural labor force begin to decrease.

In the meantime, the increasing productivity of English agriculture enabled it to feed a burgeoning population at steadily rising standards of nutrition. Indeed, for about a century, from 1660 to 1760, it produced a surplus for export before the rate of population growth overtook the rate of increase of productivity. The relatively prosperous rural population, more specialized and commercially oriented than most continental peasants, also provided a ready market for manufactured goods, ranging from agricultural implements to such consumer products as clothing, pewterware, and porcelain.

Commercialization of agriculture reflected a general process of commercialization of the entire nation. As early as the end of the seventeenth century English foreign trade per capita exceeded that of all nations except the Netherlands, and London had developed a remarkably sophisticated commercial and financial organi-

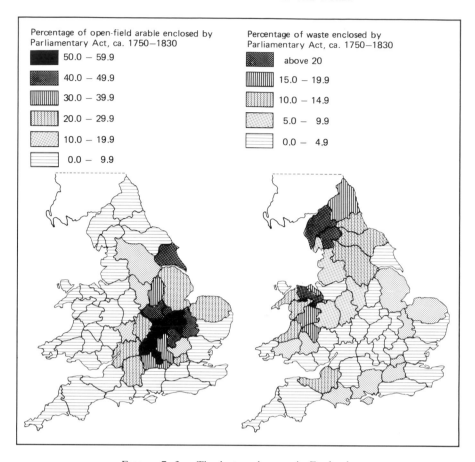

FIGURE 7–3.   The last enclosures in England.

zation that was beginning to rival that of Amsterdam. Between 1688 and 1801, while the relative share of agriculture in national income declined from 40 to 32.5 percent and that of mining, manufacturing, and building rose from 21 to only 23.6 percent, the relative share of trade and transport rose from 12 to 17.5 percent, an increase of almost 50 percent.

Already by the sixteenth century London had begun to function as a "growth pole" for the English economy. Its advantages were both geographic and political. The Romans had selected the lowest point on the Thames at which the river could be bridged for the city's site; and the network of roads the Romans built, centered on London, still served the English economy in the sixteenth and later centuries. Likewise, the "pool" of London was the highest point on the Thames accessible to ocean-going ships, and the city had already emerged in the Middle Ages as the most important port in England. The location of the nation's capital in Westminster, a short distance upriver from the City of London, with which it eventually merged, enhanced both the wealth and the population of the metropolis. Population growth

FIGURE 7–4.   Token coin. "John Wilkinson, Iron Master" minted token coins both to provide a local monetary circulation and as a form of advertisement (and no doubt also a form of ego satisfaction). (Reproduced by courtesy of the Trustees of the British Museum.)

was extremely rapid in the sixteenth and seventeenth centuries, and by 1700 London had caught up with or possibly surpassed Paris, previously the largest city in Europe.

Commercialization interacted with the developing financial organization of the nation. The origins of the English banking system are obscure, but in the years after the Restoration of 1660 a number of prominent goldsmiths in London began to function as bankers. They issued deposit receipts that circulated as banknotes, and granted loans to creditworthy entrepreneurs. The founding of the Bank of England in 1694, with its legal monopoly of joint-stock banking, forced the private bankers to give up their issues of banknotes, but they continued to function as banks of deposit, accepting drafts and discounting bills of exchange. Meanwhile, the provinces outside London remained without formal banking facilities, although "money scriveners" (brokers), attorneys, and wealthy wholesale merchants performed some of the elementary banking functions, such as discounting bills of exchange and remitting funds to London. The Bank of England established no branches, and its banknotes (of large denomination) did not circulate outside London. Moreover, the Royal Mint was extremely inefficient; the denomination of its gold coins was too large to be useful in paying wages or retail trading, and it minted very few silver or copper coins. This dearth of small change prompted private enterprise to fill the gap: industrialists, merchants, and even publicans issued scrip and tokens that served the needs of local monetary circulation (Fig. 7-4). From these diverse origins came the institution of "country banks" (i.e., any bank not located in London), whose growth was extremely rapid in the second half of the eighteenth century; by 1810 there were almost 800 of them.

The euphoria engendered by the Glorious Revolution resulted in the creation of a number of joint-stock companies in the 1690s, some of them like the Bank of

England with royal charters and grants of monopoly. (The law at that time was ambiguous on the question of business organization.) A similar euphoria suffused the country after the successful conclusion of the War of the Spanish Succession, and it culminated in the speculative financial boom known as the South Sea Bubble. The episode received its name from the South Sea Company, chartered in 1711 with a nominal monopoly of trade with the Spanish Empire, although the real reason for its creation was to raise money for the government to prosecute the war. (A similar speculative mania took place in France at the same time. Called the Mississippi Bubble, it was inspired by a Scottish financial adventurer with the unlikely name of John Law, who persuaded the Duke of Orleans, regent for the child King Louis XV, to allow him to form a bank, the Banque Royale, and also a company to exploit French possessions in North America, then called Mississippi.) The bubble burst in 1720 when Parliament, at the behest of the South Sea Company, passed the Bubble Act. The act prohibited the formation of joint-stock companies without the express authorization of Parliament, which proved loath to grant such authorizations. As a result England entered its ''industrial revolution'' with a legal barrier against the joint-stock (or corporate) form of business organization, condemning most of its industrial and other enterprises to partnerships or simple proprietorships. Whether or not that restriction hindered English industrialization has been debated extensively; but, in any case, it was not a fatal hindrance. The Bubble Act was eventually repealed in 1825.

Another major consequence of the Glorious Revolution, already noted (p. 157), was to place the public finances of the kingdom firmly in the hands of Parliament, which significantly reduced the cost of public borrowing and thereby freed capital for private investment. Although the system of taxation was highly regressive (i.e., it taxed the low-income population more heavily, proportionately, than the wealthy), that, too, permitted the accumulation of capital for investment. It is questionable whether much of that accumulation went directly into industry, since most industrial enterprises grew from small beginnings by means of reinvested profits. Indirectly, however, by means of infrastructure investments, especially in transportation, the capital contributed importantly to the process of industrialization.

The movement of large quantities of bulky, low-value goods, such as grain from the fields to the growing urban markets, timber for building, coal and ores from mines to smelters and foundries, required cheap, dependable transportation. Before the railway era water routes provided the most economical and efficient arteries of transport. Britain owed much of its early prosperity and its head start in modern industry to its island location, which not only granted it virtually costless protection from the disruption and destruction of continental warfare, but also provided it with cheap transportation. The long coastline, excellent natural harbors, and many navigable streams eliminated much of the need for overland transportation, which hindered the growth of commerce and industry on the Continent.

Even with Britain's natural advantages, the demand for improved transportation facilities increased apace. In the thirty-year period from 1660 to 1689 fifteen private acts of Parliament were passed for river and harbor improvements; from 1690 to 1719, 59 acts (including some for turnpike construction) were passed; and from

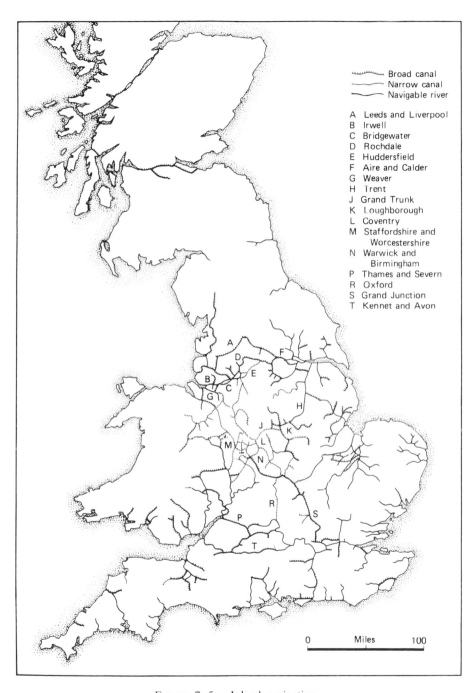

Broad canal
Narrow canal
Navigable river

A Leeds and Liverpool
B Irwell
C Bridgewater
D Rochdale
E Huddersfield
F Aire and Calder
G Weaver
H Trent
J Grand Trunk
K Loughborough
L Coventry
M Staffordshire and
     Worcestershire
N Warwick and
     Birmingham
P Thames and Severn
R Oxford
S Grand Junction
T Kennet and Avon

0   Miles   100

FIGURE 7–5.   Inland navigation.

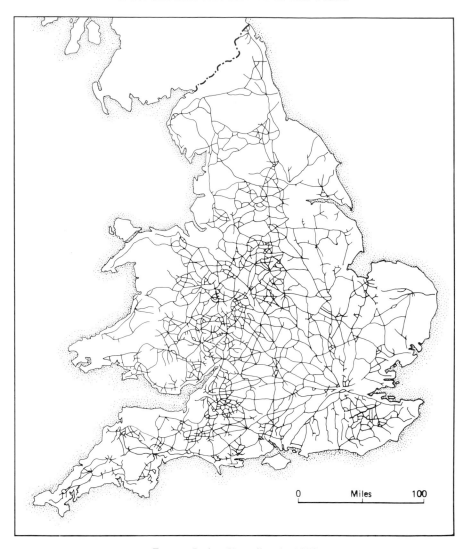

FIGURE 7–6.   Turnpikes in 1770.

1720 to 1749, 130 acts. The 1750s witnessed the advent of the canal age, during which waterways were built to connect navigable rivers with each other or mines with their markets. Sometimes canal construction entailed great ingenuity, employing aqueducts and subterranean tunnels. Altogether, between 1750 and 1820, 3000 miles of navigable waterways, mainly canals, were added to the 1000 already in existence, at a cost of £17 million (Fig. 7-5). By means of these canals and navigable rivers, all the major centers of production and consumption were connected together and with the important ports. Canal enterprises were organized as private, profit-making companies chartered by act of Parliament (a principal exception to the objective of the Bubble Act), which charged tolls for independent boat or barge operators, or sometimes operated their own barge fleets for hire.

Britain's network of canals and navigable rivers was extremely efficient for its time but still did not satisfy the demand for inland transportation. Traditionally the maintenance of roads was the responsibility of the parishes, using the forced labor of the local inhabitants. Not surprisingly, the condition of roads so maintained was deplorable. Beginning in the 1690s Parliament created, by private acts, turnpike trusts that undertook to build and maintain stretches of good roads on which users, whether traveling by wagon, by carriage, on horseback, or by foot, were charged tolls. Such trusts were not commercial companies, but were instigated and supervised by trustees, usually local landowners, farmers, merchants, and industrialists, who sought both to lower their tax liabilities for parish road upkeep and to improve road access to markets. Although most turnpikes were relatively short, about thirty miles or so, many were interconnected, and eventually formed a dense network. The 1750s and 1760s saw the greatest spurt of turnpike construction: from 3400 miles in 1750, the network grew to 15,000 miles in 1770 (Fig. 7-6) and reached a peak mileage of 22,000 in 1836, by which time the railways had begun to render them and the canals obsolete.

Industrial Technology and Innovation

Historians impressed with the revolutionary nature of industrial change point to the rapid mechanization and growth of the cotton industry in the last two decades of the eighteenth century. Almost a century earlier, however, and within only a few years of one another, two other innovations were made whose impact might be regarded as even more fundamental to industrialization, although many years passed before their importance was felt. These innovations were the process for smelting iron ore with coke, which freed the iron industry from exclusive reliance on charcoal, and the invention of the atmospheric steam engine, a new and powerful prime mover that supplemented and eventually replaced wind- and watermills as inanimate sources of power.

Many attempts had been made to replace charcoal with coal in the blast furnace, but impurities in the raw coal doomed them all to failure. In 1709 Abraham Darby, a Quaker ironmaster at Coalbrookdale in Shropshire, processed coal fuel in much the same way that other ironmasters made charcoal out of timber—that is, he heated coal in a closed container to drive off impurities in the form of gas, leaving a residue of coke, an almost pure form of carbon, which he then used as fuel in the blast furnace to make pig iron (Figs. 7-7 and 7-8).

In spite of Darby's technological breakthrough, the innovation diffused slowly; as late as 1750 only about 5 percent of British pig iron was produced with coke fuel. The continued rise in the price of charcoal after 1750, however, together with such innovations as Henry Cort's puddling and rolling process of 1783–84, finally freed iron production altogether from reliance on charcoal fuel. (Cort's process melted bars of pig iron in a reverberatory furnace, so that the iron did not come in direct

FIGURE 7–7. Coalbrookdale at night. Coalbrookdale was (and is) located in bucolic rural Shropshire, but it was the site of one of the most important events in industrial history: the first use of coke for smelting iron ore. Today it is the site of an important museum of industrial archeology. (Trustees of the Science Museum, London.)

FIGURE 7–8.   The iron bridge. This bridge over the Severn River near Coalbrookdale, with girders of cast iron weighing almost 400 tons, was built by Abraham Darby III, the grandson of the man who first smelted iron with coke, in 1779. It was assembled with interlocking joints and wedges, without bolts or welding. It still stands today, as shown in this 1986 tourist snapshot. (Rondo Cameron.)

contact with the fuel; the molten iron was then stirred or "puddled" with long rods to help burn off the excess carbon. Finally, the semimolten iron was run through grooved rollers, which both squeezed out more impurities and imparted desired shapes to the bars of wrought iron; see Fig. 7-9.) Ironmasters achieved economies of scale by integrating all of these operations in one location, usually at or near the site of coal production, and both total iron output and the proportion made with coal fuel accelerated dramatically. By the end of the century iron production had expanded to more than 200,000 tons, virtually all coke-smelted, and Britain had become a net exporter of iron and iron wares.

Steam power was first employed in the mining industries. As the demand for coal and metals expanded, so the efforts to obtain them from ever deeper mines intensified. Many ingenious devices were invented to rid the mines of water, but flooding remained a major problem, and the chief obstacle to further expansion of output. In 1698 Thomas Savery, a military engineer, obtained a patent for a steam pump, which he appropriately called "The Miner's Friend." A few of Savery's pumps were erected in the first decade of the eighteenth century, mainly in the Cornish tin mines, but the device had several practical defects—among them a tendency to explode. Thomas Newcomen, an ironmonger and tinkerer who was familiar with the problems of the mining industries, set out to remedy these defects

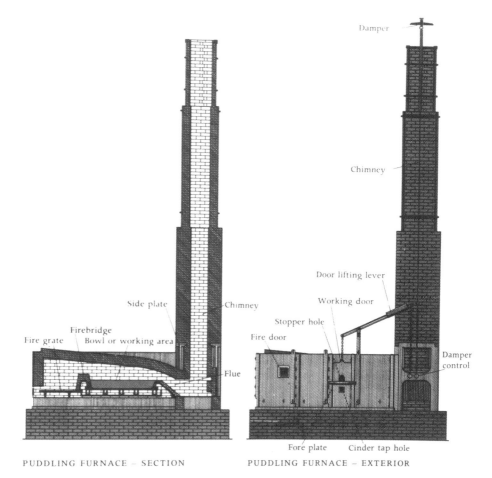

PUDDLING FURNACE – SECTION   PUDDLING FURNACE – EXTERIOR

FIGURE 7–9.   Cort's puddling furnace. The puddling furnace was used (along with rolling mills) to refine high-carbon pig iron into low-carbon bar iron. The puddling furnace and rolling mill freed the iron industry from reliance on charcoal fuel. (From *The Archaelogy of the Industrial Revolution*, by Brian Bracegirdle, London, 1974. Reprinted with permission.)

by trial-and-error experiments, and in 1712 succeeded in erecting his first atmospheric steam pump for a coal mine in Staffordshire (Fig. 7-10).

Newcomen's engine forced steam from a boiler into a cylinder containing a piston that was connected by means of a rocking T-beam to a pump. After the steam forced the piston to the top of the cylinder, a jet of cold water inside the cylinder condensed the steam and created a vacuum, allowing the weight of the atmosphere to depress the piston and actuate the pump—hence the name atmospheric steam engine. The Newcomen engine was large (it required a separate building to house it), cumbersome, and expensive; but it was also effective, if not thermally efficient. By the end of the century several hundred had been erected in Britain, and a number

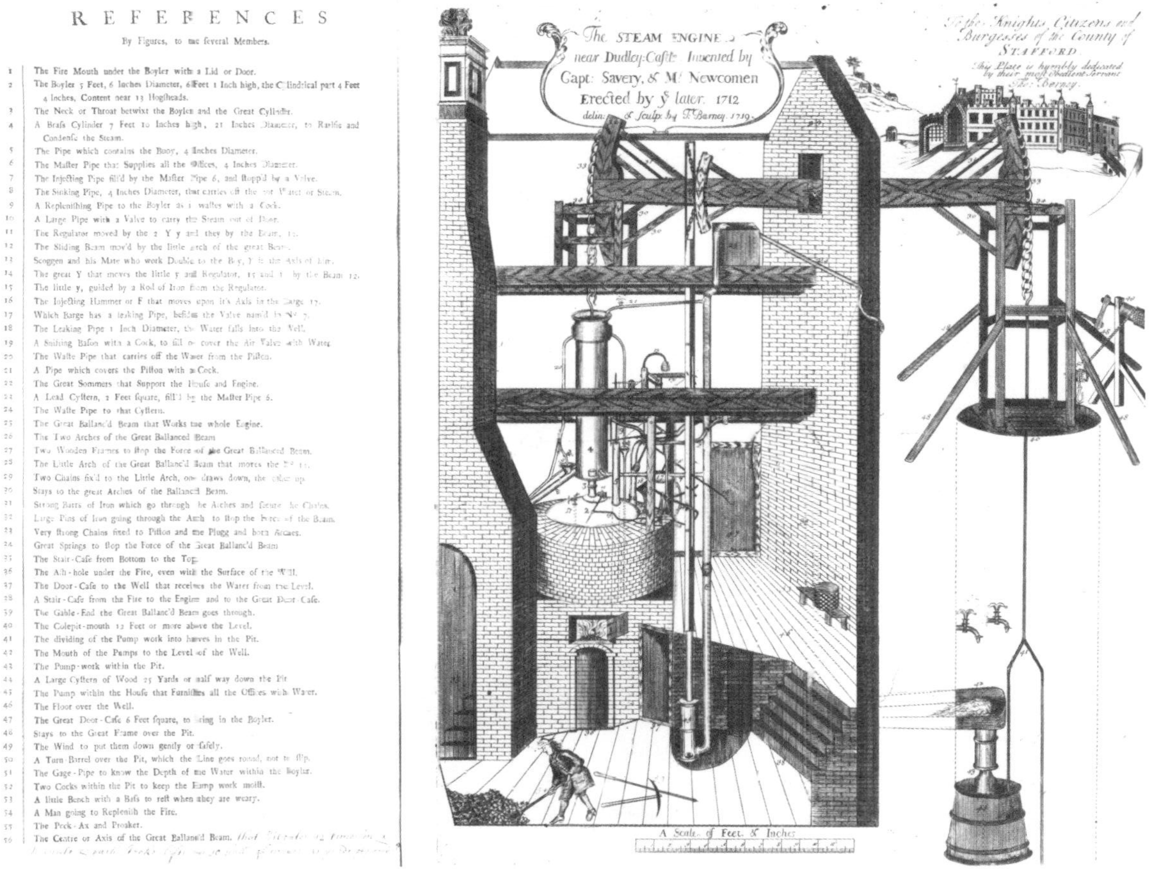

FIGURE 7–10.   The Newcomen engine has been called "the main factor in the exploitation of the mineral wealth of Britain, thereby aying the foundations of the industrial development of the country." (From H. W. Dickinson, in *A History of Technology*, IV, edited by Charles Singer et al., Oxford, 1958, 180–81. Trustees of the Science Museum, London.)

179

on the Continent as well. They were employed mainly in coal mines, where fuel was cheap, but also in other mining industries. The engines were also used to raise water to operate waterwheels when the natural fall was inadequate, and for public water supplies.

The major deficiency of the Newcomen engine was its large consumption of fuel in relation to the work produced. In the 1760s James Watt, a "mathematical instrument maker" (laboratory technician) in the University of Glasgow, was asked to repair a small working model of a Newcomen engine used for demonstration purposes in the course on natural philosophy. Intrigued, Watt began to experiment with the engine; in 1769 he took out a patent for a separate condenser, which eliminated the need for alternate heating and cooling of the cylinder. A number of technical difficulties, including that of obtaining a cylinder sufficiently smooth to prevent the steam from escaping, still plagued the engine and delayed its practical use for several years. In the interim Watt formed a partnership with Matthew Boulton, a successful hardware manufacturer near Birmingham, who provided Watt with the time and facilities for further experimentation. In 1774 John Wilkinson, a nearby ironmaster, patented a new boring machine for making cannon barrels, which also sufficed for engine cylinders. The following year Watt obtained an extension of his patent for twenty-five years, and the firm of Boulton and Watt began commercial production of steam engines. One of their first customers was John Wilkinson, who used the engine to operate the bellows for his blast furnace.

Most of Boulton and Watt's early engines were used for pumping mines, especially the tin mines of Cornwall, where coal was expensive and the savings in fuel consumption compared with the Newcomen engine were considerable. But Watt made a number of other improvements, among them a governor to regulate the speed of the engine and a device to convert the reciprocating motion of the piston to rotary motion. The latter, in particular, opened a host of new applications for the steam engine, such as flour milling and cotton spinning. The first spinning mill to be driven directly by a steam engine began production in 1785, dramatically hastening a process of change that was already underway.

The textile industries had already grown to prominence in the "preindustrial" era in Britain under the putting-out system. Manufactures of woolen and worsted goods were by far the most important (see p. 116), although linen counted for more in Scotland and Ireland than in England and Wales. (In England, corpses were required by law to be buried in woolen shrouds, whereas in Scotland that privileged status was reserved for linen.) The silk industry, introduced in the early decades of the eighteenth century, employed factories and water-powered machinery in imitation of the Italians; but the demand for silk was restricted by its high cost and continental competition.

The manufacture of cotton cloth, like that of silk, was a relatively new industry in Great Britain. Introduced into Lancashire in the seventeenth century, probably by immigrants from the Continent, it was stimulated by the Calico Acts of the early eighteenth century (see p. 160). At first the industry employed the hand processes used in the woolen and linen industries and, because of the weakness of the yarn, used linen warps to produce a type of cloth called fustians. Since it was new, cotton manufacture was less subject than other industries to restrictive legislation and guild rules and to traditional practices that obstructed technical change. Deliberate efforts to invent labor-saving machinery for both spinning and weaving were made as early as the 1730s. The early spinning machines were not successful, but in 1733 a Lancashire mechanic, John Kay, invented the flying shuttle, which enabled a single weaver to do the work of two, thereby increasing the pressure of demand for yarn. In 1760 the Society of Arts added to the incentive of the market by offering a prize for a successful spinning machine. Within a few years several devices for mechanical spinning were invented. The first was James Hargreaves's spinning jenny, invented in 1764 but not patented until 1770. The jenny was a relatively simple machine; in fact, it was little more than a spinning wheel with a battery of several spindles instead of one. It did not require mechanical power and could be operated in a spinner's cottage, but it allowed one person to do the work of several.

The water frame, a spinning machine patented in 1769 by Richard Arkwright, had more general significance. Arkwright, originally a barber and wigmaker, probably did not invent the water frame himself, and his patent was subsequently voided; but of all the early textile innovators he was the most successful as a businessman. Because the water frame operated with water power and was heavy and expensive, it led directly to the factory system on the model of the silk industry. The factories, however, were built most often near streams in the country or in small villages, so that they did not result in concentrations of workers in the cities. Moreover, because water power actuated the machinery, the early factories required relatively few adult males as skilled workers and supervisors; most of the labor force consisted of women and children, who were both cheaper and more docile.

The most important of the spinning inventions was Samuel Crompton's mule, so called because it combined elements of the jenny and the frame. Perfected between 1774 and 1779 but never patented, the mule could spin finer, stronger yarn than any other machine or hand spinner. After it was adapted for steam power, about 1790, it became the favored instrument for cotton spinning. Like the water frame, it allowed large-scale employment of women and children; but unlike the frame, it favored the construction of huge factories in cities where coal was cheap and labor plentiful. Manchester, which had only two cotton mills in 1782, had fifty-two twenty years later.

The new spinning machines reversed the pressure of demand between spinning and weaving, and led to a more insistent search for a solution to the problems of mechanical weaving. In 1785 Edmund Cartwright, a clergyman without training or experience in either mechanics or textiles, solved the basic problem through the application of sheer intelligence and took out a patent for a power loom. Many minor practical difficulties hindered the progress of mechanical weaving, however, and it was not until the 1820s, when the engineering firm of Sharp and Roberts in Manchester built an improved power loom, that machinery began to replace the handloom weavers in large numbers.

A rapid increase in the demand for cotton accompanied the technical innovations. Since Britain grew no cotton domestically, the import figures for raw cotton provide a good indication of the pace at which the industry developed. From less than 500 tons at the beginning of the century, imports rose to about 2500 tons in the 1770s, on the eve of the major innovations, and to more than 25,000 tons in 1800.

Initially India and the Levant were the chief sources of supply, but their production did not expand rapidly enough to satisfy the growing demand. Cotton production began in Britain's Caribbean Islands and in the American South, but the high cost of separating seeds by hand from the short-staple American fibers, even with slave labor, discouraged it until 1793 when Eli Whitney, a New Englander visiting the South, invented a mechanical "gin" (engine). This machine (with improvements) answered the need so well that the southern United States quickly became the leading supplier of raw material to what soon became Britain's leading industry. In 1860 Britain imported more than 500,000 tons of raw cotton.

The innovations in spinning and weaving, along with the gin, were the most important but by no means the only innovations affecting the cotton industry. A host of minor improvements took place in all stages of production, from the preparation of the fibers for spinning to bleaching, dyeing, and printing. As costs of production fell and the quantity produced increased, a large and growing percentage of the output was exported; by 1803 the value of cotton exports surpassed that of wool, and half or more of all cotton goods, yarn as well as cloth, went to markets overseas.

The drastic reductions in the price of cotton manufactures affected the demand for wool and linen cloth, and provided both incentives and models for technical innovation. Unlike cotton, however, the latter industries were encrusted with tradition and regulation, and the physical characteristics of their raw materials also made them more difficult to mechanize. Innovation in these industries had scarcely begun before 1800, and they were not fully transformed until the second half of the nineteenth century.

The technical changes involving cotton textiles, the iron industry, and the introduction of steam power constitute the heart of the so-called industrial revolution in Britain, but they were not the only industries so affected. Nor did all the changes require the use of mechanical power. At the very time that James Watt was perfecting the steam engine, his illustrious countryman Adam Smith wrote in the *Wealth of Nations* of the great increases in productivity obtained in a pin factory simply by the specialization and division of labor. In some respects, Smith's pin factory can be regarded as symbolic of the many industries engaged in the production of consumer goods, from such simple wares as pots and pans to highly sophisticated ones like clocks and watches.

Another representative industry was the manufacture of pottery. The introduction of fine porcelain from China led to a fashion among the rich to substitute it for gold and silver plate, and also provided a model for more utilitarian wares. Simultaneously, the growing popularity of tea and coffee and the increase in incomes among the middle classes led them to prefer domestically produced "chinaware" to wooden or pewter bowls and tableware. As in the iron industry, the rising price of charcoal induced the pottery industry to concentrate in areas well supplied with coal. Staffordshire became the preeminent industry site, where hundreds of small masters produced for a national market. For the most part they relied on an extensive division of labor to increase productivity, although a few of the more progressive ones, such as Josiah Wedgwood, came to employ steam engines to grind and mix raw materials.

The chemical industry also underwent significant expansion and diversification. Some of the advances resulted from the progress of chemical science, especially that associated with the French chemist Antoine Lavoisier (1743–94) and his disciples. But even more industrial advance resulted from the empirical experiments of the manufacturers of soap, paper, glass, paints, dyes, and textiles, as they sought to cope with shortages of raw materials. It is likely that in the eighteenth century chemists learned as much from the industrial users of chemicals as the latter benefited from chemical science. (Much the same can be said for other sciences.) Sulfuric acid, one of the most versatile and widely used of all chemical substances, provides an example. Although known to alchemists, its production was both expensive and dangerous because of its corrosive properties. In 1746 John Roebuck, an industrialist who had also studied chemistry, devised an economical production process using lead chambers; in partnership with another industrialist, Samuel Garbett, he began production of sulfuric acid on a commercial scale. Among other immediate uses, their product was employed as a bleaching agent in the textile industries in place of sour milk, buttermilk, urine, and other natural substances. Sulfuric acid in turn was replaced in the 1790s, when Scottish firms introduced chlorine gas and its derivatives as a bleaching agent, a discovery of the French chemist Claude Berthollet. By then, however, it had already achieved many other industrial uses.

Another group of chemicals widely used in industrial processes were the alkalis, especially caustic soda and potash. In the eighteenth century these were produced by burning vegetable matter, especially kelp and barilla, but because the supply of these seaweeds was inelastic, new methods of production were sought. It was another Frenchman, Nicholas Leblanc, who discovered in 1791 a process for producing alkalis using sodium chloride, or common salt. As with Berthollet's chlorine bleach, Leblanc's process was first applied commercially in Great Britain. This "artificial soda," as it was called, had many industrial uses in the manufacture of soap, glass, paper, paint, pottery, and other products, and it also produced valuable by-products such as hydrochloric acid.

The coal industry, whose growth had been stimulated by the scarcity of timber for fuel, and had in turn prompted the invention of the steam engine, remained for the most part a highly labor-intensive industry, although it also required much capital. Its by-products also proved useful. Coal tar, a by-product of the coking process, replaced the natural tar and pitch for naval stores when the Napoleonic Wars cut off supplies from the Baltic, and coal gas lit the streets of London as early as 1812.

The coal mines were also responsible for the first railways in Britain. As mines went deeper with long underground tunnels, the coal was brought to the main shaft for hoisting on sledges pulled by women or boys, frequently the wives and children of the miners. By the 1760s ponies were used underground in some mines, and were soon pulling wheeled carriages on tracks of metal plates, eventually on rails of cast or wrought iron. Even earlier, in the seventeenth century, tracks and rails had been used on the surface in the vicinity of the mines to facilitate haulage, with horses as the usual draft animals. In the great coalmining regions on the estuary of the Tyne River, in the vicinity of Newcastle, and in South Wales, rails were laid from the mines to wharves along the river or sea to which coal carts descended by gravity. The carts were returned to the mines by horses and, in the early years of the nineteenth century, by stationary steam engines using cables to pull up the empty carts. By the time of the first successful locomotive Britain already had several hundred miles of railways.

The steam locomotive was the product of a complex evolutionary process with many antecedents. The principal ancestor was, of course, the steam engine, as improved by James Watt, yet Watt's engines were too heavy and cumbersome and did not generate enough power per unit of weight to serve as locomotives. Moreover, Watt himself opposed the development of locomotives on grounds of their potential danger, and discouraged his assistants from working with them. As long as his patent for the separate condenser was in force (until 1800), effective progress was barred. Apart from the steam engine itself, the design and construction of locomotive engines required the development of accurate and powerful machine tools. John Wilkinson, whose boring machine made it possible for Watt to build his engine, was one of many talented engineers and machine-makers. Others included John Smeaton (1724–92), founder of the civil engineering profession, whose innovations brought the efficiency of both waterwheels and atmospheric steam engines to their peaks. Henry Maudsley (1771–1831), another in this pantheon, invented a slide-rest screw-cutting lathe about 1797, which made it possible to produce accurate metal parts.

Richard Trevithick (1771–1833), a Cornish mining engineer, deserves credit for building the first working locomotive in 1801. Trevithick used a high-pressure engine (unlike Watt), and designed his locomotive to operate on ordinary roads. Although technically operable, the locomotive was not an economic success because the roads could not bear the weight. In 1804 he built another locomotive to run on a short mine railway in the South Wales coalfield; again, although the locomotive was successful, the light cast iron rails could not bear the weight. After a few more such trials Trevithick devoted himself to building pumping engines for the Cornish mines, at which he was eminently successful.

Although many other engineers, such as John Blenkinsop, contributed to the development of the locomotive, George Stephenson (1781–1848), an autodidact, was the most conspicuously successful. Employed as an engine-wright in the Newcastle mining district, in 1813 he built a stationary steam engine with cables for hauling empty coal cars back to the mine from the loading wharves. In 1822 he persuaded the promoters of the projected Stockton and Darlington Railway, a colliery line, to use steam rather than horse traction, and on its opening in 1825 he personally drove an engine of his own design. The Liverpool and Manchester, generally regarded as the world's first common carrier railway, opened in 1830. All of its locomotives were designed and built by Stephenson, whose "Rocket" had won the famous Rainhill trials the previous year.

Regional Variation

In this summary account of the dawn of modern industry the terms *Great Britain* and *England* have been used more or less interchangeably. Most early accounts of

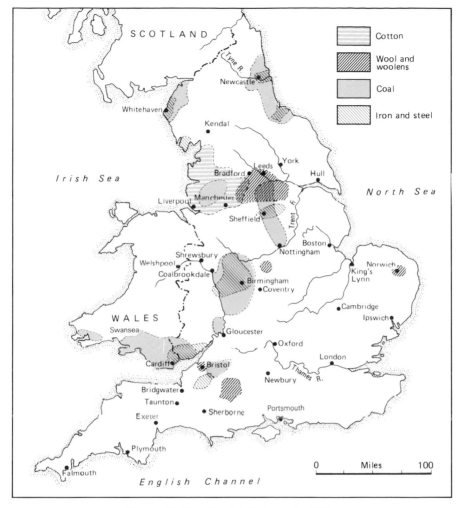

FIGURE 7–11.   English industry in 1800.

the so-called industrial revolution concentrated on England alone. It is important to recognize, however, the great regional variations in industrialization within England, as well as the very different courses of economic change within the component parts of the United Kingdom of Great Britain and Ireland.

Within England, the differential pace of change clearly emphasized the importance of the coalfields, located mainly in the northeast (especially Tyneside) and the Midlands, although Lancashire also had important deposits (see Fig. 7-11). Lancashire became almost synonymous with cotton, but it also contained important glass and chemical works, and the cotton industry had outposts in the East Midlands (Derbyshire and Nottinghamshire) as well. The iron industry and its many manufacturing branches concentrated in the West Midlands (Birmingham and the "Black Country," Shropshire), South Yorkshire (especially Sheffield), and the northeast (especially Newcastle, which was also a center of the chemical industry). The woolen industries tended to concentrate in the West Riding of Yorkshire (especially Bradford and Leeds) at the expense of the older, preindustrial centers of East Anglia and the West Country. Staffordshire almost monopolized the pottery industry, and had important ironworks as well. Cornwall remained the major source of tin and copper, but had few manufacturing industries as such. Except for the burgeoning metropolis of London, with its many consumer goods industries (especially brewing), the south remained primarily agricultural. It was not necessarily poor on that account; it had the most fertile soil and the most advanced agrarian organization, and the growing urban centers; buoyant demand for food ensured southern farmers and landlords a good return on their labor and capital. The mainly pastoral extreme north and northwest, on the other hand, lagged behind most other regions in income and wealth.

Wales, conquered by the English in the Middle Ages, had always been treated as something of a poor relation. In the latter part of the eighteenth century the extensive coalfields of South Wales provided the basis for a large iron industry, which by 1800 produced about one-fourth of British iron; but it was oriented to the export trade and spawned few subsidiary processing industries. The island of Anglesey contained important copper mines, but the ores were smelted mostly in South Wales, around Swansea. The northeastern part of the country, adjacent to Cheshire and Lancashire, benefited slightly from the spillover of their industries; yet most of the interior of the country, mountainous and infertile, remained pastoral and poor. The road to fame and fortune for ambitious Welshmen lay through England or Scotland. One who took this road was Robert Owen, who made a fortune in the cotton industries of Manchester and New Lanark before devoting the remainder of his long life to philanthropic and humanitarian causes.

Scotland, unlike Wales, maintained its independence from England until the voluntary union of the parliaments in 1707. In the mideighteenth century, however, Scotland was a poor and backward country. The majority of its sparse population still engaged in near-subsistence agriculture, and in large areas of the Highlands the tribal system of social and economic organization remained intact. Less than a century later Scotland stood with England at the forefront of the world's industrial nations. With less than one-seventh of the population of Great Britain, Scotland produced more than a fifth of the value of cotton textiles and more than a fourth of the pig iron. The Carron Company, established in 1759, was the first large-scale integrated ironworks using coke fuel anywhere in the world. The Scots also accounted for many of the leading innovators and entrepreneurs in the chemical and engineering industries. In short, Scotland's transformation from backward household economy to leading industrial economy was even more spectacular than the contemporary industrialization of England.

The reasons for Scotland's remarkable transformation have often been debated. Its only natural resource of any significance consisted of the coal measures (intermingled with "blackband" iron ore) in the narrow Lowland stretch between the Firth of Forth and the Firth of Clyde, an area that supported most of Scotland's urban population and almost all of its modern industries. Scotland's inclusion in the British Empire after 1707 gave it access not only to English markets, but to those of

England's colonies in North America and elsewhere, which undoubtedly contributed to the quickening of the tempo of economic life. The country's educational system, from parish schools to its four ancient universities (as compared with only two in England), created an unusually literate population for the time. Similarly, Scotland's precocious banking system, entirely distinct from England's and virtually free of government regulation, allowed Scottish entrepreneurs relatively easy access to credit and capital. Finally, one should not dismiss the fact that Scotland remained without a distinct political administration, apart from the local governments, from the Treaty of Union until 1885. Although this situation was deplored by those who felt that a distinctively Scottish government might have taken more vigorous and effective action to promote economic growth, it may be that the absence of a central government in Scotland was a blessing in disguise.

Ireland, in sad contrast to Scotland, failed almost entirely to industrialize. The English treated Ireland, even more than Wales, as a conquered province. Whether this was the main or even a determining reason for Ireland's fate cannot be considered here. The fact is that the Irish population, like that of Great Britain, more than doubled between the mideighteenth century and 1840, but without appreciable urbanization or industrialization. When the disastrous potato famine struck in the mid-1840s, Ireland lost a fourth of its population in less than a decade through starvation and emigration.

Social Aspects of Early Industrialization

Table 7-1 indicates the approximate population of Great Britain and of England and Wales at selected dates between 1700 and 1850. The figures show the rapid rise in population during the early stages of industrialization; a more detailed breakdown would show that population began to grow in the 1740s after remaining virtually stagnant in the first part of the century, that the growth rate accelerated in the 1780s and reached a maximum in the decade 1811–20, and then declined slightly to 1850. Britain fully participated in, and possibly even led, Europe's third logistic.

That the growth of population was not uniquely related to the process of industrialization is supported by the fact that it was a general European phenomenon, not confined to Great Britain and other industrializing nations. On the other hand, it would be incorrect to say that there was *no* relation; the contrasting fates of Britain

TABLE 7-1.   Populations of England and Wales
and Great Britain, 1700–1850 (millions)

	1700	1750	1800	1850
England and Wales	5.8	6.2	9.2	17.8
Great Britain	—	7.4	10.7	20.6

Source: B. R. Mitchell and P. Deane, *Abstract of British Historical Statistics* (Cambridge, 1962).

and Ireland in the middle decades of the nineteenth century suggest that industrialization was at least a permissive factor in the continued growth of population.

The mechanisms of the growth that occurred in the eighteenth century are imperfectly understood, in large part for want of sufficiently detailed information. It is possible that the birth rate rose somewhat because of earlier marriages, as the growth of both cottage and factory industries allowed young couples to set up households without waiting for a farmstead or completing an apprenticeship. Even more likely, the death rate declined because of several interrelated factors: the introduction of the practice of inoculation against small pox early in the century and vaccination from 1798, improvements in medical knowledge and the establishment of new hospitals, and, most important, a rise in the standard of living, which was both an effect and a cause of economic growth. Agricultural progress brought both a greater abundance and a greater variety of foodstuffs, improving nutrition; increased coal production made for warmer dwellings; soap production, which doubled in the second half of the century, indicated a greater awareness of personal hygiene and, together with the greatly increased output of cheap cotton cloth, contributed to higher standards of cleanliness.

Immigration and emigration also affected the total population. Throughout the eighteenth and early nineteenth centuries the greater economic opportunities of England and Scotland attracted Irish men and women, either temporarily or permanently, even before the huge influx following the potato famine. Political and religious refugees also came from continental Europe. On the other hand, more than a million English, Welsh, and Scots left their homelands for overseas destinations in the eighteenth century, mainly in the British colonies; most went in search of greater economic opportunity, but some—delinquent debtors and others classified as criminals—were forcibly deported to America and later to Australia. On balance, Great Britain probably lost more than it gained through international migration in the eighteenth century.

Even more important for the process of economic growth, internal migration greatly altered the geographical location of the population. Most of this migration was for relatively short distances, from the countryside to the growing industrial areas, but, together with the higher rates of natural increase, it produced two notable changes in the spatial distribution of the population: (1) a shift in density from the southeast to the northwest and (2) increasing urbanization.

At the beginning of the eighteenth century the bulk of the population of England was concentrated south of the Trent River, much of it in about a dozen counties in the southeastern corner of the country; Wales and Scotland were much less densely populated than England. By the beginning of the nineteenth century the most densely populated county outside the London metropolitan area was Lancashire, followed by the West Riding of Yorkshire and four counties including the West Midlands coalfields. The Lowland belt of Scotland between the Firths of Forth and Clyde and the Tyneside coalfield also registered notable gains. This distribution reflected the importance of coal in the industrializing economy.

In 1700 London, with a population in excess of half a million, was far and away the largest city in Britain, and probably the largest in Europe. No other British city exceeded 30,000. By the time of the first census in 1801 London had more than a million, and Liverpool, Manchester, Birmingham, Glasgow, and Edinburgh each had more than 70,000 and were growing rapidly. The census of 1851 officially classified more than half the population as urban, and by 1901 the proportion had grown to more than three-quarters.

The growth of cities was not an unmixed blessing. They contained huge ramshackle tenements and long rows of miserable cottages in which the families of the working classes crowded four and even more persons per room. Sanitary facilities were generally nonexistent, and refuse of all kinds was disposed of by being thrown into the street. Drainage facilities, where they existed, usually took the form of open ditches in the middle of the streets, but more often than not rain, waste-water, and refuse were left to stand in stagnant pools and rotting piles that filled the air with vile odors and served as breeding ground for cholera and other epidemic diseases. The streets were mostly narrow, crooked, unlighted, and unpaved.

In part, the deplorable conditions resulted from extremely rapid growth, inadequacy of the administrative machinery, lack of experience by local authorities, and the consequent absence of planning. For example, Manchester grew from a "mere village" at the beginning of the eighteenth century to a town of 25,000 in 1770, and to more than 300,000 in 1850; but it did not secure a charter of incorporation until 1838. The rapid growth of cities is even more surprising in view of the fact that it resulted entirely from migration from the countryside; because of the hideous sanitary conditions, the death rate exceeded the birth rate (infant mortality was especially high), and the rate of natural increase was actually negative. That people consented to live in such conditions is evidence of the great economic pressures forcing them to move. Although the agricultural labor force continued to grow until about 1850, the increase of the rural population was greater than could be absorbed by traditional rural pursuits, including cottage industry as well as purely agricultural labor.

An earlier textbook asserted that workers were "forced into the factories by the lure of high wages." Such a statement reveals more about the preconceptions of the author than about the economic conditions of the era. That factory workers received higher wages than either agricultural laborers or workers in domestic industry there can be no doubt. This was true not only of the adult male labor force, but of women and children, too. Many accounts of the so-called industrial revolution in Britain stress the employment of women and children in the factories, as though it were a novelty; nothing could be further from the truth. The employment of women and children in both agriculture and domestic industry was a phenomenon of long-standing, which the factory system merely adopted.

Factories developed first in textiles and spread slowly to other industries. Factories could pay higher wages because the productivity of labor was higher as a result of both technological advance and the provision of more capital per laborer. In this way factories gradually attracted more labor, and the general trend of real wages was upward. This trend may have been interrupted during the French wars, from 1795 to 1815, when the demands of government finance created an inflationary situation in which many wage earners fell behind in the growth of real income. The upward trend of real wages resumed after 1812–13 for most categories of workers, although the periodic depressions of the era brought distress to workers as a result of unemployment.

For the last century and more a major scholarly debate has raged over the question of the standard of living of British workers from the latter part of the eighteenth century to the middle of the nineteenth century. (There is no dispute that the standard rose after 1850.) No consensus has been reached, nor is it likely that one will be; the available data are not conclusive and, more important, it is difficult to assign accurate weights to the changing fortunes of various segments of the population. Some groups, such as factory workers and skilled artisans, clearly improved their lot; others, such as the unfortunate handloom weavers, disappeared as a result of technological obsolescence (but of course they went into other occupations).

On balance, it seems likely that there was a gradual improvement in the standard of living of the working classes in the century from 1750 to 1850, although some groups probably experienced a setback during the French wars. The debate is further complicated by the relative changes in the distribution of income and wealth. Most workers, including even the lowest paid, improved their situation slightly, but the incomes of those who depended mainly on rent, interest, and profit rose in greater proportion. In other words, the inequality of the distribution of income and wealth, which was already great in the preindustrial economy, became even greater in the early stages of industrialization.

8

Economic Development in the Nineteenth Century: Basic Determinants

The nineteenth century witnessed the definitive triumph of industrialism as a way of life in Europe, especially western Europe. From their beginnings in Britain, the forms of modern industry spread across the English Channel and the North Sea to Belgium, France, Germany, and the other nations of Europe, as well as across the Atlantic to the United States and, much later, to other areas of the world. In the process they greatly transformed the conditions of life and work in the affected areas. The transformations took different shapes in different regions and nations, depending on local circumstances and the timing of the onset of industrialization. These differences are sketched out in the chapters that follow. In this chapter we consider the broad general tendencies in the basic determinants—population, resources, technology, and institutions.

Population

After virtual stagnation from the early or midseventeenth century to the mideighteenth century, the population of Europe again began to grow from about 1740 (Fig. 8-1). By 1800 it had risen to almost 200 million, or slightly more than one-fifth of the estimated world total of approximately 900 million. (More precise—not necessarily more accurate—figures are given in Table 8-1.) In the nineteenth century population growth in Europe accelerated, and by 1900 the figure exceeded 400 million, or about one-quarter of the world total of approximately 1.6 billion. (These figures do not include the population of European origin overseas, notably in the United States, the British dominions, and Latin America, which would swell the proportion of European stock to more than 30 percent of the world total.) Population growth continued in the twentieth century, although the rate of growth in Europe decelerated slightly while that of the rest of the world increased. By 1950 Europe's population had increased to more than 550 million in a world total of almost 2.5 billion.

Such rates of increase, in both Europe and the world as a whole, were unprecedented.

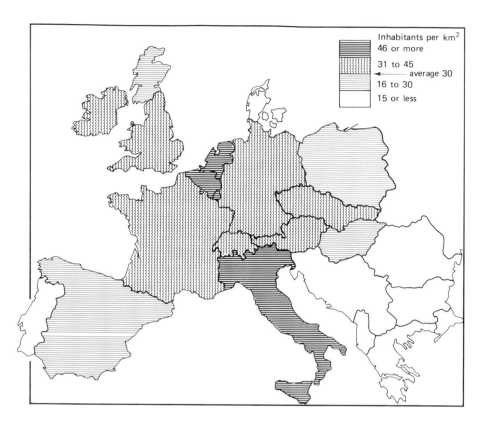

Inhabitants per km²
46 or more
31 to 45 ← average 30
16 to 30
15 or less

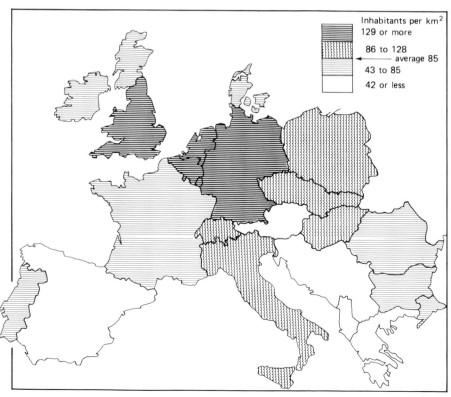

Inhabitants per km²
129 or more
86 to 128 ← average 85
43 to 85
42 or less

TABLE 8-1.   Growth of Population (millions)

	1800	1850	1900	1950
Europe	187.0	266.0	401.0	559.0
United Kingdom	16.1	27.5	41.8	50.6
Great Britain[a]	10.7	20.9	37.1	
Ireland	5.2	6.5	4.5	
Germany	24.6	35.9	56.4	69.0[b]
France	27.3	35.8	39.0	41.9
Russia	37.0	60.2	111.0	193.0[c]
Spain	10.5	na	16.6	28.3
Italy	18.1	24.3	32.5	46.3
Sweden	2.3	3.5	5.1	7.0
Belgium	na	4.3	6.7	8.6
Netherlands	na	3.1	5.1	10.0
North America	16.0	39.0	106.0	217.0
United States	5.3	23.2	76.0	151.7
South America	9.0	20.0	38.0	111.0
Asia	602.0	749.0	937.0	1302.0
Africa	90.0	95.0	120.0	198.0
Oceania	2.0	2.0	6.0	13.0
World total	906	1171	1608	2400

[a]Census dates, 1801, 1851, and 1901.
[b]West Germany.
[c]1946.

Source: W. S. Woytinsky and E. S. Woytinsky, *World Population and Production: Trends and Outlook.* (New York, 1953), pp. 34 and 44. Great Britain and Ireland from B. R. Mitchell and P. Deane, *Abstract of British Historical Statistics* (Cambridge, 1962), pp. 8–10.

Apart from short-term fluctuations (which could sometimes be severe, as during the Black Death), world population had doubled approximately every thousand years from the invention of agriculture until the end of the eighteenth century. In the nineteenth century the population of Europe doubled in less than a hundred years, and in the twentieth century even that rate has been exceeded for the world as a whole. At present rates of natural increase, world population will double in twenty-five or thirty years.

During the nineteenth century Britain and Germany, the two most important industrial nations in Europe, had growth rates in excess of 1 percent per year. (A constant rate of 1 percent would result in a doubling of the population in about seventy years.) Russia, on the other hand, one of the least industrialized countries in Europe, had the highest growth rate of any major European country—about 2 percent on average for the entire century. France, another important industrial country, which had the largest population in western Europe at the beginning of the

FIGURE 8–1 (LEFT).   **Top:** Population densities in Europe, about 1750. **Bottom:** Population densities in Europe, about 1914. (From Figures 1.9a and 1.13b, in *Atlas of World Population,* by Colin McEvedy and Richard Jones. Penguin Books, 1978. Copyright © 1978 by Colin McEvedy and Richard Jones.)

century, lagged far behind the others, especially in the second half of the century; for the century as a whole its growth rate averaged only about 0.4 percent per year.

There is thus no clear correlation between industrialization and population growth. Other causal factors must be sought. Before improvements in transportation that permitted the large-scale importation of foodstuffs from overseas in the last quarter of the nineteenth century, one major constraint on population growth was Europe's own agricultural resources. Agricultural production increased enormously during the century, for two reasons. First, the amount of land under cultivation was extended. This was especially important in the case of Russia, which had vast tracts of vacant land, and also in other parts of eastern Europe and Sweden. Even in western Europe, however, more land was made available by the abolition of fallow and the cultivation of formerly marginal lands or wastelands. Second, agricultural productivity (output per worker) increased because of the introduction of new, more scientific techniques. Better knowledge of soil chemistry and an increased use of fertilizer, at first natural, then artificial, increased yields on ordinary soils and made possible the cultivation of former wasteland. The lower cost of iron promoted the use of improved, more efficient tools and implements. Agricultural machinery, such as steam-driven threshers and mechanical reapers, debuted in the second half of the century.

Cheap transportation also facilitated the migration of the population. As in Britain, the migration was of two types: internal and international. Altogether, about 60 million left Europe between 1815 and 1914. Of these, almost 35 million went to the United States, and an additional 5 million to Canada. Some 12 or 15 million went to Latin America, chiefly Argentina and Brazil. Australia, New Zealand, and South Africa took most of the rest. The British Isles (including Ireland) supplied the largest number of emigrants, about 18 million. Large numbers also left Germany, the Scandinavian countries, and after about 1890, Italy, Austria-Hungary, and the Russian Empire (including Poland). Migration within Europe was also substantial, although in some cases it was only temporary. Large number of Poles and other Slavic and Jewish peoples moved west into Germany, France, and elsewhere. France attracted Italians, Spanish, Swiss, and Belgians, while England obtained immigrants from all of Europe. In the east the tsar settled about 1.5 million peasant families in Siberia between 1861 and 1914, in addition to many criminals and political deportees.

Except for the latter, the migrations were mostly voluntary. In a few cases the migrants fled from political persecution or oppression, but the majority moved in response to economic pressure at home and opportunities for a better life abroad. In the eight years following the Great Famine of 1845, for example, more than 1.2 million people left Ireland for the United States, and many more crossed the Irish Sea to Britain. New and nearly empty lands overseas, such as Canada, Australia, and New Zealand, attracted a steady stream of immigrants, most of them from the British Isles. Relatively large numbers of Italians and Germans migrated to what became the economically most progressive countries of South America.

Internal migration, although less dramatic, was even more basic to the process of economic development in the nineteenth century. Important regional shifts in the concentration of population took place in all countries, but the most fundamental change was the growth of the urban population, both in toto and as a percentage of the total. At the beginning of the nineteenth century England was already the most urbanized nation, with about 30 percent of its population living in agglomerations of 2000 inhabitants or more. The Low Countries, with a long urban tradition, probably had a similar proportion (the province of Holland, dominated by the metropolis of Amsterdam, was above 50 percent). Italy, also with a long urban tradition, had suffered a depopulation of its major cities in the early modern period, and by the beginning of the nineteenth century the urban population probably amounted to no more than one-fourth or one-fifth of the total. Similar proportions obtained in France and western Germany, whereas in most of the rest of Europe, and elsewhere in the world, the urban population amounted to no more than 10 percent of the total.

Urbanization, along with industrialization, proceeded apace in the nineteenth century. Britain once again led the way. By 1850 more than half the British population lived in towns and cities of more than 2000 inhabitants, and by 1900 the proportion reached three-quarters. By that time most of the other industrial nations were at least 50 percent urbanized, and even the predominantly agrarian nations showed a strong tendency to urbanization. For example, in the Russian Empire, which as a whole had no more than 12.5 percent of its population living in urban agglomerations, Moscow and St. Petersburg could nevertheless boast populations of 1 million or more.

The population of industrial countries not only lived in cities, it preferred the largest cities. For example, in England and Wales the proportion of the population living in small towns (2000 to 20,000 inhabitants) has remained roughly constant, at about 15 percent, from the beginning of the nineteenth century to the present, whereas the proportion in the large cities (over 20,000 inhabitants) has risen from 27 percent to more than 70 percent. In 1800 there were barely twenty cities in Europe with populations of as much as 100,000, and none in the Western Hemisphere; by 1900 there were more than 150 such cities in Europe and North America, and by 1950 more than 600. In the midtwentieth century there were more cities with a population in excess of 1 million (sometimes greatly in excess) than there were cities with a population of 100,000 in 1800.

There are many social and cultural reasons for people wanting to live in cities. Historically, the chief limitation on the growth of cities has been economic—the impossibility of supplying large urban populations with the necessities of life. With the technological improvements of modern industry not only were these limitations relaxed, but in some cases economic considerations required the growth of cities. In preindustrial societies most of even the nonagricultural population lived in rural areas. It was cheaper to carry the finished products of industry, such as textiles and iron, to distant markets than to carry food and raw materials to concentrations of workers. The introduction of steam power and the factory system, the transition from charcoal to coke as fuel for the iron industry, and the improvements in transportation and communications changed the situation. The rise of the factory system necessitated a concentration of the work force. Because of the new importance of coal, some of the largest centers of industry arose on or near the sites of coal deposits—the Black Country of England, the Ruhr area in Germany, the area around Lille in northern France, and the Pittsburgh area in America. These examples also underline the importance of resources in modern economic growth.

Resources

Industrial Europe did not experience any magical increase in the quantity or quality of its natural resources, compared with preindustrial Europe, but, as a result of technological change and the pressure of increased demand, resources that were formerly unknown or of little value suddenly acquired enormous, even critical, importance. This was notably the case for coal, and those regions of Europe endowed with plentiful coal deposits became the primary sites of heavy industry in the nineteenth century (Fig. 8-2). Regions without indigenous coal supplies had to import coal, although, of course, they also continued to rely on their traditional sources of water and wind power. In the late nineteenth century, with the introduction of hydroelectricity, those regions that were abundantly supplied with water power, such as Switzerland, parts of France, Italy, and Sweden-Norway, obtained a new source of comparative advantage.

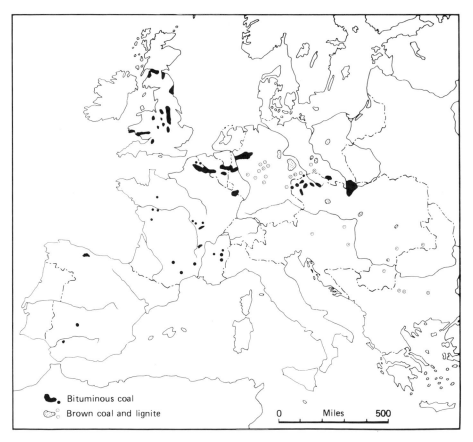

Bituminous coal
Brown coal and lignite

0   Miles   500

FIGURE 8–2.   Coalfields of Europe. (From *An Historical Geography of Europe 1800– 1914*, by Norman J. G. Pounds. Copyright © 1985 by Cambridge University Press. Reprinted with permission.)

Europe as a whole was relatively well endowed with conventional mineral resources, such as iron ore, other metallic ores, salt, and sulfur. Some of these, such as the tin of Cornwall, had been exploited from antiquity, and most of the others had been tapped to a limited extent in the Middle Ages and early modern times, but the demands of modern industry greatly intensified the pressure for their use. This resulted in a systematic search for previously unknown sources, and scientific and technological research to enhance their exploitation. In some cases, as domestic sources were exhausted, the search for new supplies extended overseas, where European capital and technology facilitated the opening of new territories, as in the American West, the British dominions, and parts of Latin America. Increasingly in the later nineteenth century the search for raw materials, in addition to other motives, led the European nations to extend political control over the poorly organized or weakly ruled areas of Africa and Asia.

The Development and Diffusion of Technology

Simon Kuznets, a Nobel Prize winner in economics, referred to the period in which we live as the "modern economic epoch."[1] According to him, an economic epoch is determined and shaped by the applications and ramifications of an "epochal innovation." For example, in his view the epochal innovation of the early modern period of European history was the development of navigational and related techniques that permitted the discovery of America and of an all-water route to the Orient, achievements that Adam Smith, writing in 1776, called "the two greatest and most important events recorded in the history of mankind."[2] According to Kuznets (and no doubt Smith would have agreed), a large part of the economic history—and even the political, social, and cultural history—of the years from 1492 to 1776 can be explained by reference to the progress of exploration and discovery, maritime commerce, the growth of navies, and related phenomena.

The present (modern) economic epoch, in Kuznet's terms, began in the latter half of the eighteenth century, and the epochal innovation he associated with it is "the extended application of science to problems of economic production."[3] As noted in the previous chapter, scientific knowledge as such had only limited applications in economically productive processes in the eighteenth, or even in the first half of the nineteenth century. That period in technological history from the early eighteenth century to approximately 1860 or 1870 is best regarded as the era of the artisan-inventor. Thereafter, however, scientific theories increasingly formed the basis of production processes, notably in such new industries as electricity, optics, and organic chemicals; but they also greatly influenced technical developments in metallurgy, power production, food processing and preservation, and agriculture, to mention only the most prominent fields.

---

[1] Simon Kuznets, *Modern Economic Growth: Rate, Structure, and Spread* (New Haven, CT, 1966), Chap. 1.

[2] Adam Smith, *An Inquiry into the Nature and Causes of the Wealth of Nations* (Glasgow ed.), R. H. Campbell and A. S. Skinner, eds. (Oxford, 1976), II, p. 626.

[3] Kuznets, *Modern Economic Growth*, p. 9.

In analyzing the process of technological change in any period of history, but especially in the modern economic epoch, it is wise to bear in mind the distinctions between three closely related but conceptually different terms: invention, innovation, and diffusion of new technology. Invention, in terms of technology, refers to a patentable novelty of a mechanical, chemical, or electrical nature. In and of itself, invention has no particular economic significance. Only when it is inserted into an economic process—that is, when it becomes an innovation—does it assume economic significance. For example, James Watt's invention of the separate condenser for the Newcomen steam engine, which he patented in 1769, played a negligible role in the economy until, in partnership with Matthew Boulton, he began to produce and market steam engines commercially in 1776. Diffusion refers to the process by which an innovation spreads, within a given industry, between industries, and internationally across geographical frontiers. Diffusion is by no means an automatic process of replicating the initial innovation; because of the different requirements of different industries, different factor proportions in different environments, and cultural differences among nations, it may face problems similar to those connected with introducing an original innovation.

The industrial superiority that Great Britain had achieved by the first quarter of the nineteenth century rested on technological advances in two major industries, cotton textiles and iron manufacture, supported by an extensive use of coal as an industrial fuel and by the growing use of the steam engine as a source of mechanical power. The mechanization of cotton spinning was virtually complete by 1820, making it the first modern factory industry, while that of weaving had scarcely begun. The other principal textile industries, wool and linen, had also just begun to mechanize, although they, along with cotton weaving, made rapid strides in the next several decades. The iron industry had completed the transition to coke smelting of iron ore and the use of the puddling process and rolling mills for refining the product of the blast furnace. Coal was used extensively not only to fire steam engines, blast furnaces, and puddling furnaces, but also as fuel in a number of other industries, such as glass manufacture, salt refining, brewing, and distilling. Steam engines provided power for textile factories and iron foundries, and for actuating pumps in coal and tin mines; they were also used less extensively in flour mills, pottery factories, and other industries.

In the next half century—that is, until about 1870—the efforts of many continental industrialists, sometimes abetted by their governments, were devoted to acquiring and naturalizing the technological gains of British industry. Some details of these efforts are related in the chapters devoted to individual countries. Meanwhile, however, the pace of technological change accelerated and spread to many industries not previously affected by science-based (or science-influenced) technology. Indeed, some previously nonexistent industries were created de novo as a result of scientific discoveries.

The textile industries, together the largest employers of labor and the most important in terms of value of output of any manufacturing industry in almost every country, underwent numerous minor technical improvements as they expanded output enormously. Many innovations were the work of continental and American industrialists, as they sought to achieve or surpass the technical efficiency of their

British rivals. Overall, however, there were no major technical breakthroughs comparable to the remarkable series of innovations in the last third of the eighteenth century. Such was not the case in other industries, however. Many of the most revolutionary technical developments occurred long after the conventional dates for the industrial revolution in Britain.

Prime Movers and Power Production

When Watt's basic patent expired in 1800 fewer than 500 of his engines were at work in Britain and only a few dozen on the Continent. As fundamental as his contributions were to the evolution of steam technology, Watt's engines had many limitations as industrial prime movers. In the first place, their thermal efficiency was quite low, generally less than 5 percent (that is, they extracted less than 5 percent of the theoretically possible work from the heat energy consumed). On average, they generated only about 15 horsepower, scarcely more than a moderately efficient windmill or waterwheel. They were also heavy, clumsy, and subject to numerous breakdowns. Finally, they worked at relatively low pressure, only a few pounds above atmospheric pressure, which greatly limited their effectiveness. Several reasons accounted for their limited usefulness, among them imperfect scientific knowledge, insufficient strength of the metals used in their construction, and lack of accurate tools.

The next fifty years witnessed many important developments in steam engine technology. Lighter, stronger metals, more accurate machine tools, and better scientific knowledge, including mechanics, metallography, calorimetry, and the theory of gases as well as the embryonic science of thermodynamics, all contributed. Although it is likely that scientists learned more from the steam engine—culminating in Helmholtz's formulation in 1847 of the first law of thermodynamics—than they contributed to it, their contributions were not negligible. The first advances, however, came from practical mechanics and engineers such as the Cornishman Richard Trevithick and the American Oliver Evans, who constructed and experimented with high-pressure engines, which Watt regarded as both unsafe and impractical. These and other experiments led to the use of steam engines to propel steamboats and locomotives, with profound consequences for the transportation industry. Many engines were used in industry as well. By 1850 France had more than 5000 fixed or stationary engines, Belgium more than 2000, Germany almost 2000, and the Austrian Empire about 1200. Although exact figures are not available, it is probable that Great Britain had more steam engines than all the continental countries combined. As early as 1838 the textile industries alone (which were, however, the largest users) had more than 3000. In comparison, the United States in 1838 had fewer than 2000 stationary steam engines in all its industries.

The power and efficiency of engines had increased greatly, as well. Engines producing 40 to 50 horsepower were common, and some produced more than 250. Thermal efficiency was three times more than that of the best Watt engines. Compound, double- and triple-acting engines were introduced. By 1860 large compound marine engines could develop more than 1000 horsepower.

Technological progress also occurred in the case of the steam engine's chief early competitor, the waterwheel. From the 1760s on, while Watt was experimenting with and improving the steam engine, other engineers and inventors turned their attention to improving the waterwheel. They introduced new, more efficient designs and, as a result of the fall in the price of iron, large all-metal wheels came into widespread use. By the early decades of the nineteenth century some very large wheels could generate more than 250 horsepower. Moreover, in the 1820s and 1830s French scientists and engineers invented and perfected the hydraulic turbine, a highly efficient device for converting the force of falling water into useful energy. It is not widely appreciated, but water power reached its peak use (apart from the generators of electricity, which came later) in the third quarter of the nineteenth century. It was only after about 1850, and more noticeably after 1870, that steam power clearly outdistanced its rival.

By the end of the nineteenth century the effective limits of the reciprocating steam engine had been reached, with some triple-expansion marine engines capable of generating 5000 horsepower. Even these huge installations, however, were inadequate for the newest use of steam power, the generation of electricity. For one thing, the maximum rotational velocity of the crankshaft that a reciprocating engine could produce was too low for the much higher velocities required by a dynamo, or electrical generator. In addition, the vibrations of the reciprocating engine were inimical to efficient generation of electricity. The solution to these problems was found in the steam turbine, developed in the 1880s by the British engineer, Charles A. Parsons, and the Swedish inventor, Gustav de Laval. Progress with this new device was rapid, and by the early decades of the twentieth century it was possible to generate more than 100,000 kilowatts from a single installation.

Electrical phenomena had been observed in early times, but as late as the eighteenth century electricity was regarded as simply a curiosity. Toward the end of that century the researches of Benjamin Franklin in America and of the Italians Luigi Galvani and Alessandro Volta, who invented the voltaic pile or battery, raised it from the status of a parlor trick to a laboratory pursuit. In 1807 Sir Humphry Davy discovered electrolysis, the phenomenon by which an electric current decomposes the chemical elements in certain aqueous solutions, which gave rise to the electroplating industry. The next phase in the study of electricity was dominated by Davy's student Michael Faraday, the Danish physicist Hans Oersted, and the French mathematician André Ampère. In 1820 Oersted observed that an electric current produces a magnetic field around the conductors, which led Ampère to formulate a quantitative relationship between electricity and magnetism. Between 1820 and 1831 Faraday discovered the phenomenon of electromagnetic induction (the generation of an electrical current by revolving a magnet inside a coil of wire) and invented a primitive, hand-operated generator. Building on these discoveries, Samuel Morse developed the electric telegraph in American between 1832 and 1844. But the industrial use of electricity was held up by the difficulties involved in devising an economically efficient generator.

Scientists and engineers experimented with a variety of devices to generate electricity, and in 1873 a papermaker in southeastern France attached his hydraulic turbine, which drew water from the Alps, to a dynamo. This apparently simple innovation had important long-range consequences, for it enabled regions poor in coal but rich in water power to supply their own energy requirements. The invention of the steam turbine in the following decade freed the generation of electricity from water power sites and shifted the energy balance back toward coal and steam. Nevertheless, the development of hydroelectric power became tremendously important for coal-short countries previously in the backwaters of industrial development.

A host of practical applications for electricity were developing contemporaneously. Electricity had been used in the new electroplating industry and in telegraphy since the 1840s. Lighthouses began to use electric arc lamps in the 1850s, and by the 1870s arc lamps were used in a number of factories, stores, theaters, and public buildings. The perfection of the incandescent electric lamp almost simultaneously between 1878 and 1880, by Joseph Swan in England and Thomas Edison in the United States, made arc lighting obsolete and inaugurated a boom in the electrical industry. For several decades electricity competed hotly with two other recently perfected illuminants, coal gas and kerosene.

Electricity has many other uses than illumination. It is one of the most versatile forms of power available. In 1879, the same year that Edison patented his electric lamp, Werner von Siemens of Germany invented the electric tram, or streetcar, with revolutionary consequences for mass transportation in the burgeoning metropolises of the time. Within a few years electric motors had found dozens of industrial applications, and inventors were even beginning to think of household appliances. Electricity can also be used to produce heat, and in this way came to be employed in the smelting of metals, notably the recently discovered aluminum.

Petroleum is another major energy source that came into prominence in the second half of the nineteenth century. Although it had been known and used earlier through accidental discoveries, its commercial exploitation began with the drilling of Drake's well at Titusville, Pennsylvania, in 1859. Like electricity, liquid petroleum and its by-product, natural gas, were at first used primarily as illuminants. Crude petroleum consists of several components, or "fractions." Of these kerosene was at first considered the most valuable because of its suitability for oil lamps. Other fractions were used as lubricants, the demand for which grew rapidly with the spread of machinery with moving parts, and for medicinal purposes. The heavier, residual fractions, at first treated as waste products, eventually were used in household and industrial heating, in competition with coal and other traditional energy sources. The highest, most volatile fractions, naptha and gasoline, were long regarded as dangerous nuisances. Meanwhile, however, a number of inventors and engineers, notably the Germans Nikolaus Otto, Karl Benz, and Gottfried Daimler, were experimenting with internal combustion engines. By 1900 a variety of such engines were available, most of which used as fuel one of the several distillates of liquid petroleum, such as gasoline and diesel oil. By far the most important use for the internal combustion engine was in light transport facilities such as automobiles, motor trucks, and buses; in the hands of such entrepreneurs as the Frenchmen Armand Peugeot, Louis Renault, and André Citroen, the Englishman William Morris, and the American Henry Ford, it gave rise to one of the twentieth century's major industries. The internal combustion engine also had industrial applications, and in the twentieth century it made possible the development of the aircraft industry.

Cheap Steel

By the beginning of the nineteenth century coke smelting and the puddling process for producing pig iron and refining it into wrought iron were virtually universal in Britain, giving British ironmasters a competitive advantage over their foreign counterparts. In the latter part of the eighteenth century attempts had been made in both France and Prussian Silesia, under royal patronage, to introduce coke smelting, but neither was economically successful, and in the turmoil of the Revolutionary and Napoleonic wars no further experiments were conducted. With the return of peace after 1815, continental ironmasters rushed to adopt the puddling and rolling method for converting pig iron to wrought iron, but, because of the different relative prices of charcoal and suitable coking coal on the Continent and in Britain, they were slower to adopt coke smelting. The first successful coke-fired blast furnaces on the Continent were built in Belgium (then a part of the kingdom of the United Netherlands) in the latter 1820s; a few French ironmasters adopted coke in the 1830s and 1840s, but the process did not become predominant until the 1850s. Germany was even slower to adopt coke smelting, with the great rush occurring in the 1850s. In the United States, with its abundant supplies of timber for charcoal, and an alternative to coke in the form of eastern Pennsylvania's anthracite coal, coke smelting was not widely adopted until after the Civil War. Elsewhere in Europe—in Sweden, the Austrian Empire, Italy, and parts of Russia—small charcoal-fired industries hung on tenaciously.

The only major technical innovation in the iron industry in the first half of the nineteenth century was the hot blast, patented by the Scottish engineer James B. Neilson in 1828. By using waste gases to preheat the air used in the blast furnace, the hot blast brought about more complete combustion of the fuel, lowered fuel consumption, and sped up the smelting process. It was quickly adopted by ironmasters in Scotland, on the Continent, and even in the United States, but more slowly in England and Wales.

The most dramatic technological innovations affecting the iron industry, occurring in the second half of the century, concerned the manufacture of steel. Steel is actually a special variety of iron; it contains less carbon than cast iron, but more than wrought iron. As a result, it is less brittle than the former, but harder and more durable than the latter. It had been made for many centuries, but in small quantities at high cost, so that its use was limited to such quality products as files, watch springs, surgical instruments, swordblades, and fine cutlery. In 1856 Henry Bessemer, an English inventor, patented a new method for producing steel directly from molten iron, eliminating the puddling process and yielding a superior product (Fig. 8-3). The output of Bessemer steel increased rapidly and soon displaced ordinary iron for a variety of uses. The Bessemer process did not always yield a uniformly high grade of steel, however, and could not be used with phosphorus-bearing iron ores. To remedy the former defect, in the 1860s a father-and-son team of French metallurgists, Pierre and Émile Martin, and the Siemens brothers, Friedrich in Germany and William in England, developed the open hearth, or Siemens-Martin, furnace. It was slower and somewhat more costly than the Bes-

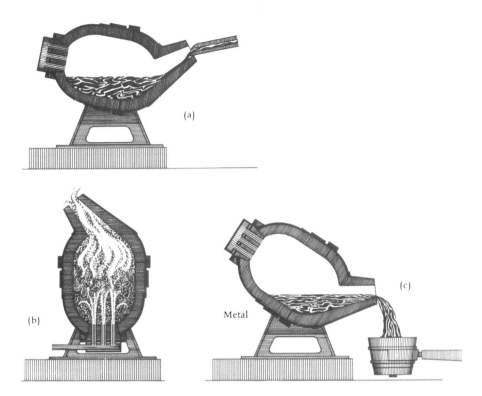

FIGURE 8–3.   The Bessemer converter, which produced steel "without fuel" by blowing air through molten iron to burn off excess carbon: (a) Turned down for charging; (b) blowing; and (c) turned down for tapping. (From *The Archaeology of the Industrial Revolution,* by Brian Bracegirdle, London, 1974. Reprinted with permission.)

semer process, but produced a higher-quality product. In 1878 two English cousins, Sidney G. Thomas and Percy C. Gilchrist, patented the "basic" process (so named because it used limestone or other basic materials to line Bessemer's converter or the open hearth furnace to neutralize the acidic phosphorus in the ore), which permitted the use of the plentiful phosphorus-bearing iron ores. As a result of these and other innovations, the annual world production of steel rose from less than half a million tons in 1865 to more than 50 million tons on the eve of World War I.

The expansion of the steel industry had a profound impact on other industries, both those that supplied the steel industry (such as coal) and those that used steel. Steel rails for railways lasted longer and were safer than iron ones. Steel plates for shipbuilding resulted in larger, lighter, faster ships and could also be used as heavy armor for warships. The use of steel beams and girders made it possible to build skyscrapers and a variety of other structures. Steel soon replaced iron and wood in tools, toys, and hundreds of other products ranging from steam engines to hairpins.

Transport and Communications

The steam locomotive and its adjuncts, iron (or steel) railways, more than any other technological innovation of the nineteenth century, epitomized the process of economic development (Fig. 8-4). They were both the symbols and the instruments of industrialization. Before the railway inadequate transportation facilities constituted a major obstacle to industrialization in both continental Europe and the United States. Lacking Britain's endowment of natural waterways and handicapped by greater distances to cover, continental and American industrialists found themselves pent up in local markets that offered little scope for extensive specialization and expensive capital equipment. The railway and, to a lesser extent, the steamship changed that state of affairs. Railways offered cheaper, faster, more dependable transportation; also, during the period of their construction, from roughly 1830 until the end of the century, their demands for iron, coal, timber, bricks, and machinery proved a potent stimulus to the industries that supplied them.

As noted in the previous chapter, by the end of the eighteenth century British coalfields had many miles of railways on which wagons were propelled by gravity, horsepower, and human beings. The opening of the Stockton and Darlington Railway in 1825 heralded the railway age, and the Liverpool–Manchester Railway, the first designed specifically for steam locomotion and as a common carrier, inaugurated it in 1830. Thereafter the British railway network developed rapidly. Britain had both the technical expertise and the reservoirs of capital needed for construction; Parliament, under the influence of liberal ideas on economic policy that had recently triumphed, readily granted charters to private joint-stock companies. Frenzies of both speculation and construction resulted ("manias," as they were called), inevitably punctuated by financial crises. Nevertheless, by 1850 Britain had constructed more than a fourth of its eventual network, almost as much as the rest of Europe combined (see Table 8-2).

France, Austria, and the United States had short, horsedrawn railways by 1830 (and France even had a few miles of steam railways), but the United States outstripped even Britain and rivaled all of Europe in the construction of railways. It drew on European capital and suppliers as well as on the abundant enthusiasm of private promoters and state and local governments to span the vast distances of the country. Many of the railways were cheaply constructed, however, and were built to widely varying standards.

Belgium made the best early showing of any continental country in railway planning and construction. Rejoicing in its newly won independence (from the United Netherlands), the middle-class government resolved to build a comprehensive network at state expense to facilitate the export of Belgian manufactures and win the transit trade of northwestern Europe. The first section, and the first wholly steam-operated railway on the Continent, opened in 1835. Ten years later the basic state-owned network was complete, after which the job of providing branch and secondary lines was turned over to private enterprise.

France and Germany were the only other continental nations to make significant railway progress by midcentury. Germany, although divided into several independent and rival states, achieved the most. Beginning with the short Nuremberg-Fürth

FIGURE 8–4. Steam locomotives. The steam locomotive epitomized nineteenth-century technology, and its evolution was rapid, from the "Rocket" (**top**) of 1829 to the huge freight locomotives (**bottom**) of the early twentieth century, so prevalent in the United States. (The Rocket from the Science Museum, London; locomotive from *The Railroad Scene*, by William D. Middleton, New York, 1927.)

TABLE 8-2.   The Growth of Railways
(length of railway open, in kilometers)

Country	1840	1870	1914
Austria-Hungary	144	6,112	22,981[a]
Belgium	334	2,897	4,676[a]
Denmark	0	770	3,951
Finland	0	483	3,683
France	410	15,544	37,400
Germany	469	18,876	61,749
Italy	20	6,429	19,125
Netherlands	17	1,419	3,339
Norway	0	359	3,165
Russia	27	10,731	62,300
Spain	0	5,295	15,256
Sweden	0	1,727	14,360
Great Britain	2,390	21,558[b]	32,623
United States	4,510	84,675	410,475

[a]1913.
[b]1871.

Source: B. R. Mitchell, *European Historical Statistics, 1750–1970* (New York, 1975), pp. 582–84. United States figures from *Historical Statistics of the United States, Colonial Times to 1957* (Washington, 1960), pp. 427, 429.

line in 1835, construction took place at varying but generally rapid rates in several of the states. Some followed a policy of state ownership and operations; others left railways to private enterprise, though usually with subsidies. Still others allowed both state and private enterprise. Although France had a centralized government and by 1842 a comprehensive railway plan centered on Paris, it built more slowly. Parliamentary wrangling over the question of state or private enterprise and sectional conflicts over the location of the main lines held up the railway era in France until the coming of the Second Empire. After 1852 construction proceeded rapidly.

Elsewhere progress was minimal before midcentury. The first railway in the Austrian Empire, a horse-powered line in Bohemia from Budweiss to Linz, dated from the 1820s. In 1836 the government granted a concession for the first steam railway to a private company sponsored by the Rothschild family, but in 1842 the state undertook to construct railways for its own account, a policy that continued until financial difficulties in the following decade forced the state to dispose of its railways to private companies. By midcentury only some 1700 kilometers of state and private railways were in operation, and those almost exclusively in Bohemia and the German-speaking portions of the empire.

The Netherlands had a flurry of activity in the late 1830s and early 1840s, which linked the major cities, but the financial results were poor and the railways fell into disfavor. The excellent Dutch canals and the few brick highways across the flat countryside were adequate for the needs of internal commerce. The Netherlands still lived by the sea and maintained communication with the interior by the Rhine and

Meuse rivers. The railway network did not connect with that of the rest of Europe until 1856.

A few short railways had been constructed in the Italian peninsula in the 1830s and 1840s, but, divided as the country was among several petty, impoverished principalities, railways made little progress until the advent of the statesman Camillo de Cavour in the kingdom of Sardinia in the 1850s. Switzerland and Spain had both inaugurated short lines in the 1840s, but, as in Italy, serious construction did not begin until the 1850s.

The government of the tsar, after connecting the city of St. Petersburg with the imperial summer palace outside the city by rail in 1838, did not again venture into the field until the middle 1840s. It then undertook, primarily for military purposes and by means of loans raised abroad, the important lines from St. Petersburg to Moscow and from St. Petersburg to the Austrian and Prussian frontiers. (Nicholas I is reputed to have settled a dispute among his engineers as to the route of the Moscow–St. Petersburg railway by drawing a straight line between the two cities on the map with a ruler and stating, ''There, gentlemen, is the line I want you to build.'') The midcentury mark passed, however, with only the relatively short section from Warsaw to the Austrian frontier in operation.

Elsewhere in eastern and southeastern Europe, whether in areas dominated by emperor, tsar, or sultan, there were no thoughts of railways in 1850. Even in the West, Denmark had just begun to make plans, and three countries—Sweden, Norway, and Portugal—had neither railways nor plans.

The second half of the nineteenth century was the great age of railway construction, in Europe and elsewhere, as is evident from Table 8-2. British engineers, with their head start in experience and numerous foundries and machine shops, built some of the first railways on the Continent; subsequently they were responsible for the greater part of construction in India, Latin America, and southern Africa. Americans built their own railways from the beginning, although with the assistance of European (mainly British) capital and some equipment. The French, after some early lessons from the British, not only built their own railways, but most of those throughout southern and eastern Europe, including Russia. The Germans also built most of their own railways, and some in eastern Europe and Asia, and in the process strengthened their huge metallurgical and engineering establishments.

The first locomotives, although marvels in their day, were actually quite puny (see Fig. 8-4, top). Continual improvements in locomotive design created the huge machines of the late nineteenth and early twentieth centuries, by which time electric traction and diesel engines had begun to challenge the primacy of steam locomotives. Tunnels pierced the Alps as early as the 1870s. Sleeping cars, although introduced in the United States in 1837, did not beome common in Europe until the 1870s, by which time continuous networks of rails crossed political boundaries with ease. The famed Orient Express, from London and Paris to Constantinople, made its first run in 1888.

The steamship, although developed earlier than the locomotive, played a less vital role in the expansion of commerce and industry until late in the century. In fact, for ocean commerce the wooden sailing ship reached its peak development, both technically and in the tonnage of goods carried, after 1850. In the first half of

FIGURE 8–5. Steamboat on the Rhone River. In the first half of the nineteenth century steamboats made their greatest contributions in inland navigation. Ocean steamers came later. (From *La Navigation à vapeur sur la Saone et le Rhone*, by Felix Rivet, Paris, 1962. Reprinted courtesy of Chamber of Commerce, Lyon, France.)

the century steamers made their greatest contribution in the development of inland commerce (Fig. 8-5). Credit for inventing the steamboat is usually given to the American, Robert Fulton, whose ship, the *Clermont,* made its first successful run on the Hudson in 1807, although there are earlier claimants for the distinction. Within a few years steamers appeared on the Great Lakes and the rivers of the Mississippi system as well as in coastal waters. Prior to 1850 steamboats probably contributed more than railroads to opening the trans-Allegheny West. In Europe they could be seen on such broad rivers as the Rhine, the Danube, the Rhone, and the Seine, as well as on the Mediterranean and Baltic seas and the English Channel. Steam came to the North Atlantic with the voyage of the auxiliary steamer *Savannah* in 1820, but regular trans-Atlantic service did not begin until 1838, when the *Sirius* and *Great Western* made simultaneous voyages from England to New York. Samuel Cunard, an Englishman, inaugurated his famous line in 1840, but soon ran into stiff competition from other companies. Until the end of the American Civil War ocean steamers carried chiefly mail, passengers, and expensive, lightweight cargo. The true age of the ocean steamer did not arrive until the development of the screw propeller (1840s), the compound engine (1850s), steel hulls (1860s), and the opening of the Suez Canal in 1869. Thereafter its progress was rapid, and by 1900 use of the large sailing ship was limited to bulky, cheap, nonperishable commodities.

Perhaps no single invention in the nineteenth century compared with that of printing in the fifteenth century in its effect on the field of communication. Nevertheless, the cumulative effects of nineteenth-century innovations were comparable. Papermaking machinery, invented about 1800, and the cylindrical printing press, first used by the London *Times* in 1812, greatly reduced the cost of books and newspapers. Wood pulp replaced rags as the raw material for paper in the 1860s. Together with the reductions in stamp and excise taxes on paper and printing, they brought reading material within reach of the masses and contributed to their increasing literacy. Improvements in printing and typesetting culminated in the linotype machine, invented by the German-American Ottmar Mergenthaler in 1885, further extending the influence of the daily newspaper. By 1900 several newpapers in the largest cities had daily circulations of more than a million copies, compared with 50,000 for the London *Times* during the 1860s, the largest circulation of that period.

The invention of lithography in 1819 and the development of photography after 1827 made possible the cheap reproduction and wide dissemination of visual images. Great Britain introduced the penny post in 1840; in that year the number of letters delivered by the Royal Mail was more than twice the number delivered in 1839. Within a few years most Western countries had adopted a system of flat-rate, prepaid postal charges.

Even more significant was the invention in 1832 of the electric telegraph by the American Samuel Morse. By 1850 most major cities in Europe and America had been linked by telegraph wires, and in 1851 the first successful submarine telegraph cable was laid under the English Channel. In 1866, after ten years of trying and several failures, the American Cyrus W. Field succeeded in laying a telegraph cable under the North Atlantic Ocean, providing nearly instantaneous communication between Europe and North America. Other submarine telegraph cables followed. The telephone, patented by Alexander Graham Bell in 1876, made distant communication even more personal, but its principal use in the beginning was in facilitating local communications.

The Italian inventor-entrepreneur Guglielmo Marconi, building on the scientific discoveries of the Englishman James Clerk Maxwell and the German Heinrich Hertz, invented wireless telegraphy (or radio) in 1895. As early as 1901 a wireless message was transmitted across the Atlantic, and by the time of the *Titanic* disaster in 1912 radio had come to play a significant role in ocean navigation. In the field of business communications the invention of the typewriter (Scholes's patent, 1868; "Model I Remington," 1874) and other rudimentary business machines helped busy executives keep up with and contribute to the increasing flow of information that their large-scale operations and worldwide activities made necessary. The typewriter also played a role in bringing women into the office work force.

The Application of Science

All of these developments relied much more than earlier technological innovations on the application of science to industrial processes. The electrical industry, in particular, required a high degree of scientific knowledge and training. In other industries scientific advance became more and more the prerequisite of technological advance. This did not mean, however, that scientists abandoned their laboratories for boardrooms, or conversely, that businessmen became scientists. It did mean, however, an increasing interaction between scientists, engineers, and entrepreneurs. Marconi, although he had a smattering of scientific knowledge, was primarily a businessman. Bessemer and Edison were prototypes of a new occupational category, the professional inventor. Edison, who invented the phonograph and the moving picture camera as well as the incandescent lightbulb and a host of lesser novelties, actually devoted a large part of his time to business matters setting up large-scale generating and transmitting equipment for electricity. Increasingly, technological development required the cooperation of numbers of scientific and engineering specialists whose work was coordinated by business executives who realized the potentialities of, though they did not possess special expertise in, the new technology.

The science of chemistry proved especially prolific in the birth of new products and processes. It had already created artificial soda, sulfuric acid, chlorine, and a number of heavy chemicals of particular importance in the textile industry. While seeking a synthetic substitute for quinine in 1856 William Perkin, an English chemist, accidentally synthesized mauve, a highly prized purple dye. This was the beginning of the synthetic dyestuffs industry, which within two decades virtually drove natural dyestuffs from the market. Synthetic dyestuffs proved to be the opening wedge of a much larger complex of organic chemical industries, whose output included such diverse products as drugs and pharmaceuticals, explosives, photographic reagents, and synthetic fibers. Coal tar, a by-product of the coking process previously regarded as a costly nuisance, served as the principal raw material for these industries, thus turning a bane into a blessing.

Chemistry also played a vital role in metallurgy. In the early nineteenth century the only economically important metals were those known from antiquity: iron, copper, lead, tin, mercury, gold, and silver. After the chemical revolution associated with Antoine Lavoisier, the great French chemist of the eighteenth century, many new metals were discovered, including zinc, aluminum, nickel, magnesium, and chromium. In addition to discovering these metals, scientists and industrialists found uses for them and devised economical production methods. One major use was in making alloys, a mixture of two or more metals with characteristics different from its components. Brass and bronze are examples of natural alloys (alloys that occur in nature). Steel is actually an alloy of iron with a small amount of carbon and sometimes other metals. In the second half of the nineteenth century metallurgists devised many special alloy steels by adding small amounts of chromium, manganese, tungsten, and other metals to impart specially desired qualities to ordinary steel. They also developed a number of nonferrous alloys.

Chemistry likewise came to the aid of such old, established industries as food production, processing, and preservation. The scientific study of soil that was initiated in Germany in the 1830s and 1840s, notably by the agricultural chemist Justus von Liebig, led to greatly improved agricultural practices and the introduction of artificial fertilizers. Scientific agriculture thus developed along with scientific industry. Canning and artificial refrigeration produced a revolution in dietary habits and, by permitting the importation of otherwise perishable foodstuffs from the New World and Australasia, allowed Europe's population to grow far beyond what its own agricultural resources would support.

The Institutional Framework

Economic development may take place in a variety of institutional contexts, as the previous chapters have shown. Clearly, however, some legal and social environments, just as some natural environments, are more conducive to material advance than others. The institutional setting for economic activity in nineteenth-century Europe, which produced the first industrial civilization, gave wide scope to individual initiative and enterprise, permitted freedom of occupational choice and geographical and social mobility, relied on private property and the rule of law, and emphasized the use of rationality and science in the pursuit of material ends. None of these elements was wholly new in the nineteenth century, but their juxtaposition and the explicit recognition granted to them made them powerful contributors to the process of economic development.

Legal Foundations

Great Britain, as we have seen, had already acquired a substantially modern framework for economic development, adapted to both social and material innovation and change. One of the key institutions of that framework was the legal system known as common law ("common" because, from at least the time of the Norman Conquest, it was common to the entire kingdom of England, superseding purely local laws and customs). The distinctive features of common law were its evolutionary character, its reliance on custom and precedent as set forth in written legal decisions, and its flexibility. It provided protection for private property and interests against the depredations of the state ("an Englishman's home is his castle") and at the same time protected the public interest from private exactions (e.g., by prohibiting combinations in restraint of trade). It also incorporated the customs of merchants (the "law merchant") as they had developed in specialized commercial courts. Transmitted to the English colonies in the process of settlement, common law became the basis for the legal systems of the United States and the British dominions when they achieved independence or autonomy.

On the Continent, meanwhile, the outmoded institutions of the past had ossified in the face of the eroding forces of change to the point that gradual, peaceful transition to the new order was no longer possible. The French Revolution, by shattering the Old Regime, opened new vistas and new opportunities for enterprise and ambition. It abolished outright the decaying remnants of the feudal order, and instituted a more rational legal system that was eventually enshrined in the Napoleonic Codes.

The charter of the new order is to be found in the Declaration of the Rights of Man and the Citizen (borrowing heavily from the American Declaration of Independence, which, in turn, had borrowed from the writings of the French *philosophes*).

The first article proclaimed that "men are born and remain free and equal in their rights," rights specified as liberty, property ("inviolable and sacred"), security, and resistance to oppression. The Declaration also spelled out the guarantees necessary to preserve those rights: uniformity of laws, freedom of speech and of the press, equitable taxation imposed by the citizens themselves or their representatives, and responsibility of public officials. All citizens were to be "equally admissible to all dignities, offices, and public employments, according to . . . their virtues and their talents."

The revolutionary assemblies went beyond mere declarations to the specifics of the legal foundations of the new order. In addition to abolishing the feudal regime and establishing private property in land, they did away with all internal customs duties and tariffs, abolished craft guilds and the whole apparatus of state regulation of industry, prohibited monopolies, chartered companies, and other privileged enterprises, and replaced the arbitrary and inequitable levies of the Old Regime with a rational and uniform system of taxation. In 1791 the Assembly went so far as to pass the drastic Le Chapelier Law prohibiting organizations or associations of both workers and employers.

Naturally, the French carried their revolutionary reforms to the lands they conquered in the course of the Revolutionary and Napoleonic wars. Belguim, the left bank of the Rhine in Germany, much of Italy, and for a short time Holland and parts of northern Germany were all incorporated in the French Empire. With few exceptions the whole body of reform was applied to these territories directly. The Confederation of the Rhine, the Swiss Confederation, the Grand Duchy of Warsaw, the kingdom of Naples, and Spain, all under French "protection," accepted the greater part of the revolutionary legislation. The influence of the reforms extended even to countries not directly dominated by the French. Prussia was the most profoundly affected. After the humiliation of Jena in 1806 a group of intelligent and patriotic officials came to the fore in the Prussian administration, determined to regenerate the country by administrative and social reforms so that it might withstand the conqueror and assume the leadership of a German nation.

The purgative work of the revolution should not be regarded as mere negative acts of demolition. On the contrary, these acts represented the essential first steps toward a positive, constructive, and fairly consistent policy. In the end, however, modern French institutions—and those of several other nations influenced by the French—received their definitive stamps not from the revolution itself but from Napoleon. The reaction in public opinion that made the dictatorship of Napoleon possible was a reaction to the excesses of the revolution and to the corruption and license that flourished under the Directory. As such, it favored a compromise with some, though by no means all, of the institutions and traditions of the Old Regime. Napoleon's genius and good fortune lay in his ability to synthesize the highly rational achievements of the revolution with the deeply ingrained habits and traditions built up over a thousand years of history. His policies were further influenced by his military cast of mind, which was impressed by hierarchical order and called for strict discipline, and by the exigencies of continual warfare.

The Napoleonic synthesis is perhaps best seen in the great work of legal codification begun during the revolution but completed under the empire. A classic compromise between the received Roman law, as adopted to local needs and customs, and the new revolutionary legislation, the Codes nevertheless preserved the fundamental principles of the revolution: equality before the law, a secular state, freedom of conscience, and economic freedom. The *Code civile*, promulgated in 1804, is the most fundamental and important. Written by middle-class lawyers and jurists, it clearly reflected the preoccupations and interests of the propertied classes. It treated property as an absolute, sacred, and inviolable right. It also specifically sanctioned freedom of contract and gave valid contracts the force of law. It recognized the bill of exchange and other forms of commercial paper, and expressly authorized loans at interest—a provision of signal importance for the development of industry in Roman Catholic countries.

As the French abolished the institutions of the Old Regime in the territories they conquered, they laid the foundations of the new. The *Code civile*, which accompanied the French armies of occupation, remained when the latter had departed. Throughout Europe and beyond, including Louisiana and Quebec as well as practically all of Latin America, the *Code civile* was either adopted outright or formed the basis of national codes.

Another of the Napoleonic Codes of particular importance for economic development was the *Code de Commerce*, promulgated in 1807. Before this, no single comprehensive rule had governed the forms of business enterprise. In Britain the Bubble Act of 1720 prohibited joint-stock companies unless they possessed charters from Parliament (see p. 172); similar prohibitions had long been the rule on the Continent. The enlarged scale of enterprise brought about by the new technology required new legal forms to facilitate the accumulation of capital and spread the risks of investment. Britain repealed its Bubble Act in 1825, but incorporation continued to require a special charter until 1844, when associations of twenty-five or more persons were allowed to form joint-stock companies by simple registration. Even then, limited liability for shareholders was not normally available until a series of acts in the 1850s granted limited liability on registration, under certain conditions. A new, comprehensive law of 1862 made limited liability generally available.

The *Code de Commerce* distinguished three main types of business organizations: (1) simple partnerships, in which the partners were individually and collectively liable for all debts of the business; (2) *sociétés en commandite*, limited partnerships in which the active partner or partners assumed unlimited liability for the affairs of the concern, whereas the silent or limited partners risked only the amounts they actually subscribed; and finally (3) *sociétés anonymes*, corporations in the American sense, with limited liability for all owners. These were ''anonymous'' companies in the sense that the names of individuals could not figure in the official designation of the company. Because of their privileges, each *anonyme* had to be explicitly chartered by the government, which, in the first half of the century, was extremely loath to grant such charters. A *commandite*, however, could be established simply by registering with a notary public, and it quickly became the favored form of enterprise. Eventually, a law of 1863 allowed free incorporation with limited liability to firms whose capital stock did not exceed 20 million francs, and in 1867 another law removed even that restriction.

The *commandite* form was adopted in most continental nations and performed a vital function in gathering capital for commerce and industry in the transitional period before free incorporation, at a time when most governments proved even more conservative than the French in granting charters for *anonymes*. After France adopted free incorporation in 1867, other countries soon followed. By 1900 only Russia and the Ottoman Empire, among the major nations, still required specific authorization for incorporation. In the United States, on the other hand, where egalitarian sentiments and hostile attitudes toward special privilege were stronger than in Europe, and where the individual states as well as the federal government could charter corporations, free incorporation became the rule as early as the 1840s.

Economic Thought and Policy

The period of the Napoleonic Wars witnessed what was in some respects the culmination of the economic nationalism and imperialism of earlier centuries, with the attempted British blockade of the Continent and Napoleon's Continental System as a response. Neither was entirely effective in its principal purpose, which was to limit or destroy the war-making potential of the adversary's economy, but they both represented the extremes of the policies of economic nationalism. Even earlier, however, intellectual currents that condemned those policies had begun to flow.

In the 1760s and 1770s the Physiocrats (called *les économistes* in France) had begun to advocate the merits of economic freedom and competition. In 1776, the year of the American Declaration of Independence, Adam Smith published in *The Wealth of Nations* what was to become a declaration of individual economic independence. Smith has sometimes been portrayed as an apologist for businessmen or "the bourgeoisie," but that results from a misreading (or a nonreading) of his text. His criticisms of merchants are no less scathing than his condemnation of foolish or misguided governments. Concerning their penchant for monopoly, for example, he wrote, "People of the same trade seldom meet together, even for merriment and diversion, but the conversation ends in a conspiracy against the publick, or in some contrivance to raise prices."[4] Smith's major concern throughout the book, however, was to show that the abolition of vexatious and "unreasonable" restrictions and restraints on individual enterprise would promote competition within the economy, and this, in turn, would maximize the "wealth of nations." Smith's book was quite popular for a philosophical treatise. It went through five editions before his death in 1790, and was subsequently translated into all major languages. Statesmen and politicans on both sides of the Atlantic quoted him in support of or opposition to particular pieces of legislation during his lifetime, and he acquired a number of disciples on the Continent. It was not until well after his death, however, and until a number of other writers such as the Reverend T. R. Malthus and David Ricardo had contributed to the body of literature known as "classical political economy," that Smith's ideas began to be implemented in legislation. This occurred first in the United Kingdom in the 1820s and 1830s. Some of the reforms, it is true, such as the revision of the criminal and penal laws in a humanitarian direction, the reduction in the number of capital offenses, and the creation of a metropolitan police force,

---

[4] Smith, *Wealth of Nations* (Glasgow ed.), I, p. 145.

owed more to Jeremy Bentham and the Utilitarians than to Smith and the classical economists (although there was some overlap in the two schools, particularly in the person of John Stuart Mill). The foremost achievement of the latter group was the repeal of the Corn Laws, which ushered in a long period of free trade in Great Britain (see later, p. 277).

In addition to free trade, the tenets of economic liberalism (as the new doctrine was known) called for a reduction of the role of government in the economy. In its name the system of taxation was overhauled and simplified, and the Combination Acts, the Navigation Acts, the Usury Laws, and other legislative symbols of the Old Regime in economic life were all repealed. According to Smith and his "system of natural liberty," the government had only three functions to perform: "first, the duty of protecting society from the violence and invasion of other independent societies; secondly, the duty of protecting, as far as possible, every member of the society from the injustice or oppression of every other member of it, or the duty of establishing an exact administration of justice; and thirdly, the duty of erecting and maintaining certain publick works and certain publick institutions, which it can never be for the interest of any individual, or small number of individuals, to erect and maintain. . . ."[5]

This idealized description of the role of government according to the classical economists gave rise to a myth, namely, the myth of laissez faire. The phrase first appeared in English usage in 1825, and is literally translated by the imperative "let do." The popular understanding of it was that individuals, especially businesspersons, should be free of all governmental restraint (except criminal laws) to pursue their own selfish interests. Thomas Carlyle satirized it as "anarchy plus a constable."

Laissez faire in practice, however, was by no means as heartless, as selfishly motivated, or as inexorable as extremist statements indicated. The main target of the classical economists was the old apparatus of economic regulation, which in the name of national interest frequently erected pockets of special privilege and monopoly, and in other ways interfered with individual liberty and the pursuit of wealth. At the same time that Parliament was dismantling the old system of regulation and special privilege, moreover, it was enacting a new series of regulations concerned with the general welfare, especially of those least able to protect themselves. The measures included the Factory Acts, new health and sanitary laws, and the reform of local government. These were not the work of any one class or segment of the population, although they drew on the intellectual capital of the Utilitarians. Humanitarian reformers of both aristocratic and middle-class backgrounds joined forces with leaders of the working classes to agitate for them, and they were voted for by Whigs and Tories as well as Radicals.

Economic liberalism had its advocates on the Continent as well, but they never achieved the same degree of success as their British counterparts. One reason for this was that the tradition of state paternalism was much more deeply ingrained on the Continent than in Britain. Another was that, since Britain was the acknowledged technological leader, many individuals looked to the government to help close the

---

[5] Smith, *Wealth of Nations* (Glasgow ed.), II, pp. 687–88.

gap. Free trade won some adherents, and there was some reduction in government interference in the economy, but on the whole government played a more active role than in Great Britain.

Overseas, the United States had a unique blend of government and private enterprise. The classical economists had few purist adherents in the United States. Varied as actual economic policies were in the numerous burgeoning states, they achieved a pragmatic and workable compromise between the demands of individual liberty and the requirements of society. Because of rival sectional interests and the triumph of the Jeffersonian and Jacksonian Democrats, the federal government played the minimal role assigned it by classical theory and, until the Civil War, generally followed a liberal or low-tariff commercial policy. State and local governments, on the other hand, took an active role in promoting economic development. The "American System," as Henry Clay called it, regarded government as an agency to assist individuals and private enterprise to hasten the development of the nation's material resources.

Class Structure and Class Struggles

Socially, Old Regime Europe was organized into three "orders," the nobility, the clergy, and all others—the common people or "commons" (see Chapter 3, p. 48). A modern functional analysis in terms of social class would revise the classification slightly. At the top of the social pyramid was a ruling class of landowners, which included some non-nobles as well as the higher clergy and the nobility per se. The economic basis of their political power and social status was ownership of the land, which enabled them to live "nobly" without working. Next on the social scale was an upper middle class, or *haute bourgeoisie,* of great merchants, high government officials, and professionals such as attorneys and notaries; although these frequently owned some real property as well, the principal bases of their positions were their special knowledge and skills, their stock in trade (for merchants), and their personal contacts with the aristocracy. Still lower on the social scale stood a lower middle class, or *petite bourgeoisie,* consisting of artisans and handicraftsmen, retail tradesmen and others engaged in service occupations, and independent small holders. At the bottom lay the peasants, domestic workers in cottage industries, and agricultural laborers, whose numbers included many indigents and paupers.

The shift from agriculture to the new forms of industry and the growth of cities brought about the rise of new social classes. It is readily apparent that an individual's place in the social hierarchy depends in part on how he or she makes a living, and individuals in the same occupation are likely to share common values and a common outlook, different from and perhaps conflicting with the values and outlooks of those engaged in other occupations. The nineteenth century sometimes saw bitter struggles between rival groups for social and political recognition and dominance.

At the beginning of the century the peasants were by far the most numerous group. At the end of the century they still constituted a majority in Europe as a whole, but in the more industrialized areas their relative numbers had drastically decreased. Isolated by poor communications and bound by a traditionalist mentality, their greatest desire was to obtain land. Their participation in broad social movements was generally sporadic and limited to their immediate economic interests.

In the years immediately after Waterloo the landed aristocracy continued to enjoy social prestige and political power, despite the effects of the French Revolution. Its position of leadership was sharply challenged, however, by the rapidly growing middle classes. By the middle of the century the latter had succeeded in establishing themselves in the seats of power in most of western Europe, and during the second half of the century they made deep inroads into the exclusive position of the aristocracy in central Europe.

At the beginning of the nineteenth century urban workers constituted a small minority of the population, but with the spread of the industrial system they began to gain numerical superiority. To speak of "the" working class is misleading, however, for there were many gradations and differences within the laboring population. Factory workers proper, although among the objects of greatest attention for historians of industrialization, were only one element of it, and not the largest at that. Moreover, within that one element were many different attitudes and circumstances among, for instance, textile workers, iron workers, pottery workers, and others. Miners, although they resembled factory workers in some respects, differed in others. Domestic servants, artisans, and handicraftsmen had all existed before the rise of modern industry. Many of the skilled workers sank to the status of the unskilled as machines replaced them in their work. Others, however, including carpenters, masons, machinists, and typesetters, found demand for their services increasing with the growth of industries and cities. Casual laborers, such as dockers and porters, constituted another important group, as did transport workers, clerical workers, and others. Their common characteristic that enables us to treat them as one for some purposes (although even this was not precise or universal) was their dependence for a living on the sale of their labor for a daily or weekly wage.

Karl Marx prophesied in the middle of the nineteenth century that the polarization he thought he observed in the then-advanced industrial societies would continue until, ultimately, only two classes would be left, the ruling class of capitalists (who in his opinion would absorb and replace the aristocracy) and the industrial proletariat. Gradually, all of the intervening classes would be forced down into the proletariat until the latter, with its overwhelming weight of numbers would rise up in revolution and overthrow the ruling class of capitalists. As a prediction, this prophesy has been falsified by the facts of history. Rather than polarize two mutually antagonistic classes, the spread of industrialization has greatly swollen the middle classes of white-collar workers, skilled artisans, and independent entrepreneurs. Successful revolutions, when they occurred, as in Russia in 1917, were the work of small bands of militant professional revolutionaries preying on the weaknesses of societies debilitated by war.

The more usual forms of working-class solidarity and self-help were trade unions and eventually, in some countries, working-class political parties. Although trade unions have a long history, running back to the journeyman's associations of the later Middle Ages, the modern movement dates from the rise of modern industry. In the first half of the nineteenth century unions were weak, localized, and usually short-lived in the face of opposition by antagonistic employers and un-favorable or repressive legislation. Most Western nations have passed through at least three phases in their official attitudes toward trade unions. The first phase, that of outright prohibition or suppression, was typified by the Le Chapelier Law of 1791 in France, the Combination Acts of 1799–1800 in Britain, and similar legisla-tion in other countries. In the second phase, marked in Britain by the repeal of the Combination Acts in 1824–25, governments granted limited toleration to trade unions, allowing their formation but frequently prosecuting them for engaging in overt action such as strikes. A third phase, not achieved until the twentieth century in some countries and not at all in others, accorded full legal rights to working men and women to organize and engage in collective activities.

In Britain in the 1830s the trade union movement became involved in a broader political movement known as Chartism, whose purpose was to achieve suffrage and other political rights for the disenfranchised. After the movement's failure in 1848, trade union organization fell off until 1851. Then the Amalgamated Society of Engineers (machinists and mechanics) was formed, the first of the so-called New Model unions. The distinctive feature of the New Model union was that it organized skilled workers only, and on a craft basis; it represented the "aristocracy" of labor. Unskilled workers and workers in the new factory industries remained unorganized until near the end of the century. New Model unions aimed modestly at improving the wages and working conditions of their own members, already the best paid in British industry, through peaceful negotiations with employers and mutual self-help. They eschewed political activities and rarely resorted to strikes except in desperation. As a result, they grew in strength, but membership remained low. Attempts to organize the large mass of semiskilled and unskilled workers resulted in successful strikes by the "match girls" (young female workers in the match indus-try) in 1888 and the London dockworkers in 1889. By 1900 trade union membership surpassed 2 million, and in 1913 it had reached 4 million, or more than one-fifth of the total work force.

On the Continent trade unions made slower progress. From the beginning French unions were closely associated with socialism and similar political ide-ologies. The varying and mutually antagonistic forms taken by French socialism badly splintered the movement, resulted in a fickle and fluctuating membership, and made it nearly impossible to agree on nationwide collective action. In 1895 French unions succeeded in forming a national nonpolitical General Confederation of Labor (CGT), but even it did not include all active unions, and it frequently had difficulty in commanding local obedience to its directives. The French labor movement re-mained decentralized, highly individualistic, and generally ineffective.

The German labor movement dated from the 1860s. Like that of the French, it was associated from the beginning with political parties and political action; unlike the French movement, it was more centralized and cohesive. The German labor movement had three main divisions: the Hirsch-Dunker or liberal trade unions, appealing mainly to skilled craftsmen; the socialist or "free" trade unions, with a far larger membership; and developing somewhat later, the Catholic or Christian trade unions, founded with the blessing of the Pope in opposition to the "godless" socialist unions. By 1914 the German trade union movement had 3 million members, five-sixths belonging to the socialist unions, making it the second largest in Europe.

In the economically backward countries of southern Europe, and to some extent in Latin America, French influence predominated in working-class organizations. Trade unions were fragmented and ideologically oriented. They were savagely repressed by employers and the state and were mostly without consequence. Trade unions in the Low Countries, Switzerland, and the Austro-Hungarian Empire followed the German model. They achieved moderate success at the local level, but religious and ethnic differences as well as opposition from government hindered their effectiveness as national movements. In the Scandinavian countries the labor movement developed its own distinctive traditions. It allied itself with the cooperative movement as well as with the Social Democratic political parties and by 1914 had done more than any other trade union movement to alleviate the living and working conditions of its membership. In Russia and elsewhere in eastern Europe trade unions remained illegal until after World War I.

Early attempts to form mass working-class organizations in the United States had limited effectiveness in the face of governmental and employer opposition, and the difficulty in securing cooperation among working men of different skills, occupations, religions, and ethnic backgrounds. In the 1880s Samuel Gompers took the lead in organizing closely knit local unions of skilled workers only, and in 1886 he united them into the American Federation of Labor (AFL). Like the New Model unions of Britain, the AFL followed bread-and-butter tactics, concentrating on the welfare of its own members, steering clear of ideological entanglements, and avoiding overt political action. Consequently, it succeeded in achieving many of its limited goals but left the majority of American industrial workers unorganized. In the British dominions trade unions developed in traditional British form, but with greater commitment to socialist programs. The first Trades Union Congress in Australia took place in 1879, only eleven years after the first of its kind in Britain.

Education and Literacy

Another feature of economic development in the nineteenth century, less remarked on but scarcely less remarkable than the growth of cities, of industrial workers, and of incomes, was the growth of literacy and education. Tables 8-3 and 8-4 present some rough data for selected countries at selected dates. In all cases it is virtually certain that the rates were lower in 1800 than in 1830 or 1850. The tables show a rough (not a precise) correlation between levels and rates of industrialization, on the one hand, and educational effort and attainment on the other. It is significant that Great Britain (or the United Kingdom), the first industrial nation, ranks high in both tables, but not at the top. In general, the countries of northwestern Europe (and the United States) make the best showing, in terms of both effort and attainment, whereas those of southern and eastern Europe (for which Spain, Italy, and Russia are proxies) are least impressive. This corresponds with both levels and rates of industrialization.

Perhaps the most surprising aspect of Table 8-3 is the high standing of Sweden in both 1850 and 1900; Sweden was a poor country in the midnineteenth century,

TABLE 8-3.   Adult Literacy of Selected Countries
(percentage)

Country	ca. 1850	ca. 1900
Sweden	90	(99)[a]
United States (white only)	85–90	94
Scotland	80	(97)
Prussia	80	88
England and Wales	67–70	(96)
France	55–60	83
Austria (excluding Hungary)	55–60	77
Belgium	55–60	81
Italy	20–25	52
Spain	25	44
Russia	5–10	28

[a]Figures in parentheses are almost certainly exaggerated.

Source: Calculated from Carlo M. Cipolla, Literacy and Development in the West (Harmondsworth, 1969), Tables 21, 24, and 31; figures in parentheses are from Michael G. Mulhall, Dictionary of Statistics (London, 1899; reprinted 1969), p. 693.

but in the second half of the century had one of the highest rates of growth of any country in Europe. Its high initial level of literacy is attributable to religious, cultural, and political factors antedating the onset of industrialization, but the large stock of human capital so acquired stood it in good stead once industrialization had begun. The same broad generalization applies, perhaps to a lesser degree, to the other Scandinavian countries, the United States, Germany (Prussia), and (within the United Kingdom) Scotland.

Beyond the quantitative data one must inquire into the nature and scope of education. Before the nineteenth century publicly supported educational institutions scarcely existed. The well-to-do hired private tutors for their children. Religious and charitable institutions, and in a few cases fee-charging proprietary schools, provided elementary education for a fraction of the population, mainly in the towns.

TABLE 8-4.   Primary School Enrollment Rates
from Selected Countries (per 10,000 population)

Country	1830	1850	1900
United States	1500	1800	1969
Germany	1700	1600	1576
United Kingdom	900	1045	1407
France	700	930	1412
Spain	400	663	1038
Italy	300	463 (1860)	881
Russia	—	98 (1870)	348

Source: Richard A. Easterlin, "Why Isn't the Whole World Developed?" Journal of Economic History, 41 (March 1981).

No one dreamed of universal literacy; indeed, much influential opinion opposed literacy for the "laboring poor" as being incompatible with their "stations" in life. Technical education was provided almost exclusively through the apprenticeship system. Secondary and higher education was reserved largely for the children (mainly sons) of the privileged classes, except for aspiring members of the clergy. With few exceptions (notably in Scotland and the Netherlands), the ancient universities had long since ceased to be centers for the advancement of knowledge; mired in a traditional curriculum that emphasized the classics, they trained bureaucrats for church and state, and gave the semblance of a liberal education to the sons of the ruling classes.

The French Revolution introduced the principle of free publicly supported education, but in France itself the principle was disregarded by the Restoration governments until after 1840. Meanwhile, several of the German, Scandinavian, and American states, which already benefited from a tradition of widespread primary education, established publicly supported systems, although they did not become compulsory or universal until later in the century. In England the Factory Act of 1802 required owners of textile mills to provide elementary instruction for their apprentices, but the law was poorly enforced; another law of 1833 required instruction for all child workers. In the first half of the century many artisans and skilled workers attended "mechanics' institutes," evening schools supported by either fees or charitable institutions, but Britain lagged notably in the provision of public education. Southern and eastern Europe lagged even further than Britain.

The French Revolution occasioned other educational innovations of particular significance for the industrial age. These were specialized schools for science and engineering, of which the École Polytechnique and the École Normale Supérieure are the most famous. Established at university level, but outside the university system (which Napoleon reorganized to train professionals and bureaucrats), these institutions not only provided advanced instruction but also engaged in research. They were widely imitated throughout Europe, except in Great Britain, and a graduate of the Polytechnique organized instruction at the U.S. Military Academy at West Point, the first engineering school in America.

The post-Napoleonic reform era in Germany resulted in the revitalization of its old universities and the creation of several new ones. Scientific training borrowed heavily from the curriculum and methods of the École Polytechnique, but was made available to a far larger number of students than in the French system. Thus, as science became more and more the foundation of industry, Germany stood ready to take advantage of the situation. When American educators in the 1870s began to be concerned with the need to remodel their system of higher education, they turned to Germany rather than France or England for a model. Subsequently, French and British universities, and those of other countries as well, fell in line.

International Relations

At the Congress of Vienna in 1814–15 Napoleon's victors tried to reestablish the Old Regime, politically, socially, and economically, but their efforts proved in vain. The ideological forces of democracy and nationalism unleashed by the French Revolution, together with the economic forces of incipient industrialization, undermined their efforts. Moreover, a divergence of interest among the victors, notably between Great Britain and the restored rulers of continental Europe, hastened the breakdown of the restored old order. The final decay of the Old Regime, except in Russia and the Ottoman Empire, became evident in the revolutions of 1830 and 1848 on the Continent.

The revolutions were not predominantly economic manifestations, but they did have significant economic consequences, mainly as a result of the realignment of political forces. In France, for example, the revolution of 1830 replaced a resolutely backward-looking government with one more amenable to commercial and industrial interests, whereas in that of 1848 the urban working classes made a determined bid for political power before they were crushed by the forces of repression. The revolution of 1830 in the southern Netherlands resulted in the creation of a new nation, Belgium, which soon showed itself to be one of the most economically progressive on the Continent. The revolutions of 1848 in central Europe finally brought about the extinction of the remnants of the feudal regime.

In all of these revolutions nationalism was a potent force. Nationalism as an ideology did not belong to any social class as such. It was principally espoused by members of the educated middle classes, but it also reflected the aspirations of the divided peoples of Italy and Germany for unified nations, and the aspirations of the subject nationalities in the Austrian, Russian, and Ottoman empires, the Belgian Netherlands, Norway, and Ireland for autonomy and freedom. In Germany the achievement of economic unification under the Prussian-dominated Zollverein in the 1830s preceded the achievement of political unification in 1871, and helped lay the foundations for German industrial power. The failure to achieve a similar economic unification before the creation of the kingdom of Italy in 1861 (in spite of an attempt in 1848) hindered that country's rise to great power status. The achievement of independence by Greece, Serbia, Rumania, and Bulgaria from the Ottoman Empire, unaccompanied by any notable economic progress, made those countries pawns in the chess games of power politics.

The nineteenth century witnessed no massive, devastating wars such as the Napoleonic Wars that began it or World War I which ended it. The relatively brief, limited wars that occurred in between sometimes had significant political outcomes, with implications for economic policy, but they did not seriously hinder the accumulation of capital or the process of technical change. Toward the end of the century, it is true, political tensions, sometimes exacerbated by economic rivalry, became more acute and overflowed into the revival of European imperialism. The economic aspects of that imperialism will be analyzed in a subsequent chapter. For the present it is sufficient to note that this revival of imperialism greatly enlarged the world market system, with Europe at its core.

9

Patterns of Development: The Early Industrializers

From one point of view the process of industrialization in the nineteenth century was a European-wide phenomenon. (The fact that by the century's end the United States had become the leading industrial nation does not alter the matter, because the United States was basically European in its culture.) One hardy scholar has even estimated Europe's gross "national" product for the nineteenth century (see Fig. 9-1). Although such estimates are easy to criticize in detail because of deficient sources, the two outstanding features of Figure 9-1 are undoubtedly correct in general: (1) the numerous short-term fluctuations and (2) the steady long-term growth.

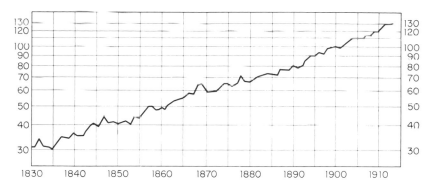

FIGURE 9–1.   Index of Europe's gross national product (1899–1901 = 100). [From Paul Bairoch, "Europe's Gross National Product: 1800–1975," in *Journal of European Economic History*, 5 (fall 1976), 288.]

From a different point of view, however, industrialization was basically a regional phenomenon, as discussed previously (p. 184). The regions in question might lie wholly within a single nation, as in the case of southern Lancashire and its immediately adjacent areas; or they might overlap national boundaries, as with the Austrasian coalfield, which runs from the English Channel in northern France through Belgium and into western Germany, terminating with the Ruhr area. For many scholars regional analysis offers the most satisfying means of understanding the process of industrialization.

Yet a third way to view the process of industrialization, however, is the more conventional method of looking at it in terms of national economies. Such a procedure has the disadvantages of possibly overlooking the international and supranational ramifications of the process and of ignoring or slighting its regional dynamics; but it has two powerful offsetting advantages. The first is the purely technical advantage that most quantitative descriptive measures of economic activity are collected and aggregated in terms of national economies. Second, and more fundamentally, the institutional framework of economic activity, and the policies intended to influence the direction and character of that activity, are most often set within national boundaries.

Fortunately, the three approaches are not mutually exclusive. In the preceding chapter the international and supranational aspects of the process of industrialization were stressed, particularly with regard to population and technology; the international dimensions of commerce and finance are again stressed in Chapter 11. In this and the following chapter, we look at the varied national patterns of growth, glancing at their regional manifestation as well, when important.

Great Britain

We begin with Great Britain, "the first industrial nation." At the end of the Napoleonic Wars Britain was clearly the world's leading manufacturing nation, producing about one-quarter of the total *world* industrial production according to some estimates.[1] Moreover, as a result of both its lead in manufacturing and its role as the world's overwhelmingly superior sea power, achieved during the late wars, it emerged as the world's leading commercial nation as well, accounting for between one-fourth and one-third of total international commerce—well over twice that of its leading rivals. Britain retained its dominance as both an industrial and trading nation for most of the nineteenth century. After some slippage in the middle decades of the century, it still accounted for about one-quarter of the total international commerce in 1870, and actually increased its share of total industrial production to more than 30 percent. After 1870, even as total output and trade continued to increase (e.g., industrial production increased by 250 percent between 1870 and 1913), it gradually lost its lead to other rapidly industrializing nations. The United States overtook it in total industrial production in the 1880s and Germany in the first decade of the twentieth century. On the eve of World War I it was still the leading commercial nation, but it then commanded only about one-sixth of total trade and was closely followed by Germany and the United States.

Textiles, coal, iron, and engineering, the bases of Britain's early prosperity, remained its standbys. As late as 1880 its production of cotton yarn and cloth surpassed that of the rest of Europe combined; by 1913, although its relative

---

[1] That is, the output of modern, market-oriented industries, there is no way of computing the value of traditional household industries in India and China (and elsewhere), much of which was destined for family consumption.

position had declined, it still accounted for one-third of total European production, more than twice as much as its closest competitors. Similarly, in the iron industry Britain reached its relative peak around 1870, producing more than half the world's pig iron. By 1890, however, the United States had snared the lead, and in the early years of the twentieth century Germany also forged ahead. In the coal industry, on the other hand, Britain maintained its lead in Europe (although the United States overtook it at the beginning of the twentieth century) and produced a surplus for export. On a per capita basis, Britain produced almost twice as much coal throughout the century as its leading European rivals, Belgium and Germany. (See Fig. 9-3, later.) The northeastern coalfield (Northumberland and Durham) and South Wales had exported to the Continent from early in the century, and even before; in 1870 that trade was valued at 3 percent of total British exports. The rapid industrialization of Britain's coal-poor European neighbors resulted in a remarkable rise; in 1913 exports of coal, a raw material, accounted for more than 10 percent of the value of all exports from the world's most highly industrialized nation.

The engineering industry, a creation of the late eighteenth century, can trace its roots in all three of the industries just mentioned. The textile industry needed machine builders and machine menders, the iron industry produced its own; and the coal industry's need for efficient pumps and cheap transport resulted in the development of both the steam engine and the railway. Railways, as suggested in the previous chapter, constituted the most important new industry of the nineteenth century. They were especially important for both their forward and backward linkages to other industries. Moreover, as a result of Britain's pioneering role in the development of railways, the foreign demand both in Europe and overseas for British expertise, materiel, and capital provided a strong stimulus for the entire economy.

Similarly, the evolution of the shipbuilding industry from sail to steam propulsion, and from wood to iron to steel construction, was another potent stimulus. Newly constructed steam tonnage did not surpass new sailing tonnage until 1870, but its predominance increased rapidly thereafter; by 1900 the output of new sailing ships had fallen to less than 5 percent of the total. Iron began to replace wood on a large scale in the construction of both steam and sailing ships in the 1850s, and steel to replace iron in the 1880s. In the early years of the twentieth century the British shipbuilding industry produced, on average, more than 1 million tons a year, virtually all steel steamships. That accounted for more than 60 percent of world construction. (For a few years in the 1880s and 1890s Britain produced more than 80 percent of the total.) A substantial fraction of this output, between one-sixth and one-third, was exported.

These impressive achievements notwithstanding, the pace and extent of British industrialization should not be exaggerated, as they often have been. Recent research has demonstrated that the rate of industrial growth in the century 1750–1850 was considerably slower than earlier, impressionistic estimates had implied, and that,

even as late as 1870 about half the total steam horsepower in manufacturing was in textiles, and that in many trades power-driven mechanization had as yet made comparatively little impact. The great majority of industrial workers in 1851 and perhaps in 1871 were not in large-scale factory industry but were still craftsmen in small workshops. The massive application of steam power did not occur until after 1870, rising from a total of around 2 million h.p. by that date to nearly 10 million h.p. in 1907.[2]

The census of 1851 further confirms these generalizations. For example, agriculture was still the largest single employer of labor—until as late as 1921—with domestic service second. The textile industries accounted for less than 8 percent of the total labor force (the cotton industry alone for about 4 percent). Blacksmiths outnumbered workers in the primary iron industry by 112,500 to 79,500; shoemakers (274,000) were more numerous than coalminers (219,000).

Britain reached its peak of industrial supremacy in relation to other nations in the two decades from 1850 to 1870. The growth rate of gross national product from 1856 to 1873—both peak years of the business cycle—averaged 2.5 percent, the highest of the entire century. Between 1873 and 1913 it fell to 1.9 percent, lower than its century-long average and substantially lower than those of the United States and Germany during the same period. On a per capita basis it was even lower than that of France, traditionally regarded as the laggard among the great powers. How should this lackluster performance be evaluated?

In the first place, growth rates are to some extent misleading, as units with a small statistical base can post high growth rates with very modest absolute increments of increase. More fundamentally, Britain could not retain its preeminence indefinitely, as other less-developed but well-endowed nations began to industrialize. In that sense, Britain's *relative* decline was inevitable. Moreover, in view of the vast resources and rapid population growth of the United States and Russia, it is not surprising that they would eventually overtake the small island nation in total output. More difficult to explain is the low rate of growth of output per capita; from 1873 to 1913 the rate of growth of total factor productivity (output per unit of all inputs) was zero.

Many explanations for this disappointing performance have been offered. Some are highly technical, involving the relative prices of primary products and manufactured goods, the terms of trade, investment ratios and patterns, and so on. At the risk of some oversimplification, for present purposes these can be ignored. Some have seen availability of natural resources and access to raw materials as a problem, but it was certainly a minor one. The cotton industry, of course, had always depended on imported raw cotton, but that did not prevent Britain from becoming the world's leading manufacturer of cotton goods; and in any case all other European cotton producers also obtained their raw material from abroad, frequently by way of Britain. Native ores of nonferrous metals—copper, lead, and tin—were gradually depleted or could not compete with cheaper supplies from overseas; but in most cases those cheaper supplies were mined and imported by British firms operating abroad. By the beginning of the twentieth century the iron industry imported about one-third of its ore, mainly from Spain; but that was largely because of the

[2] A. E. Musson, "Technological Change and Manpower," *History*, 67 (June 1982): 240–41.

failure of the industry to switch completely to the Thomas-Gilchrist basic process of steelmaking, which would have allowed it to use domestic phosphoric ore.

The latter instance points to another possible cause of Britain's relative decline: entrepreneurial failure. The question has been (and still is) hotly debated by scholars, with no definitive resolution in sight. Without question, Victorian Britain had some dynamic, aggressive individual entrepreneurs; William Lever (of Lever Brothers, later Unilever) and Thomas Lipton (tea), among others, became household names. On the other hand, there is abundant evidence that late Victorian entrepreneurs generally did not exhibit the dynamism of their forebears, as the sons and grandsons of the founders of family firms adopted the life-style of leisured gentlemen and left the day-to-day operation of their firms to hired managers. The tardy, almost half-hearted introduction of new, high-technology industries (in those days), such as organic chemicals, electricity, optics, and aluminum, even though many of the inventors were British, is one sign of entrepreneurial lethargy. Even more telling is the tardy and partial response of British entrepreneurs to new technology in those staple industries in which they were, or had been, leaders. The slow and incomplete adoption of the Thomas-Gilchrist process is a case in point and, in the same industry, the relatively slow adoption of the Siemens-Martin furnace. The textile industries long resisted the introduction of superior spinning and weaving machinery invented in the United States and on the Continent, and the Leblanc soda producers fought a thirty-year losing rearguard action against the Solvay ammonia-soda process, introduced from Belgium.

In part, the backwardness of the British educational system may be blamed for both the industrial slowdown and the entrepreneurial shortcomings. Britain was the last major Western nation to adopt universal public elementary schooling, important for training a skilled labor force. The few great English universities paid slight attention to scientific and engineering education (the Scottish universities did, however). Although they had revived somewhat from their eighteenth-century torpor, they were still primarily engaged in educating the sons of the leisured classes in the classics. This was part of the perpetuation of aristocratic values, with their disdain for commercial and industrial achievements. The contrast with the eighteenth century is striking, and ironic; at that time British society was widely regarded as more fluid and open than those of the *ancien régime* on the Continent. A century later the perceptions, if not the reality, were reversed.

This discussion of the triumphs and tribulations of British industry in the nineteenth century has proceeded with only incidental mention of the international context—a glaring omission that will be remedied to some extent in Chapter 11, but a few remarks need to be made here to put the discussion in proper perspective.

Of all large nations, Great Britain was most dependent on both imports and exports for its material well-being. Thus, the commercial policies, especially tariffs, of other nations had important repercussions. More than that, Britain depended to a greater extent than even smaller nations on the international economy for its sustenance. It had by far the largest merchant marine and the largest foreign investments of any nation—both important earners of foreign exchange. From the beginning of the nineteenth century, if not before, in spite of its important export industries, Britain had an "unfavorable," or negative, balance of commodity trade. The deficit was covered (more than covered) by the earnings of the merchant marine and foreign investments, which enabled the latter to increase almost continually throughout the century. Furthermore, in the latter part of the century the central role of London in the international insurance and banking industries added still more to these invisible earnings. The importance of these international sources of income can be judged by a brief comparison: earlier we compared the growth rate in gross national product from 1856 to 1873 with that for 1873 to 1913: 2.5 versus 1.9. The comparable figures for gross *domestic* product (i.e., GNP less foreign earnings) were 2.2 and 1.8.

To conclude this all-too-brief discussion of Great Britain's pattern of industrialization in the nineteenth century, it should be said that, for all its vicissitudes, the per capita real income of Britons increased by roughly 2.5 times between 1850 and 1914, income distribution became *slightly* more equal, the proportion of the population in dire poverty fell, and the average Briton in 1914 enjoyed the highest standard of living in Europe.

The United States

The most spectacular example of rapid national economic growth in the nineteenth century was the United States. The first federal census of 1790 recorded fewer than 4 million inhabitants. In 1870, after the limits of continental expansion had been reached, the population had risen to almost 40 million, larger than that of any European nation except Russia. In 1915 the population surpassed 100 million. Although the United States received the bulk of emigration from Europe, the largest element of population growth resulted from the extremely high rate of natural increase. At no time did the foreign-born population surpass one-sixth of the total. Nevertheless, the American policy of almost unrestricted immigration until after World War I placed a definite stamp on national life, and America became known as the melting pot of Europe.

The numbers of immigrants entering the country annally rose rapidly though unsteadily from fewer than 10,000 in 1820–25 to more than 1 million in the early years of the twentieth century. Until the 1890s the great majority came from northwestern Europe; immigrant stock from those countries continued to constitute the largest part of the foreign-born population. By 1900, however, new immigrants from Italy and eastern Europe dominated the listings. In 1910 the foreign-born population numbered 13,500,000 or about 15 percent of the total. Of these, about 17 percent came from Germany; 10 percent from Ireland; almost as many from Italy and the Austro-Hungarian Monarchy; about 9 percent each from Great Britain, Scandinavia, Canada (many of British origin), and Russia; almost 7 percent from Russian, Austrian, and German Poland; and a scattering from other countries.

Income and wealth grew even more rapidly than population. From colonial times the scarcity of labor in relation to land and other resources had meant higher wages and a higher standard of living than in Europe. It was this fact, together with the related opportunities for individual achievement and the religious and political liberties enjoyed by American citizens, that drew the immigrants from Europe.

Although the statistics are imperfect, it is probable that the average per capita income at least doubled between the adoption of the Constitution and the outbreak of the Civil War. Almost surely, it more than doubled between the end of that war and the outbreak of World War I. What were the sources of this enormous increase?

Abundant land and rich natural resources help explain why the United States had higher per capita incomes than Europe, but they do not in themselves explain the higher rate of growth. The reasons for that are to be found mainly in the same forces that were operating in western Europe, namely, the rapid progress of technology and the increasing regional specialization, although there were also special factors at work in the United States. For example, the continued scarcity and high cost of labor placed a premium on labor-saving machinery, in agriculture as well as industry. In agriculture the best European practices yielded consistently higher returns per acre than in the United States, but American farmers using relatively inexpensive machinery (even before the introduction of tractors) obtained far larger yields per worker (Fig. 9-2). A similar situation prevailed in manufacturing.

The huge physical dimensions of the United States, with varied climates and resources, permitted an even greater degree of regional specialization than was possible in individual countries in Europe. Although at the time it won its independence almost 90 percent of its labor force was engaged primarily in agriculture, and much of the remainder in commerce, the new nation soon began to diversify. In

FIGURE 9–2. Harvesting wheat in Nebraska. Labor scarcity and the introduction of labor-saving machinery characterized both industry and agriculture in the United States. Here one man with a reaper-binder and a team of four horses does work that would have occupied a dozen European workers. (From *The American Land*, by William R. Van Dersal. Copyright 1943 by Oxford University Press. Reprinted with permission.)

1789, the year the Constitution took effect, Samuel Slater arrived from England and the following year, in partnership with Rhode Island merchants, established America's first factory industry. Soon afterward, in 1793, Eli Whitney's invention of the cotton gin set the American South on its course as the major supplier of the raw material for the world's largest factory industry (see Chapter 7, p. 182).

This dichotomy led to one of the first major debates on economic policy in the new nation. Alexander Hamilton, the first secretary of the treasury, wished to sponsor manufactures by means of protective tariffs and other measures (see his *Report on Manufactures*, 1791). Thomas Jefferson, the first secretary of state and third president, on the other hand, preferred the "encouragement of agriculture, and of commerce as its handmaid" (from his first inaugural address, in 1801). The Jeffersonians won the political struggle, but the Hamiltonians (after Hamilton's tragic and untimely death) saw his ideas triumph. The New England cotton industry, after weathering some severe ups and downs before 1815, emerged in the 1820s, and remained until 1860, as America's leading factory industry and one of the world's most productive. In its shadow a number of other industries, notably the manufacture of guns by means of interchangeable parts (another Eli Whitney innovation) developed, and laid the basis for subsequent mass production industries.

Another advantage of the size of the United States was its potential for a large domestic market, virtually free of artificial trade barriers. But to realize that potential required a vast transportation network. At the beginning of the nineteenth century the sparse population was scattered along the Atlantic seacoast; communication was maintained by coastal shipping supplemented by a few post roads. Rivers provided the only practical access to the interior, and that was severely limited by falls and rapids. To remedy this deficiency the states and municipalities, in cooperation with private interests (the federal government was scarcely involved), engaged in an extensive program of "internal improvements," meaning primarily the construction of turnpikes and canals. By 1830 more than 11,000 miles of turnpikes had been built, mainly in southern New England and the mid-Atlantic states. Canal construction got seriously underway after 1815 and reached a peak in the 1820s and 1830s. By 1844 more than 3000 miles had been constructed and more than 4000 by 1860. Public funds accounted for almost three-quarters of the total of 188 million dollars invested. A few of the enterprises—notably New York State's Erie Canal—were spectacularly successful, but the majority were not; many did not even recoup the capital invested.

A major reason for the disappointing economic performance of the canals was the advent of a new competitor, the railway. The railway age began almost simultaneously in the United States and Great Britain, although for many years the United States depended heavily on British technology, equipment, and capital. Nevertheless, American promotors quickly seized the opportunity of this new means of transportation. By 1840 the length of completed railways exceeded not only that in Britain, but that in all of Europe, and continued to do so for most of the century (see Table 8-2).

As in Britain, railways in America were important not only as producers of transportation services but also for their backward links to other industries, especially iron and steel. Although this importance has sometimes been exaggerated, it should not be ignored. Before the Civil War, it is true, the iron industry was largely scattered, small-scale, and dependent on charcoal technology, and much railway material, especially rails, was imported from Britain. Even so, in 1860 iron ranked fourth in value added by manufacture, after cotton, lumber, and boots and shoes. After the war, with the widespread adoption of coke smelting, the introduction of the Bessemer and open hearth processes of steelmaking, and the enormous expansion of demand by the transcontinental railways, it quickly became the largest American industry in terms of value added by manufacture.

In spite of the rapid growth of manufactures, the United States remained a predominantly rural nation throughout the nineteenth century. The urban population did not pull abreast of the rural until after World War I. In part this was because much manufacturing took place in what were essentially rural areas. As noted earlier, the iron industry was mainly rural-based until after the Civil War. Other industries, employing cheap and efficient water power, remained so even longer. Although steam engines gradually encroached on water power, it was the advent of central electricity-generating stations that caused the decline of rural-based industries. The westward movement continued after the Civil War, encouraged by the Homestead Act and the opening of the trans-Mississippi West by the railways. Agricultural products continued to dominate American exports, although the non-agricultural labor force surpassed workers in agriculture in the 1880s, and the income from manufacturing began to exceed that from agriculture in the same decade. By 1890 the United States had become the world's foremost industrial nation.

Belgium

The first region of continental Europe to adopt fully the British model of industrialization was the area that became the kingdom of Belgium in 1830. In the eighteenth century it had been (with the exception of the Prince-Bishopric of Liege) a possession of the Austrian Habsburgs. From 1795 until 1814 it was incorporated in the French Republic-Empire, and from 1814 to 1830 it formed a part of the kingdom of the United Netherlands. In spite of these frequent and, in the short run, upsetting political changes, it exhibited a remarkable degree of continuity in its pattern of economic development.

Proximity to Britain was not a negligible factor in its early and successful imitation of British industrialization, but there were other, more fundamental reasons. In the first place, the region had a long industrial tradition. Flanders was an important center of cloth production in the Middle Ages, and in the east the Sambre-Meuse valley was famous for its metalwares (see Chapter 5, p. 118). Bruges and Antwerp were the first northern cities to assimilate Italian commercial and financial techniques in the late Middle Ages. Although the region's economy suffered from Spanish rule and other misfortunes after the revolt of the Dutch (see Chapter 5, p. 97), it recovered somewhat under the more benign rule of Austria in the eighteenth century. An important hand-powered rural linen industry grew up in Flanders, and coalmining developed in the Hainaut basin and the Sambre-Meuse valley.

Second, Belgium's natural resource endowment resembled that of Britain. It had easily accessible coal deposits and, despite its small size, produced the largest outputs of any continental country until after 1850. It also had iron ore in proximity to the coal deposits, and ores of lead and zinc as well. In fact, a Belgian entrepreneur, Dominique Mosselman, played a leading role in founding the modern zinc industry, and the firm he created, the Société de la Vieille Montagne, virtually monopolized the industry for many years.

Third, in part because of its location, traditions, and political connections, the region that became Belgium received important infusions of foreign technology, entrepreneurship, and capital, and enjoyed a favored position in certain foreign markets, especially those of France. The process began under the Old Regime, and accelerated during the period of French domination. The Biolley family, natives of Savoy, settled at Verviers early in the eighteenth century and entered the woolen industry. By the end of the century they had by far the most important establishments in that industry. The Biolleys attracted still other migrants who came to work for them and eventually set up on their own. Among these was William Cockerill, a skilled mechanic from the Leeds woolen industry, who came to Verviers by way of Sweden and in 1799 set up his shop for the construction of spinning machinery. Louis Ternaux, a native of Sedan who fled France in 1792 and traveled in Britain studying British industrial processes, returned to France under the Directory and set up several woolen factories in both France and the annexed Belgian provinces. In 1807 one of his factories near Verviers, equipped with water-powered spinning machinery built by Cockerill, employed 1400 workers.

In 1720 an Irishman, O'Kelly, erected the first Newcomen steam pump on the continent for a coal mine near Liege. Ten years later an Englishman, George Sanders, built another for a lead mine at Vedrin. Before the end of the Old Regime almost sixty Newcomen engines were at work in the area that became Belgium. In 1791 the Périer brothers of Chaillot, near Paris, installed the first Watt-type engine in the same area, and by 1814 had built eighteen or more of a total of twenty-four of that type on the future Belgian territory. They were employed in textile factories, ironworks, and the cannon foundry in Liege, which the Périers operated themselves, in addition to coal mines; but their small total number is indicative of the relatively poor performance of the engines. Mineowners, in particular, preferred the older Newcomen-type engines, which continued to be erected as late as the 1830s.

Coal mines were the largest users of steam engines, of both the Newcomen and Watt varieties, and also attracted the greatest amount of French entrepreneurship and capital. During the French domination a traffic of great importance to both the Belgian coal industry and French industry in general developed and survived the various political transformations after 1814. In 1788 the Austrian Netherlands exported 58,000 tons of coal to France, whereas Britain supplied 185,000 tons; in 1821 the southern Netherlands exported 252,000, Britain 27,000 tons; and in 1830 Belgium sent more than 500,000 tons, Britain about 50,000 tons. The network of canals and other waterways connecting northern France with the Belgian coalfields, begun during the Old Regime but continued by the succeeding regimes, greatly facilitated this traffic. French capitalists found Belgian coal an attractive investment. During the great industrial booms of the 1830s, 1840s, and as late as the

1870s, when coal production spurted, new mines were sunk in Belgium with French capital.

The cotton industry grew up in and around the city of Ghent, which in effect became the Belgian Manchester. Already the principal market for the rural linen industry of Flanders, the city saw the establishment, from the 1770s, of several calico printing works—which, however, did not use mechanical power. At the beginning of the nineteenth century a local entrepreneur, Lievin Bauwens, not previously associated with the textile industry, went to England at great personal risk, with France and Britain at war, as an industrial spy. He managed to smuggle out some Crompton mule-spinning machines, a steam engine, and even skilled English workers to tend the machines and build others on their model. He installed the machines in an abandoned convent in Ghent in 1801; thus began the modern Belgian cotton industry. Bauwens soon had local competition, but the industry grew rapidly, especially with the protection of Napoleon's Continental System. By 1810 it employed 10,000 workers, mainly women and children. The vagaries of war and, even more, the peace that followed subjected the industry to violent fluctuations that bankrupted many entrepreneurs, including Bauwens, but the industry itself survived and grew. Power looms for weaving appeared in the 1830s, and by the end of the decade the introduction of mechanical spinning of flax, also in Ghent, spelled the doom of the rural linen industry.

A traditional iron industry, fueled by charcoal, had long existed in the Sambre-Meuse valley and in the Ardennes Mountains to the east. It played a significant part in the military-industrial effort of the Revolutionary and Napoleonic wars, but remained wedded to traditional techniques. In 1821 Paul Huart-Chapel introduced puddling and rolling in his ironworks near Charleroi. In 1824 he began building a coke-fired blast furnace, which finally became operational in 1827—the first commercially successful one on the Continent. Others soon followed including, in 1829, that of John Cockerill, whose partner was none other than the Dutch government of King Willem I.

In 1807 William Cockerill moved his textile machine works from Verviers to Liège and took two of his sons, James and John, into his partnership. William retired in 1813, and John bought out his brother in 1822. Meanwhile, about 1815, the firm began to manufacture steam engines as well as textile machinery; for this purpose they employed a large number of skilled English workers, some of whom later went into business for themselves or to work for other Belgian firms. The Cockerills announced plans to build coke blast furnaces as early as 1820, and in 1823 John obtained a subsidized loan from the Dutch government for that purpose. He also hired as a consultant David Mushet, a well-known Scottish ironmaster. But difficulties, both financial and technical, continued to plague the enterprise. In 1825 the government purchased a half interest in it for 1 million florins; but even this further infusion of government funds was insufficient to enable it to achieve its goal, and before it did so in 1829 the government had invested an additional 1,325,000 florins.

On the eve of the Belgian Revolution of 1830 (which, ironically, dispossessed the Dutch government of its investment) the Cockerill firm was unquestionably the largest industrial enterprise in the Low Countries, and probably the largest on the

Continent (Fig. 9-3). It employed almost 2000 workers and represented a capital investment of more than 3 million florins (about 1.5 million dollars), a huge sum for the times. With its coal and iron mines, blast furnaces, refineries, rolling mills, and machine shops, it was also one of the first vertically integrated metallurgical firms. As such, it served as a model for other firms in the burgeoning industry.

The Belgian Revolution, quite mild in terms of loss of life and property, nevertheless produced an economic depression because of the uncertainty over the new state's character and future. The depression ran its course shortly, however, and the middle years of the decade witnessed a vigorous industrial boom. Apart from international economic conditions, which were also favorable, two special factors were primarily responsible for the character and extent of the boom in Belgium: (1) the government's decision to build a comprehensive railway network at state expense (see Chapter 8, p. 204), a boon in particular to the coal, iron, and engineering industries; and (2) a remarkable institutional innovation in the field of banking and finance.

In 1822 King Willem I authorized the creation of a joint-stock bank, the Société Générale pour favoriser l'Industrie Nationale des Pays-Bas (known after 1830 as the Société Générale de Belgique), with its headquarters in Brussels; he endowed it with state properties valued at 20 million florins, and invested a considerable portion of his private fortune in its shares. The bank had the broadest possible attributes for an enterprise of its nature, but its performance for the first decade was lackluster at best. After the revolution, however, with a new governor appointed by the new government, it stimulated an investment boom unprecedented on the Continent. Between 1835 and 1838 it created thirty-one new *sociétés anonymes* with a combined capital of more than 100 million francs, including blast furnaces and ironworks, coalmining companies, the Phénix machine works in Ghent, the Antwerp Steamship Company, a textile factory, sugar refineries, and glassworks. In all of these promotions it had the cooperation of James de Rothschild of Paris, the most influential investment banker of his day, who facilitated access to the French capital market.

In 1835 a rival group of financiers obtained a charter for another joint-stock bank, the Banque de Belgique. Modeled on the Société Générale in every important respect (although substantially smaller), the new bank lost no time in imitating its forerunner in the field of investment banking. In less than four years it established twenty-four industrial and financial enterprises with a combined capital of 54 million francs. These included coal mines, metallurgical establishments, the St. Léonard machine works in Liege, textile mills, and sugar refineries, and of special importance, the company that became the world's largest producer of nonferrous metals, the Société de le Vieille Montagne, which it purchased from the founder, Mosselman. Like the Société Générale, the Banque de Belgique had a French connection through the private Parisian bank of Hottinguer et Cie. It was said that more than nine-tenths of its capital was French.

By 1840, if not before, Belgium was clearly the most highly industrialized country on the Continent and, in per capita terms, a very close follower of Great Britain. Although, like other early industrializers, its rate of industrial growth eventually declined slightly and fell behind that of the newly industrializing nations,

FIGURE 9–3. The Cockerill works at Seraing. The Cockerill factories on the Meuse River at Seraing, near Liege, was the first large-scale integrated industrial establishment on the Continent. This photo dates from the early twentieth century, but the enterprise was already large in the 1830s. (From *Toute la Belgique*, Monmarche.)

by 1914 it remained the most highly industrialized nation on the Continent in per capita output, second only to Great Britain in Europe. Throughout the century the bases of its prosperity continued to be the industries that had initiated its growth: coal, iron (and steel), nonferrous metals, engineering, and, to a lesser degree than in Britain, textiles. In the chemical industry the introduction of the Solvay ammonia-soda process boosted an otherwise slow-growing industry, and Belgian engineering firms excelled in the installation abroad (as well as at home) of narrow-gauge railways and, after 1880, electric street railways and interurbans. Throughout the century as well, Belgian industry depended heavily on the international economy; ultimately, 50 percent or more of its gross national product derived from exports. France was especially important. Indeed, under plausible counterfactual assumptions, if Belgium had been incorporated with France throughout the century, we would have "lost" important statistics on a regional economy, but those of France would have shown a much more impressive total growth. As it was, France in 1844 imported 30 percent of all Belgian pig iron production; nor was that an exceptional year. For the century as a whole, France imported more than 30 percent of its coal supplies, of which more than half came from Belgium, mainly from French-owned mines.

France

Of all the early industrializers, France had the most aberrant pattern of growth. That fact gave rise to a large literature, both in the nineteenth century and more recently, devoted to explaining the supposed "backwardness" or "retardation" of the French economy. Still more recently, however, new empirical research and theoretical insights have shown that the earlier debates were based on a false premise. In fact, although the *pattern* of industrialization in France differed from that of Great Britain and the other early industrializers, the *outcome* was no less efficient and, in terms of human welfare, may have been more so. Moreover, looking at the patterns of growth of successful late industrializers, it appears the French pattern may have been more "typical" than the British.

In seeking a solution to this paradox it is worth looking at the basic determinants of economic growth. The most striking feature of the nineteenth century, in the case of France, was its low rate of demographic growth (see Chapter 8, p. 194). When all relevant measures of growth (GNP, industrial production, etc.) are reduced to a per capita basis, it appears that France did very well indeed. Second is the matter of resources. British, Belgian, and eventually American and German industrialization were based to a large extent on abundant coal resources. France, although not totally deprived of coal, was much less well endowed and, moreover, the character of its deposits rendered their exploitation more expensive. These facts had important implications for other, coal-connected industries, such as iron and steel, that we will explore later. In technology, France was no laggard—far from it. French scientists, inventors, and innovators took the lead in several industries, including hydropower (turbines and electricity), steel (the open hearth process), aluminum, automobiles, and, in the twentieth century, aviation. The institutional factor is much more complex and difficult to evaluate; as noted in Chapter 8, the Revolutionary and Napoleonic regimes provided the basic institutional context for most of continental Europe, but many important changes took place in the course of the nineteenth century, the analysis of which is most conveniently postponed until another chapter.

It is now well established that modern economic growth in France began in the eighteenth century. For the century as a whole the *rates* of growth of both total output and output per capita were roughly the same in France and Britain, perhaps even slightly higher in France, although France began (and ended) with a lower per capita output. But the century ended with Britain undergoing an "industrial revolution" (in cotton) while France was caught up in the throes of a great political upheaval, the French Revolution. Therein lies an important difference that affected the relative performances of the two economies for much of the nineteenth century. For a quarter of a century, from 1790 until 1815, except for the brief truce of Amiens (1802–3) France was almost continually involved in what has been called the first "modern" war, involving mass conscription of manpower. Under wartime demand the output of the economy expanded, but mainly along established lines, with little technological progress. Some spinning machinery was established in the cotton industry, and a few steam engines were erected, but the important iron and chemical industries experienced technological stasis. Britain also went to war in 1793, but it experienced much less drain on its manpower, leaving most of the land warfare, except in the Iberian peninsula, to its continental allies. With its control of the seas (and, by the same token, with France cut off from overseas markets), its exports expanded dramatically, hastening the technological modernization of its principal industries.

After a rather severe postwar depression, which affected all of continental western Europe and even touched Great Britain, the French economy resumed its growth at even higher rates than during the eighteenth century. For the century as a whole gross national product probably grew at an average rate of between 1.5 and 2.0 percent per year, although these figures are subject to some uncertainty, especially for the first half of the century. For the period 1871–1914, for which statistics are both more numerous and more reliable, the French gross national product grew at an average annual rate of approximately 1.6 percent, whereas that of the United Kingdom grew at approximately 2.1 percent and that of Germany at 2.8 percent. These figures appear to indicate that the German economy grew almost twice as fast as the French, and the British a third again as fast. The figures can be misleading as a guide to the overall performance of the economy, however, because when all growth rates are reduced to a per capita basis the comparable rates are 1.4 percent for France against 1.7 percent for Germany and only 1.2 percent for the United Kingdom. In other words, the slow demographic growth of France accounts in large measure for the apparently slow growth of the economy as a whole. Moreover, even the per capita growth rates can be misleading because Germany, a relatively backward economy in the midnineteenth century, began with much lower per capita incomes and thus a smaller statistical base. Furthermore, as a result of the outcome of the Franco-Prussian War, two of France's most economically dynamic provinces, Alsace and Lorraine, became part of the new German Empire in 1871.

Industrial production, the leading edge of modern economic growth in France as in most other industrializing nations, grew even more rapidly than total product. It has been variously estimated at between 2.0 and 2.8 percent. The variations arise not only with different methods of estimation (and different estimators), but also according to the number of industries included in the estimates. Throughout the first half of the century—even as late as the Second Empire—handicrafts, artisan, and domestic industry accounted for three-quarters or more of total "industrial" production. The output of these activities grew more slowly than that of modern factories and other new industries, and in some cases declined absolutely; thus their exclusion from the indexes shows apparently higher growth rates. Nevertheless, their importance should not be underestimated, for in large measure they gave French industry its distinctive characteristics.

Although the overall performance of the economy was quite respectable, it experienced variations in the rate of growth (quite apart from short-term fluctuations, to which all industrializing economies were subject). Between 1820 and 1848 the economy grew at a moderate or even rapid rate, punctuated by occasional minor fluctuations. Coal production, which averaged less than 1 million tons from 1816 to 1820, exceeded 5 million tons in 1847, and coal consumption rose even more rapidly. The iron industry adopted the puddling process and began the transition to coke smelting. By midcentury more than one hundred coke furnaces produced more pig iron than 350 charcoal furnaces. The foundations of an important machinery and engineering industry were laid; by midcentury the value of machinery exports exceeded that of imports by more than 3 to 1. Many of the new machines went to the domestic textile industry, woolens and cottons in particular, which were the largest users of steam engines and other mechanical equipment, as well as the most important industries in terms of employment and value added. Consumption of raw cotton rose fivefold from 1815 to 1845, and imports of raw wool (in addition to domestic production) increased sixfold from 1830. Beet sugar refineries grew from one in 1812 to more than one hundred in 1827. The chemical, glass, porcelain, and paper industries, which also grew rapidly, were unexcelled for the variety and quality of their products. A number of new industries originated or were quickly domesticated in France in this period, including gas lighting, matches, photography, electroplating and galvanization, and the manufacture of vulcanized rubber. Improvements in transportation and communication, including extensive canal building, the introduction of steam navigation, the first railways, and the electric telegraph, facilitated the growth of both domestic and foreign commerce. The latter, valued in current prices, increased by 4.5 percent per year from 1815 to 1847, and, because prices were falling for most of that period, the real value was even greater. Moreover, France had a sizable export surplus in commodity trade throughout the period, by means of which it obtained the resources for substantial foreign investments.

The political and economic crises of 1848–51 inserted a hiatus in the rhythm of economic development. The crisis in both public and private finance paralyzed railway construction and other public works. Coal production dropped abruptly by 20 percent; iron output declined more slowly but in 1850 amounted to less than 70 percent of the 1847 figure. Commodity imports fell by half in 1848 and did not fully recover until 1851; exports dipped slightly in 1848, but recovered the following year.

With the coup d'état of 1851 and the proclamation of the Second Empire the following year, French economic growth resumed its former course at an accelerated rate. The rate of growth slackened somewhat after the mild recession of 1857, but the economic reforms of the 1860s, notably the free trade treaties (see Chapter 11) and the liberalized incorporation laws of 1863 and 1867, provided fresh stimuli. The war of 1870–71 brought economic as well as military disaster, but France recovered economically in a manner that astounded the world. It suffered less from the depression of 1873 than other industrializing nations, and recovered more quickly. A new boom developed that carried through to the end of 1881. During this period the railway network grew from about 3000 kilometers to more than 27,000, the telegraph network from 2000 to 88,000. Railway construction provided a powerful stimulus to the remainder of the economy, both directly and indirectly. The iron industry completed the transition to coke smelting in the 1850s, and in the 1860s and 1870s adopted the Bessemer and Martin processes for cheap steel. Both coal and iron production registered a fourfold increase during the period, coal production reaching 20 million tons and iron 2 million. Foreign commerce, profiting from the continued improvements in transportation and communication, increased by more than 5 percent annually, and France, still the world's second trading nation, increased its share of total world trade slightly from 10 to 11 percent. Over the period 1851 to 1881 as a whole, French wealth and income grew at their most rapid rates for the entire century, averaging 2 to 4 percent annually.

The depression that began in 1882 lasted longer and probably cost France more than any other in the nineteenth century. At its inception it resembled many other minor recessions, beginning with a financial panic, but a number of factors arose to complicate and prolong it: disastrous diseases, which seriously affected the wine and silk industries for almost two decades; large losses on foreign investments from defaulting governments and bankrupt railways; the worldwide return to protectionism in general and the new French tariffs in particular; and a bitter commercial war with Italy from 1887 to 1898. Foreign trade as a whole fell off and remained virtually stationary for more than fifteen years, and with the loss of foreign markets, domestic industry also stagnated. Capital accumulation fell to its lowest point in the second half of the century.

Prosperity returned at last, just before the end of the century, with the extension of the Lorraine ore fields and the advent of such new industries as electricity, aluminum, nickel, and automobiles. France once more enjoyed a rate of growth comparable with that of 1815–48, if not with 1851–81. *La belle époque*, as the French call the years immediately preceding World War I, was thus a period of material prosperity as well as cultural efflorescence. Although precise comparisons are not possible, it is likely that the average French person in 1913 enjoyed a material standard of living as high as or higher than that of the citizens of any other continental nation.

Certain key features of the French pattern of growth remain to be analyzed: the low rate of urbanization, the scale and structure of enterprise, and the sources of industrial power. All are interrelated, and closely related to two other features that have already been emphasized, the low rate of demographic growth and the relative scarcity of coal.

Of all major industrial nations, France had the lowest rate of urbanization. The slow growth of its total population was mainly responsible, but the proportion of the labor force in agriculture and the structure and location of industrial enterprise are also implicated. France also had the largest proportion of its labor force in agriculture among major industrial nations—about 40 percent in 1913. This fact has frequently been pointed to as primary evidence of the "retardation" of the French economy, but the correct interpretation is not so simple. A number of factors have been invoked to account for the relatively high proportion of population in agriculture—including the low rates of population growth and urbanization!—but it is less often observed that at the beginning of this century France was the only industrial nation in Europe that was self-sufficient in foodstuffs, and indeed had a surplus for export.

With respect to the scale and structure of enterprise, France was famous (or notorious) for the small scale of its firms. According to the census of 1906, fully 71 percent of all industrial firms had no wage earners; their workers—owners and family members—constituted 27 percent of the industrial labor force. At the other extreme, 574 large firms employed more than 500 workers each; their workers accounted for about 10 percent of the industrial labor force, or 18.5 percent of industrial wage workers. Significantly, these firms were concentrated in mining, metallurgy, and textiles, the same industries in which large-scale, capital-intensive enterprises prevailed in other major industrial countries, except that there were more of them. Between these two extremes lay large numbers of small and medium-sized firms employing the vast majority of wage earners. At the lower end of the scales, those employing fewer than ten workers each were in the traditional artisanal industries, such as food processing, clothing, and woodworking, whereas those with more than 100 workers were mainly in modern industries—chemicals, glass, paper, and rubber, as well as textiles, mining, and metallurgy. Two other characteristics of the relatively small scale of French enterprises should not pass notice: high value added (luxury articles) and geographical dispersion. Rather than having just a few huge conurbations of heavy industry, as Britain and Germany had, France had widely dispersed, highly diverse industries in small towns, villages, and even the countryside. In part, the dispersion was determined by the nature of the power sources available.

As previously pointed out, and as graphically depicted in Figure 9-4, France was the least well endowed with coal of all early industrializers. At the beginning of the twentieth century coal production per capita in France was about one-third that of Belgium and Germany, and about one-seventh that of Great Britain, even though France was exploiting its known reserves at a higher rate than the other countries. In the early part of the nineteenth century the most important mines, with one exception, were located in the hilly central and southern portions of the country, distant from markets and difficult of access, especially before the coming of the railway. Nevertheless, it was on the basis of these resources that France established its early coke-smelted iron industry. From the 1840s on the great northern coalfield, an

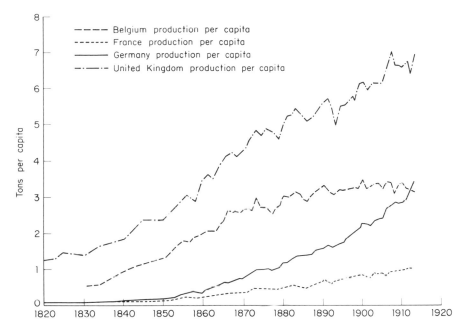

FIGURE 9–4.   Per capita production of coal, 1820–1913. (Belgium—*L'Annuaire Statistique de la Belgique,* 1871 and 1914 editions; France—*Annuaire Statistique de la France,* 1965 edition; Germany—Walther G. Hoffmann, *Das Wachstum der Deutschen Wirtschaft seit der mitte des 19. Jahrhunderts,* New York, 1965; United Kingdom— B. R. Mitchell and Phyllis Deane, *Abstract of British Historical Statistics,* Cambridge, 1962.)

extension of those in Belgium and Germany, came into use and served to fuel the growth of the modern steel industry. Still, for the century as a whole France depended on imports for about one-third of its coal consumption; and even with that French consumption per capita was only a fraction of that of its neighbors (Fig. 9-5).

To offset the scarcity and high cost of coal France relied to a much greater extent than its coal-rich neighbors on water power. It has already been pointed out that, thanks in part to improved technology, including the introduction of the hydraulic turbine, water power remained competitive with steam until near the middle of the century, even in Great Britain. On the Continent, especially in France and other coal-poor countries, it retained its importance for much longer. In France in the early 1860s falling water supplied almost twice the horsepower of steam engines, and in terms of total horsepower it continued to increase until the 1930s (quite apart from its use in generating electricity, which became increasingly important from the 1890s). But the characteristics of water as a source of power imposed restraints on its use. The best locations were generally remote from the centers of population; the number of users in any given location was limited to one or a very few, and the size of installations was similarly limited. Thus, water power, important though it was for French industrialization, helped impose a pattern: small firm size, geographical

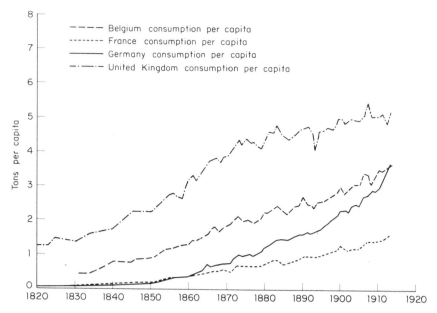

FIGURE 9–5.   Per capita consumption of coal, 1820–1913. (Belgium—*L'Annuaire Statistique de la Belgique*, 1871 and 1914 editions; France—*Annuaire Statistique de la France*, 1965 edition; Germany—Walther G. Hoffmann, *Das Wachstum der Deutschen Wirtschaft seit der mitte des 19. Jahrhunderts*, New York, 1961; United Kingdom—B. R. Mitchell and Phyllis Deane, *Abstract of British Historical Statistics*, Cambridge, 1962.)

dispersion, and low urbanization. As we will see, these characteristics came to be shared by other coal-poor countries.

Germany

Germany was the last of the early industrializers. Indeed, a case can be made that it was something of a laggard. Poor and backward in the first half of the nineteenth century, the politically divided nation was also predominantly rural and agrarian. Small concentrations of industry existed in the Rhineland, Saxony, Silesia, and the city of Berlin, but they were mostly of the handicraft or proto-industrial variety. Poor transportation and communications facilities held back economic development, and the numerous political divisions with their separate monetary systems, commercial policies, and other obstacles to commercial exchange further retarded progress.

On the eve of the First World War, in contrast, the unified German Empire was the most powerful industrial nation in Europe. It had the largest and most modern industries for the production of iron and steel and their products, electrical power and machinery, and chemicals. Its coal output was second only to that of Great Britain, and it was a leading producer of glass, optical instruments, nonferrous metals, textiles, and several other manufactured goods. It had one of the densest railway networks, and a high degree of urbanization. How did this remarkable transformation come about?

With only slight oversimplification, German economic history in the nineteenth century can be divided into three fairly distinct, almost symmetrical periods. The first, extending from the beginning of the century to the formation of the Zollverein in 1833, witnessed a gradual awakening to the economic changes taking place in Britain, France, and Belgium, and the creation of the legal and intellectual conditions essential for transition to the modern industrial order. In the second, a period of conscious imitation and borrowing that lasted until about 1870, the actual material foundations of modern industry, transportation, and finance took shape. Finally, Germany rose rapidly to the position of industrial supremacy in continental western Europe that it continues to occupy. In each of these periods foreign influences played an important role. In the beginning the influences, like the changes themselves, were primarily legal and intellectual, emanating from the French Revolution and the Napoleonic reorganization of Europe. A brisk inflow of foreign capital, technology, and enterprise, reaching a crescendo in the 1850s, marked the second period. In the final period the expansion of German industry into foreign markets dominated the picture.

The left bank of the Rhine, united politically and economically with France under Napoleon, adopted the French legal system and economic institutions, most of which were retained after 1815. Under Napoleon French influence was quite strong in the Confederation of the Rhine (most of central Germany). Even Prussia adopted in modified form many French legal and economic institutions. An edict of 1807 abolished serfdom, permitted the nobility to engage in "bourgeois occupations [commerce and industry] without derogation to their status," and abolished the distinction between noble and non noble property, thus effectively creating "free trade" in land. Subsequent edicts abolished the guilds and removed other restrictions on commercial and industrial activity, ameliorated the legal status of the Jews, reformed the fiscal system, and streamlined the central administration. Still other reforms gave Germany the first modern educational system (see Chapter 8, p. 221).

One of the most important economic reforms instigated by Prussian officials led to the formation of the Zollverein (literally, toll or tariff union). They laid the foundations in 1818 by enacting a common tariff for all of Prussia, primarily in the interests of administrative efficiency and a higher fiscal yield. Several small states, some of them completely surrounded by Prussian territory, joined the Prussian tariff system, and in 1833 a treaty with the larger states of South Germany, except for Austria, resulted in the creation of the Zollverein itself. The Zollverein did two things: in the first place, it abolished all internal tolls and customs barriers, creating a German "common market"; second, it created a common external tariff determined by Prussia. In general, the Zollverein followed a "liberal" (i.e., low-tariff) commercial policy, not on economic principle but because Prussian officials wanted to exclude protectionist Austria from participation.

If the Zollverein made a unified German economy possible, the railway made it a reality. The rivalry among the various German states, which contributed to the number and quality of German universities, also hastened railway construction. As a result the German rail network expanded more rapidly than that of France, for example, which had a unified government but was divided over the question of state versus private enterprise. Railway construction also required the states to get together to agree on routes, rates, and other technical matters, resulting in greater interstate cooperation.

Important as they were for unifying the country and stimulating the growth of both domestic and international commerce, the railways' role in the growth of industry, by means of both forward and backward linkages, was no less important. Until the 1840s Germany produced less coal than France or even tiny Belgium. It also produced less iron than France until the 1860s. Thereafter progress in both industries was extremely rapid; this progress owed much (though not everything) to the extension of the railway network, because of both the direct demand of railways for their output and the lower cost of transportation that they provided to other users.

The key to the rapid industrialization of Germany was the rapid growth of the coal industry, and the key to the rapid growth of the coal industry was the Ruhr coalfield. (The Ruhr River and valley, from which the coalfield and the industrial region, the largest in the world, get their name, actually form the southern boundary of the region. The larger part of the area lies to the north.) Just before World War I the Ruhr produced about two-thirds of Germany's coal. Prior to 1850, however, the region was much less important than Silesia, the Saar, Saxony, and even the Aachen region. Commercial production began in the Ruhr valley proper in the 1780s, under the direction of the Prussian state mining administration (Fig. 9-6). The mines were shallow, the techniques simple, and the output insignificant. In the late 1830s the "hidden" (deep) seams north of the Ruhr valley were discovered. Their exploitation, although extremely profitable, required greater capital, more sophisticated techniques (use of steam pumps, etc.), and greater freedom of enterprise. All of these were eventually supplied, although not without bureaucratic delays, largely by foreign firms (French, Belgian, British). From about 1850 coal production in the Ruhr rose very rapidly, and with it the production of iron and steel, chemicals, and other coal-based industries (Fig. 9-7).

The German iron industry as late as 1840 had a primitive aspect. The first puddling furnace began production in 1824, but it was financed by foreign capital. Medieval bloomeries were still in use in the 1840s. Coke smelting began in Silesia, but development of the West German industry is almost synonymous with development of the Ruhr, and that came largely after 1850. By 1855 there were about twenty-five coke furnaces in the Ruhr and a similar number in Silesia; these and a scattering of other coke furnaces produced almost 50 percent of the total German output of pig iron, although charcoal furnaces still outnumbered them 5 to 1.

Production of Bessemer steel began in 1863, and the Siemens-Martin process was adopted soon afterward. But not until after the Gilchrist-Thomas process was introduced, in 1881, permitting the use of the phosphoric iron ore of Lorraine, did German steel production accelerate dramatically. For the period 1870–1913 as a whole steel production increased at an average annual rate of more than 6 percent, but the most rapid growth came after 1880. German steel production surpassed that

FIGURE 9–6. The Ruhr River and valley. This picture shows the Ruhr at the beginning of the nineteenth century when it was still predominantly rural. (Note, however, the mine entrance on the left, the horses carrying bags of coal, and the ship awaiting coal at the pier.) A century later the same area was the site of the greatest concentration of heavy industry in the world. (From *Sozialgeschichte der Bergarbeiterschaft an der Ruhr im 19. Jahrhundert,* by Klaus Tenfelde. Copyright 1981 by Verlag Neue Gesellschaft GmbH, Bonn, Germany.)

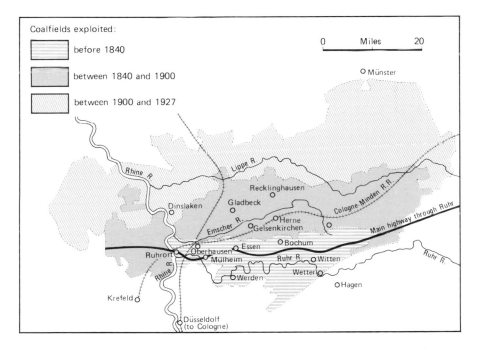

FIGURE 9–7. The Ruhr industrial area. [Used with permission based on material published in *The Times Atlas of World History* (1978, 1984).]

of Great Britain in 1895, and by 1914 it amounted to more than twice the British output. The German industry was large not only in its total output, but also in its individual units of production. In the early years of the twentieth century average output per firm was almost twice as great as that in Britain. German firms quickly adopted the strategy of vertical integration, acquiring their own coal and ore mines, coking plants, blast furnaces, foundries and rolling mills, machine shops, and so on.

The year 1870–71, so dramatic in political history with the Franco-Prussian War, the overthrow of the Second Empire in France, and the creation of a new Second Empire in Germany, was less dramatic in economic history. Economic unification had already been achieved, and a new cyclical upswing in investment, trade, and industrial production had begun in 1869. But the successful outcome of the war, including an unprecedented 5 billion franc indemnity, and the proclamation of the empire added euphoria to the boom. In 1871 alone 207 new joint-stock companies came into existence (aided, to be sure, by the new free incorporation law of the North German Confederation of 1869), and an additional 479 in 1872. In the process German investors, aided and encouraged by the banks (see p. 313), began buying back foreign holdings of German firms, and even began investing abroad. This hyperactivity came to a sudden halt with the financial crisis of June 1873 that heralded a severe depression. Nevertheless, after the depression had run its course growth resumed more strongly than before. From 1883 to 1913 net domestic product increased at an annual rate in excess of 3 percent; in per capita terms, the increase was almost 2 percent annually.

The most dynamic sectors of German industry were those producing capital goods or intermediate products for industrial consumption. Coal, iron, and steel production were notable, as we have seen. Even more notable were two relatively new industries, chemicals and electricity, as shown in Table 9-1. The table also shows that consumer goods industries, such as textiles, clothing and leather, and food processing had growth rates substantially below average. The emphasis given to capital and intermediate goods in Germany, and the relative neglect of consumer goods, contrasts markedly with the situation in France, and helps to explain their differing patterns of growth.

Prior to 1860 the chemical industry scarcely existed in Germany, but the rapid growth of other industries created a demand for industrial chemicals, especially alkalis and sulfuric acid. Stimulated by the new literature on agricultural chemistry, a German invention, farmers also demanded artificial fertilizers. Unburdened by obsolete plants and equipment, chemical entrepreneurs could use the latest technology in a rapidly changing industry. This was most strikingly exemplified by the advent of organic chemicals. As previously noted (in Chapter 8, p. 210), the first synthetic dye was discovered accidentally by an English chemist, Perkin; but Perkin had studied under A. W. Hofmann, a German chemist brought to the new Royal College of Chemistry in 1845 at the suggestion of Prince Albert. In 1864 Hofmann returned to Germany as a distinguished professor and consultant to the fledgling dyestuffs industry. Within a few years the industry, drawing on the personnel and resources of the universities, established its dominance in Europe and the world. The organic chemical industry was also the first in the world to establish its own

TABLE 9-1.   Rates of Growth of Output and Labor Productivity
in Germany, 1870–1913

Industrial Sector	% Growth Rate of Output, 1870–1913	% Growth Rate of Labor Productivity 1875–1913
Stones and earth	3.7	1.2
Metal producing	5.7	2.4
Iron manufacture	5.9	na
Steel manufacture	6.3	na
Metalworking	5.3	2.2
Chemicals	6.2	2.3
Textiles	2.7	2.1
Clothing and leather working	2.5	1.6
Foodstuffs and drinks	2.7	0.9
Gas, water, and electricity	9.7	3.6
Average for all industry and handwork	3.7	1.6

Source: Alan S. Milward and S. B. Saul, *The Development of the Economies of Continental Europe, 1850–1914* (Cambridge, MA, 1977), p. 26; derived from W. G. Hoffmann, *Das Wachstum der deutschen Wirtschaft seit der Mitte des 19. Jahrhunderts* (Berlin, 1965).

research facilities and personnel. As a result it brought about the introduction of many new products, and also dominated the production of pharmaceuticals.

The electrical industry grew even more rapidly than the chemical. Science-based, it drew on the university system for personnel and ideas, as did the chemical industry. On the side of demand, the extremely rapid urbanization of Germany that occurred just as the industry was growing up gave it an extra fillip; the German industry did not have to struggle against a well-entrenched gas light industry, as did the British industry. Illumination and urban transport were the two most important early uses for electricity, but engineers and entrepreneurs soon developed other uses. By the beginning of the twentieth century electric motors were competing with and displacing steam engines as prime movers.

A notable characteristic of the chemical and electrical industries, as indeed of coal, iron, and steel, was the large size of firms. Employees for most firms in these industries numbered in the thousands; at the extreme, the electrical firm of Siemens and Schuckert on the eve of World War I had more than 80,000 employees. To some degree the large size of firms was dictated by technical economies of scale. Deep mining, for example, required expensive pumps, hoists, and other equipment; it was more economical, therefore, to employ the machinery with a large volume of output to spread the cost. Not all instances of large firms can be explained by such logic, however. In some cases pecuniary economies of scale—arrangements that provided extra profits or rents to promoters or entrepreneurs without reducing the real cost to society—furnish a better explanation of large-scale enterprise. The close connections between the banking system and manufacturing industries in Germany are frequently held responsible; this possibility is discussed in greater detail in Chapter 12.

Yet another notable characteristic of the German industrial structure was the prevalence of cartels. A cartel is an agreement or contract among nominally independent firms to fix prices, limit output, divide markets, or otherwise engage in monopolistic, anticompetitive practices. Such contracts or agreements were contrary to the common law prohibition of combinations in restraint of trade in Britain and the United States, and to the Sherman Anti-Trust Act in the United States, but they were perfectly legal and indeed enforceable by law in Germany. Their number grew rapidly from 4 in 1875 to more than 100 in 1890 and almost 1000 by 1914. Elementary economic theory teaches that cartel behavior restricts output in order to increase profits, but such a prediction is scarcely compatible with Germany's record of rapid growth of output even—or especially—in cartelized industries. The resolution of this paradox is to be found in the combination of cartels with protective tariffs after Bismarck's conversion to protection in 1879. By means of protective tariffs, cartels could maintain artificially high prices in the domestic market (which also implied restrictions on domestic sales or other market-sharing devices), while engaging in virtually unlimited exports to foreign markets, even at prices below the average cost of production if the markup on domestic sales could offset the nominal losses on exports. The profitability of this type of activity was enhanced by the practice of the state-owned or regulated railways of charging a lower rate for shipments to the country's borders than for intracountry shipment.

As a result of these various devices, German exports increased rapidly in the world market—so much so that even free trade Britain adopted retaliatory measures, as is recounted in Chapter 11.

10

Patterns of Development: Latecomers and No-Shows

Elsewhere in Europe, before 1850, scatterings of modern industry existed—and in Bohemia more than that—but it could scarcely be said that a process of industrialization was underway. Such a process did begin in the second half of the century, markedly in Switzerland, the Netherlands, Scandinavia, and the Austro-Hungarian Empire; much more weakly in Italy, the Iberian countries, and the Russian Empire; and hardly at all in the new nations of the Balkans and the declining Ottoman Empire. When and where it did take place, it did so under vastly different circumstances than in the early industrializers, and consequently with different patterns.

To the extent that early industrialization was associated with coal—as it clearly was in Britain, Belgium, and Germany—the association shows up in the figures on per capita consumption (see Fig. 9-5). The late industrializers, on the other hand, had little or no coal within their own borders. Production in Spain, Austria, and Hungary barely sufficed to satisfy the meager domestic demand, if that. Russia had huge deposits (in the midtwentieth century the Soviet Union was the world's most important producer), but these had scarcely begun to be developed before 1914. The other countries in question had negligible coal resources, depending almost entirely on imports for their consumption.

Figure 10-1 shows the per capita coal consumption of some of the late industrializers. Two of its features should be stressed. First, as late as the beginning of the twentieth century per capita consumption in even the most successful of the late industrializers amounted to less than one-fifth that of Great Britain and less than one-third that of Belgium and Germany. Second, given the limited consumption in all of the late industrializers, the consumption of the most successful rose much more rapidly than that of the others. Since the principal use of coal in the coal-scarce countries was to fuel locomotives, steamships, and stationary steam engines, and since virtually all of the coal of the most successful late industrializers had to be imported, it appears that demand was the dominant force in promoting their greater relative consumption. In other words, their greater consumption was a result, not a cause, of successful industrialization.

To appreciate the significance of this proposition it is necessary to consider individual cases.

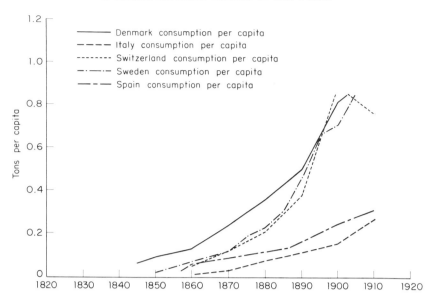

FIGURE 10–1.   Coal consumption per capita, 1820–1913. (From *European Historical Statistics, 1750–1970*, by B. R. Mitchell, New York, 1975.)

Switzerland

As Germany was the last of the early industrializers, Switzerland was the earliest of the latecomers. Some scholars dispute this statement, claiming that Switzerland was more highly industrialized than Germany, and at an earlier date—indeed, that Switzerland had undergone an "industrial revolution" or "take-off" in the first half of the nineteenth century. Such controversies are largely semantic and of little consequence; when the facts are clearly stated and the patterns exposed, the question of priority becomes one of definition. Although Switzerland had already acquired, in the first half of the century or earlier, some important assets that played an important role in its rapid industrialization after 1850—notably a high level of adult literacy—its economic structure was still largely preindustrial. In 1850 more than 57 percent of the labor force was engaged primarily in agricultural pursuits; less than 4 percent worked in factories. The great majority of industrial workers labored at home or in small workshops without machinery. Switzerland had barely entered the railway age, having fewer than thirty kilometers of recently opened track. Most important, the country lacked an appropriate institutional structure for economic development. Not until 1850 did it obtain a customs union (unlike Germany, which had a Zollverein but no central government), an effective monetary union, a centralized postal system, or a uniform standard of weights and measures.

A small country in both territory and population, Switzerland is also poor in conventional natural resources other than water power and timber; it has virtually no coal. Because of its mountains fully 25 percent of its land area is uncultivatable and, in fact, practically uninhabitable. In spite of these handicaps, the Swiss achieved one of the highest living standards in Europe by the beginning of the twentieth century, and, by the last quarter of that century, the highest in the world. How did they do it?

The population grew from just under 2 million in the early years of the nineteenth century to just under 4 million in 1914. The average growth rate was thus slightly less than those of Britain, Belgium, and Germany, but substantially higher than that of France. The population density was below that of all four countries, but this is largely explained by the nature of the terrain. Because of the scarcity of arable land, the Swiss had long practiced the combination of domestic industry with farming and dairying. They did so largely with imported raw materials and, by the latter part of the nineteenth century, with imported foodstuffs as well. Thus Switzerland, like Belgium and to an even greater extent than Britain, depended on international markets.

Swiss success in international markets resulted from an unusual, if not unique, combination of advanced technology with labor-intensive industries. This combination produced high-quality, high-priced, and high value-added products, such as traditional Swiss clocks and watches, fancy textiles, intricate specialized machinery, and exquisite cheese and chocolate candy. It is worth emphasizing that the labor-intensive industries were primarily *skilled* labor intensive. If this appears paradoxical, the resolution lies in the high rate of literacy in most of the Swiss cantons (for noneconomic reasons) and in the elaborate systems of apprenticeship that prevailed. This provided a skilled, adaptable labor force willing to work for relatively low wages. To this was added the justly famous Swiss Institute of Technology, founded in 1851, which provided trained brainpower and ingenious solutions for difficult technical problems that arose in the late nineteenth century.

Switzerland had an important cotton textile industry in the eighteenth century—the largest after England— but it was based on handicraft processes and part-time labor. In the last decade of the eighteenth century the cotton spinning industry, in particular, was virtually wiped out by competition from the more advanced British industry. After ups and downs during and immediately after the Napoleonic period, the industry revived and even prospered. It had an unusual combination of technologies: mechanized spinning (using water power for the most part, not steam), employing the cheap labor of women and children, but handloom weavers, who persisted long after their counterparts in Britain had disappeared from the scene (Fig. 10-2). This was possible because they concentrated on high-quality fabrics, including embroidery, and improved the handloom itself, incorporating elements of the Jacquard loom, invented earlier in the century for the silk industry. Eventually the improvements included mechanization, but still with special designs for high quality. By 1900 handlooms were rarities.

Although more traditional than the cotton industry, the silk industry actually contributed more to Swiss economic growth in the nineteenth century, both in terms of employment and exports, than the former. It also underwent technological modernization. Switzerland had rather small woolen and linen industries, likewise concentrating on quality products, and manufactured some clothing, shoes, and other leather goods. Overall, textiles and related products dominated Swiss exports throughout the century. In current values they rose from about 150 million francs in

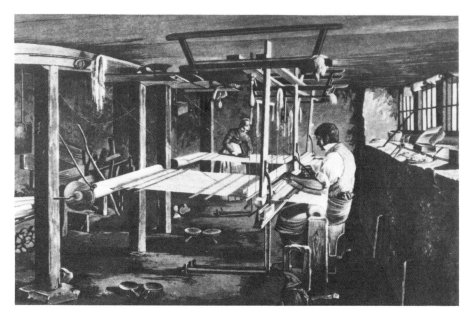

FIGURE 10–2.   Swiss handloom weaver. The Swiss specialized in high-quality, hand-produced fabrics. Here a weaver is at work in his basement workshop, assisted by his wife, around 1850. (Swiss National Museum, Zurich.)

the 1830s to more than 600 million in 1912–13. As a proportion of total exports, however, they fell from about three-quarters to slightly less than half during the same period.

The industries that gained on textiles in supplying export markets included both traditional industries and some that were the creations of the process of industrialization itself. These were, in order of importance on the eve of World War I, machinery and specialized metal products, foods and beverages, clocks and watches, and chemicals and pharmaceuticals. In view of Switzerland's lack of coal and small iron ore deposits, it wisely did not attempt to nurture a primary iron industry (the small charcoal iron industry of the Jura Mountains disappeared in the first half of the century); but relying on imported raw materials, it did develop an important industry for the transformation of metals. It began in the 1820s with the manufacture of machinery for the cotton spinning industry and, given the importance of water power to the economy, it is not surprising that it expanded to include the production of waterwheels, turbines, gears, pumps, valves, and a host of other highly specialized, high-value products. When the age of electricity dawned the industry quickly turned to the manufacture of electrical machinery; in fact, Swiss engineers contributed many important innovations to the new industry, particularly in the area of hydroelectricity. The decline in coal consumption per capita after 1900, primarily as a result of the electrification of the railways (see Fig. 10-1), offers vivid testimony.

The dairy industry, renowned for its cheese, converted its production from a handicraft to a factory process, thereby greatly expanding output and exports. The industry also developed the production of condensed milk (based on an American patent) and engendered two sibling industries, the production of chocolate and of prepared baby foods. The other traditional industry, the manufacture of clocks and watches, continued to be characterized by handwork by highly skilled (though frequently part-time) artisans and by a minute division of labor. Some specialized machinery was developed for the industry, particularly for the production of standardized, interchangeable parts; but the final assembly remained a hand process.

Finally, the chemical industry developed in response to the process of industrialization itself. Lacking natural resources, Switzerland had no heavy, or inorganic, chemical industry worth mentioning. In 1859 and 1860, after the discovery of artificial dyestuffs, two small firms began manufacturing them in Basle to supply the local ribbon industry. Subsequently two other firms joined them. Significantly, although all four began as suppliers to local industry, they soon discovered they could not compete with German firms in supplying standard bulk dyes; as a result they began to specialize in exotic, high-priced items in which they soon had a virtual world monopoly. Before the end of the century they sold more than 90 percent of their production outside Switzerland. They also developed through their own research various pharmaceutical specialties. In the early twentieth century the industry, employing fewer than 10,000 workers, accounted for 5 percent of total Swiss exports. Its exports amounted to more than 7500 francs per employee, more than twice as much as the watch industry and four times as much as the textile industry. In global terms, it was the world's second largest; although it produced only a fifth as much as Germany, that was as much as the rest of the world combined.

Probably no other country in Europe was more radically transformed by the advent of railways than Switzerland, but, paradoxically, no other railways were, as a whole, less profitable. Swiss investors apparently foresaw at least the latter possibility, for they proved extremely reluctant to invest in them, preferring those of the United States, and left their own largely to foreign (mainly French) capitalists. Construction began in earnest in the 1850s; 1882 saw completion of the first of the Alpine tunnels, that of Gotthard. By the 1890s, as a result of high construction and operating costs and low traffic density, most of the railways were in or on the verge of bankruptcy. In 1898 the Swiss government purchased the railways from their (largely foreign) owners at a fraction of their actual cost. Shortly afterward it undertook their electrification.

The trends established in the second half of the nineteenth century continued in the twentieth: the decline in the relative importance of agriculture, the growth in that of industry and (even more so) services, and the continued dependence on international demand, especially tourism (from the 1870s) and financial services (from World War I). In the 1960s the machinery and metallurgical industries accounted for about 40 percent of export earnings, chemicals and pharmaceuticals for 20 percent, clocks and watches for 15 percent, textiles for 12 percent, and food and beverages for 5 percent.

The Netherlands and Scandinavia

It may appear incongruous to include the Netherlands with the Scandinavian countries in a discussion of the pattern of industrialization; actually, it is quite logical. The common features of the Scandinavian countries that frequently cause them to be lumped together in discussions are cultural, not economic. In terms of economic structure, the Netherlands have more in common with Denmark than either do with Norway and Sweden. The usual pairing of the Netherlands with Belgium reveals that the latter was an early industrializer and the former was not; that Belgium had coal and developed heavy industry, the Netherlands did not; beyond that, the comparison is not very helpful. The comparison with other successful latecomers, on the other hand, despite differences in resources, reveals more about the process of industrialization, especially late industrialization.

All four of these countries, after lagging considerably behind the leaders in the first half of the century, spurted rapidly in the second half, and especially in the last two or three decades. For the period 1870 to 1913 Sweden had the highest rate of growth of output per capita of any country in Europe, at 2.3 percent. Denmark was second, with 2.1 percent. Norway, at 1.4, had about the same rate of per capita growth as France. Comparable figures are not available for the Netherlands, but other data indicate that they, too, experienced a high rate of growth. By 1914 these four countries, along with Switzerland, had achieved standards of living comparable with those of the early continental industrializers. In view of their later start, and their lack of coal, it is important to understand the sources of their success.

All of them, like Belgium and Switzerland, had small populations. At the beginning of the nineteenth century Denmark and Norway had fewer than 1 million, while Sweden and the Netherlands had fewer than 2.5 million inhabitants. All exhibited moderate growth rates in the course of the century, Denmark the highest and Sweden the lowest; but all more than doubled in population by 1900. Density varied greatly. The Netherlands had one of the highest densities in Europe, whereas Norway and Sweden had the lowest, even lower than Russia. Denmark was in between, but closer to the Netherlands.

Considering human capital as a characteristic of the population, we can say that all four countries were extremely well endowed. In both 1850 and 1914 the Scandinavian countries had the highest literacy rates in Europe, or in the world, and the Netherlands was well above the European average. This fact was of inestimable value in helping the national economies to find their niches in the evolving, ever-changing currents of the international economy.

With respect to resources, the most significant fact is that all four, like Switzerland but unlike Belgium, lacked coal. This is undoubtedly the main reason they were not among the early industrializers, and why they did not develop an appreciable heavy industrial sector. As for other natural resources, Sweden had the best endowment with abundant iron ore deposits, both phosphoric and non-phosphoric (also nonferrous metal ores, but these were of lesser significance), vast tracts of virgin timber, and water power. Norway also had timber, some metallic ores, and a huge water-power potential. Water power in Sweden and Norway was a significant factor in their development early in the nineteenth century—in 1820

Norway had between 20,000 and 30,000 watermills—but it became extremely important with the harnessing of hydroelectric power after 1890. Denmark and the Netherlands were almost as devoid of water power as of coal. They did have some wind power, which was not negligible, but it could scarcely serve as the basis of a major industrial development.

Location was an important factor for all four countries. Unlike landlocked Switzerland, all had immediate access to the sea. This had important implications for a significant international natural resource, fish, as well as for cheap transport, merchant marines, and the shipbuilding industry. Each took advantage of these opportunities in its own way. The Dutch, with a long tradition of fisheries and mercantile shipping, but lately somewhat moribund, had difficulty in developing good harbors suitable for steamships; eventually they did so at Rotterdam and Amsterdam, with spectacular results for transit trade with Germany and central Europe and for the processing of overseas foodstuffs and raw materials (sugar, tobacco, chocolate, grain, and eventually oil). Denmark also had a venerable commercial history, particularly with respect to traffic through the Öresund. In 1857, in return for a payment of 63 million kronor from other commercial nations, Denmark abolished the Sound Toll dues, which it had collected since 1497, along with other policy shifts toward free trade. This resulted in a significant increase in traffic through the Sound, and in the port of Copenhagen. Norway became a major supplier of fish and timber in the European market in the first half of the century, and boasted the second largest merchant marine (after Britain) in the latter half. Sweden, although slower to develop its own merchant marine, benefited from the removal of restrictions on international trade in general and from the reduced transport charges on its bulk exports of timber, iron, and oats.

The political institutions of the four countries posed no significant barriers to industrialization or economic growth. The post-Napoleonic settlement detached Norway from the crown of Denmark and attached it to that of Sweden, from which it seceded peacefully in 1905, but Sweden lost Finland to Russia in 1809. The Congress of Vienna created the kingdom of the United Netherlands, grouping the provinces of the old Dutch Republic with those to the south, which seceded, not very peacefully, to form modern Belgium in 1830. Prussia and Austria seized the duchies of Schleswig and Holstein from Denmark in 1864. Otherwise, the century passed relatively peaceably, with a progressive democratization taking place in all the countries. They were reasonably well governed, without notable corruption or grandiose state projects, although in all the government gave some aid to the railways, and in Sweden, like Belgium, the state built the main lines. As small countries dependent on foreign markets, they followed a liberal trade policy in the main, although a protectionist movement developed in Sweden. In Denmark and Sweden, the two countries whose agrarian structure most resembled those of the Old Regime, agrarian reforms took place gradually from the late eighteenth century through the first half of the nineteenth. The reforms resulted in the complete abolition of the last vestiges of serfdom and the creation of a new class of independent peasant proprietors with a pronounced market orientation.

The key factor in the success of these countries (along with high literacy, which contributed to it), as in Switzerland and in contrast to other late industrializers, was their ability to adapt to the international division of labor determined by the early industrializers, and to stake out areas of specialization in international markets for which they were especially well suited. This meant, of course, a great dependence on international commerce, which had notorious fluctuations; but it also meant high returns to those factors of production that were fortunate enough to be well placed in times of prosperity. In Sweden exports accounted for 18 percent of national income in 1870, and in 1913, 22 percent of a much larger national income. In the early twentieth century Denmark exported 63 percent of its agricultural production: butter, pork products, and eggs. It exported 80 percent of its butter, almost all to Great Britain, where it accounted for 40 percent of British butter imports. Norway's exports of timber, fish, and shipping services accounted for 90 percent of total exports—about 25 percent of national income—as early as the 1870s; by the early twentieth century those exports accounted for more than 30 percent of national income, with shipping services alone being responsible for 40 percent of foreign earnings. The Netherlands also depended heavily on the service occupations for foreign earnings. In 1909 11 percent of the labor force was employed in commerce and 7 percent in transport. The service sector as a whole employed 38 percent of the labor force and produced 57 percent of the national income.

Although these countries entered the world market in a big way in the middle of the nineteenth century, with exports of raw materials and lightly refined consumer goods, they had all developed highly sophisticated industries by the beginning of the twentieth century. This has been called "upstream industrialization;" that is, a country that once exported raw materials begins to process them and exports them in the form of semimanufactures and finished goods. Sweden's and Norway's timber trade is a prime example. In the beginning the timber was exported as logs, to be sawed into boards in the importing country (Britain); in the 1840s Swedish entrepreneurs built water-powered (later steam-powered) sawmills to convert the timber into lumber in Sweden (Fig. 10-3). In the 1860s and 1870s processes for making paper from wood pulp, at first by mechanical and subsequently by chemical means (the latter a Swedish invention), were introduced; output of wood pulp grew rapidly for the remainder of the century. Well over half of total output was exported, mainly to Britain and Germany, but the Swedes consumed an increasing amount themselves, and exported the higher value-added paper. The iron industry followed a similar pattern. Although Sweden's charcoal-smelted iron could not compete in price with coke-smelted iron or Bessemer steel, its higher quality made it especially valuable for such products as ball bearings, in the production of which Sweden specialized (and still does).

Scholars in each of the four countries have debated the timing of their respective industrial revolutions or take-offs. The 1850s, the 1860s, the 1870s—and even earlier and later decades—have their partisans, but what the debates show mainly is the artificiality and irrelevance of those two concepts. In fact, all four countries experienced very satisfactory rates of growth, cyclical fluctuations notwithstanding, from at least the middle of the century until the 1890s. Then, in the two decades immediately preceding World War I, even those satisfactory growth rates accelerated, especially in the Scandinavian countries, quickly bringing their levels of per capita income to the top rank on the Continent. No doubt the reasons for this

FIGURE 10–3.   Swedish sawmill. Timber was the leading Swedish export in the midnineteenth century. Swedish entrepreneurs built sawmills at river mouths, such as this one at Skutskär in the 1860s, to gain the higher-valued sawn timber. (From *An Economic History of Sweden*, by E. F. Heckscher, Cambridge, MA, 1954. Drawing by Robert Haglund.)

acceleration are numerous and complex, but three of them stand out immediately. First, the period was one of general prosperity, with rising prices and buoyant demand. Second, it was marked in Scandinavia by large-scale imports of capital (the Netherlands, on the other hand, was an exporter of capital in this period); more will be said about this in Chapter 11. Finally, the period coincided with the rapid spread of the electrical industry.

Electricity was a great boon to the economies of all four countries. Norway and Sweden, with their vast hydroelectric potential, were especially favored; but even Denmark and the Netherlands, who could import coal relatively cheaply from Britain's northeastern coalfield (and the Netherlands from the Ruhr, by way of the Rhine), also benefited greatly from steam-generated electricity. The Dutch had the highest per capita consumption of the coal poor countries throughout the century, whereas Denmark, with the second highest per capita consumption, spurted markedly after 1890. All four countries quickly developed important industries for the manufacture of electrical machinery and products (e.g., light bulbs in the Netherlands). Swedish and to a lesser extent Norwegian and Danish engineers became pioneers of the electrical industry. (For example, Sweden was the first country to smelt iron on a large scale by means of electricity, with no coal required; by 1918 it produced 100,000 tons by this method, about one-eighth of its total pig iron production.) No less important, electricity allowed the countries to develop metal-fabricating and machinery and machine tool industries (including shipbuilding) without coal or primary metals industries.

In short, the experience of the Scandinavian countries, like that of Switzerland, shows that it was possible to develop sophisticated industries and a high standard of living without indigenous coal supplies or heavy industries, and that there is no single model for successful industrialization.

The Austro-Hungarian Empire

Austria-Hungary, or the lands ruled before 1918 by the Habsburg Monarchy, has suffered from a somewhat unjustified reputation for economic backwardness in the nineteenth century. This stigma resulted in part from the fact that some portions of the empire definitely *were* backward, and in part from the (mistaken) association of economic performance with political failure—the break-up of the empire in the aftermath of World War I. But above all, the misperception of the actual economic performance has been a result of the absence, until recent years, of informed research. The recent efforts of a number of competent scholars of several nationalities have made it possible to present with some confidence a more balanced and nuanced account of the progress of industrialization in the Habsburg domains.

Two points need to be stressed at the outset. First, to an even greater extent than France or Germany, the Habsburg Empire was characterized by regional diversity and disparity, with the western provinces (especially Bohemia, Moravia, and Austria proper) far more advanced economically than those of the east. Second, within the western provinces some stirrings of modern economic growth could be observed as early as the second half of the eighteenth century. Two other factors, to be elaborated subsequently, deserve brief mention here: the topography, which made both internal and international transportation and communication difficult and expensive, and the paucity and poor location of natural resources, especially coal.

The eighteenth-century beginnings of industrialization within the empire are now well established. Textile, iron, glass, and paper industries grew up in both Austria proper and the Czech lands. Collectively, the textile industries were by far the largest; linens and woolens predominated, but a fledgling cotton industry existed from at least 1763. In the beginning the technology was traditional; although a few "proto-factories"—large workshops without mechanical power—existed in the woolen industry, most production took place under the putting-out system. Mechanization began in the cotton industry at the end of the century, spreading to the woolen industry in the early decades of the next and more slowly to the linen industry. By the 1840s the empire was second only to France on the Continent for production of cotton goods.

It used to be thought that the revolution of 1848 marked a great watershed in the economic as well as the political history of the empire, but that notion has now been discounted. As noted, modern industries were already well established in the western provinces before the revolution; they continued to expand at a gradual but fairly steady rate thereafter. To be sure, in Austria as elsewhere the business cycle produced short-term fluctuations in the rate of growth, and much effort has been devoted to trying to discern which of the several cyclical upturns in the nineteenth century represented the beginning of the industrial revolution (or take-off); but such efforts are now seen to be fruitless.

Impressed by the gradual but cumulative character of Austrian industrialization from the eighteenth century until World War I, one scholar characterized it as a case of "leisurely" economic growth, but the word *labored* would seem to be more appropriate. Whereas the former term evokes a picture of a man floating slowly down a gentle stream in a boat, the latter suggests one climbing a steep hill on an ill-marked road bestrewn with obstacles and impediments—surely a more apt meta-phor. Some of the obstacles—the difficult terrain and the lack of natural re-sources—were imposed by nature; others, such as social institutions inimical to growth, were the work of men.

Among the latter the persistence of legalized serfdom until 1848 was the most anachronistic. In fact, however, serfdom was less an impediment than might be thought. The reforms of Joseph II in the 1780s allowed peasants to leave the estates of their lords without penalty and to market their crops as they chose. As long as they remained on their holdings they paid rent and taxes to their feudal lords, but otherwise the feudal system had little sway. The main consequence of the abolition of serfdom in 1848 was to grant the peasants freehold tenures, and to substitute taxes paid to the state for those formerly paid to their feudal lords. Although some improvement may have occurred in agricultural productivity as a result, improve-ments undertaken by noble landowners were already working in that direction.

The abolition of the customs frontier between the Austrian and Hungarian halves of the empire in 1850 (or, more positively, the creation of an empire-wide customs union in that year) has been seen by some as a progressive achievement and by others as a perpetuation of the "colonial" status of the eastern half. Although the customs union probably did facilitate the territorial division of labor, the pattern of Austrian exports of manufactures to Hungary, and Hungarian exports of agricultural products to Austria, was already well established before 1850. The alleged de-leterious effects of the customs union for the eastern half of the empire are no longer viewed as such.

Another institutional obstacle to more rapid economic growth was the mon-archy's foreign trade policy. Throughout the century it remained staunchly protec-tionist, which facilitated Prussia's goal of excluding it from the Zollverein. High tariffs limited not only imports but exports as well, because the high-cost protected industries were unable to compete in world markets. At the beginning of the twen-tieth century the foreign commerce of tiny Belgium exceeded that of Austria-Hungary in absolute value; in per capita terms it exceeded it many times over. To be sure, the empire's geographical position and topography contributed to its poor showing in international commerce, and its internal customs union embracing both industrial and agricultural areas offset its limited access to foreign markets and sources of supply to some extent; but commercial policy must be regarded as one reason, if a minor one, for the empire's relatively poor performance.

A major reason for both slow growth and uneven diffusion of modern industry was the levels of education and literacy, major components of human capital. Although literacy levels for the Austrian half of the monarchy were about the same as for France and Belgium in the midnineteenth century, they were extremely unevenly distributed. In 1900 the percentage of adults classified as literate ranged from 99 in Vorarlberg to 27 in Dalmatia; literacy rates in the Hungarian half were even lower and displayed the same west–east gradient. Within the empire as a whole, a high correlation existed between literacy levels and levels of industrializa-tion and per capita incomes.

In spite of the obstacles, both natural and institutional, industrialization and economic growth did occur in Austria throughout the century, and also in Hungary in the latter part of it. Estimates of the growth rate of industrial production per capita in Austria in the first half of the century range from 1.7 to 3.6 percent annually, and the rate accelerated somewhat in the second half of the century. In Hungary, after that portion of the monarchy obtained autonomy and a government of its own in the Compromise of 1867, even higher rates of industrial production growth occurred. (One should bear in mind, however, the small size of the statistical base to avoid overemphasizing the rapidity of growth.)

Transportation played a crucial role in the economic development of the empire. Since much of the country was mountainous, or surrounded by mountains, land transport was expensive and water transport nonexistent in the mountainous areas. Unlike the early industrializers, Austria-Hungary had few canals. The Danube and a few other large rivers flowed southward and eastward, away from markets and industrial centers. Not until the 1830s, with the advent of river steamboats, could they be used for upstream navigation.

As noted previously (see p. 206), the earliest railways were located largely in Austria proper and the Czech lands. After midcentury, and especially after the Compromise of 1867, Hungary obtained more. The effect consolidated the already established geographical division of labor within the empire. In the 1860s more than half the goods transported on Hungarian railways consisted of grain and flour. The traffic in flour, however, enabled Hungary to begin to industrialize. In the latter part of the century Budapest became the largest milling center in Europe, second only to Minneapolis worldwide. It also manufactured and even exported milling machinery, and at the end of the century began to manufacture electrical machinery as well. For the most part, however, Hungarian industrial production consisted of consumer goods, especially food products. These included, in addition to flour, refined sugar (from beets), preserved fruits, beer, and spirits. These were Hungary's counterpart to Austria's and Bohemia's emphasis on textiles.

The empire did have some heavy industry. A charcoal-fired iron industry had existed in the Alpine regions for centuries, and Bohemia also had a long tradition of metalworking in both ferrous and nonferrous metals. With the advent of coke smelting of iron ore the charcoal industries gradually declined, but in Bohemia and Austrian Silesia, somewhat better endowed with coal than the remainder of the empire, modern metallurgical industries developed from the 1830s onward. These industries included not only the primary production of pig iron but refining and fabricating as well, along with some machinery and machine-tool factories. Some heavy chemical industries also came into existence. On the eve of World War I the Czech lands accounted for more than half of "Austrian" industrial production, including about 85 percent of coal and lignite, three-fourths of chemical output, and more than half of all iron production. Some rather sophisticated industries also grew up in Lower Austria, especially in Vienna and its suburbs. Wiener Neustadt was the site of a locomotive factory as early as the 1840s.

Some of the problems attending Austrian heavy industry are illuminated by Figure 10-4, which shows the evolution of per capita coal production and consumption of Germany, France, Austria, and Russia. From about 1880 on Austrian and French production were roughly even—both far behind that of Germany but well

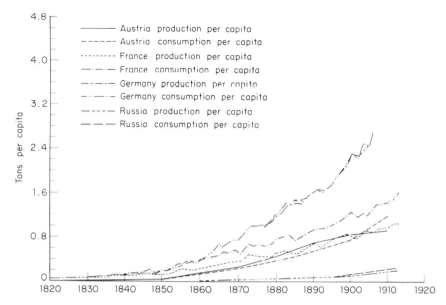

FIGURE 10-4.   Production and consumption of coal, 1820–1913. (From *European Historical Statistics, 1750–1970*, by B. R. Mitchell, New York, 1975.)

ahead of Russian production—but French consumption was somewhat higher because of imports. (Austria actually had a small export surplus, across the border to neighboring Germany, for a few decades at the end of the century.) What the figure does not reveal is the fact that about two-thirds of Austrian production was inferior lignite (or brown coal), unsuitable for metallurgical uses. Nor does it reveal the location of the deposits; mostly they were in the northern part of the country (the Czech lands), especially along the northern border with Germany, which accounts for the fact that coal-rich Germany could import coal from coal-poor Austria along the Elbe waterway. Hungarian coal production (not included in Fig. 10-4) was less than one-fourth that of Austrian and even more heavily tilted to lignite. Even so, the kingdom supported a small (and subsidized) iron and steel industry from the late 1860s.

In summary, the Habsburg Monarchy, which in industrial terms had been at parity with or even in advance of the disunited German states in the first half of the nineteenth century, steadily fell behind the industrial growth of the united German Empire after 1871. Nevertheless, the picture is not as bleak as it used to be painted. Industry in the western (Austrian) half of the monarchy continued to grow, steadily if not spectacularly, whereas that of the eastern (Hungarian) half spurted rapidly after about 1867. At the beginning of the twentieth century the western portion was at about the same level of development as the average for western Europe; the eastern portion, although lagging behind the western, was nevertheless well in advance of the rest of eastern Europe.

Southern and Eastern Europe

The patterns of industrialization of the remaining countries of Europe—the Mediterranean counties, southeastern Europe, and imperial Russia—can be dealt with more summarily. One common characteristic is failure to industrialize significantly before 1914, with resulting low levels of per capita income and a high incidence of poverty. To be sure, if one looks not at national aggregates but at individual regions, as we will do briefly later, one discovers marked regional variation, as with France, Germany, the Habsburg Monarchy, and even Great Britain. Nevertheless, the few "islands of modernity" remained surrounded by seas of backwardness.

One reason for this is a second common characteristic: abysmally low levels of human capital. Tables 8-3 and 8-4 illustrate this. Among the larger nations Italy, Spain, and Russia ranked at the bottom in both adult literacy rates and primary school enrollment rates, and the smaller countries of southeastern Europe fared no better. In primary school enrollments Rumania and Serbia came in above Russia but below Spain and Italy.

The laggard countries shared a third common characteristic that had an important bearing on their possibilities for economic development: the lack of any meaningful agrarian reform, with consequent low levels of agricultural productivity. Discussions of the patterns of industrialization of the other countries in this and the previous chapter made scant reference to the agrarian sector, since all had achieved relatively high levels of agricultural productivity. As Chapter 7 pointed out with reference to Great Britain, high agricultural productivity is necessary for any extended process of industrialization, both to supply the urban, industrial portion of the population with food and raw materials and, especially, to release labor for industrial (and other nonagricultural) occupations. In the midnineteenth century the proportion of the labor force engaged in agriculture ranged from a low of 20 percent in Great Britain through 50 to 60 percent in the other early industrializers to about 60 percent in Italy, more than 70 percent in Spain, and more than 80 percent in Russia and southeastern Europe. By the beginning of the twentieth century the proportions had fallen to less than 10 percent in Great Britain, about 20 percent in Belgium, Switzerland, and the Netherlands, and 30 to 40 percent in France and Germany, but they remained above 50 percent in Italy, at about 60 percent in the Iberian peninsula, and above 70 percent in Russia and the Balkans.

Finally, one can even mention a fourth common characteristic of the laggard nations: all suffered to varying degrees from autocratic, authoritarian, and corrupt, inefficient governments. Although the industrialized countries also experienced periods of authoritarian government from time to time, the relationship of this to the other common characteristics, especially the low levels of human capital, bears further investigation.

So much for the common characteristics. The countries also differed in significant respects. We now turn to the distinctive features of their response, or nonresponse, to the opportunity for industrialization and economic development.

The Iberian Peninsula

The economic histories of Spain and Portugal in the nineteenth century are sufficiently similar that it is convenient to deal with them as one. Both emerged from the Napoleonic Wars with primitive, even archaic, economic systems and reactionary political regimes. The latter feature sparked revolutionary outbursts in both countries in 1820; although the revolutions were ultimately unsuccessful, they led to endemic civil wars that interfered with normal economic activity and rendered impossible any coherent economic policy. Deplorable public finances plagued both countries. During the civil wars both sides (in both countries) borrowed abroad to support their military efforts. The losers, of course, defaulted, but even the winners were hard-pressed to pay their debts, and in the end they also defaulted in part. In Spain, following the damage and destruction of the Napoleonic Wars, the loss of the American colonies (except for Cuba, Puerto Rico, and the Philippines, which were lost in the aftermath of the Spanish-American War of 1898) resulted in a drastic reduction of the public revenue from 1800 to 1830. Chronic government deficits led to manipulations of the banking system, monetary inflation, and recourse to foreign borrowing, but the credit rating of the government was so poor that the terms on which it could borrow were extremely onerous. One loan of 1833 raised only 27 percent of the nominal capital. Before the century was over both countries repudiated at least a portion of their debts on more than one occasion.

Low agricultural productivity remained a fundamental weakness of both economies. As late as 1910 the primary sector, mainly agriculture, employed about 60 percent of the labor force in Spain, and at least as much in Portugal. But it was not, for the most part, commercial agriculture. One scholar has characterized the Spanish economy of the nineteenth century as a "dual economy," with a large subsistence agricultural sector, on the one hand, and a small commercial agricultural sector interacting with an even smaller urban industrial, commercial, and service sector, on the other. In the 1840s a government decree requiring the payment of taxes in cash rather than in kind sparked a peasant revolt, as no markets existed in which they could sell their produce.

Spain attempted an agrarian reform, but it was a complete fiasco. Like the government of revolutionary France, it confiscated the lands of the church, the municipalities, and the aristocrats who opposed it in the civil wars, with the intention of selling them to the peasants; but the exigencies of public finance were so great that the government ended up selling at auction to the highest bidders (who could pay in depreciated government bonds at face value); the result was that most of the land ended up in the possession of those already wealthy, both aristocrats and the urban bourgeoisie. The peasants merely endured the replacement of one set of absentee landlords by another, with no improvement of technology or increase in capital equipment. Portugal attempted no land reform at all. Meanwhile the increase in population in both countries resulted in more grain cultivation, the means of subsistence, on inferior soils and less pasture for livestock, producing a further fall in productivity.

Despite this generally depressing picture, a few bright spots existed—regional variations on a theme of backwardness. A modern cotton industry developed in Catalonia, in Barcelona and its environs, in the 1790s and, thanks to protective tariffs and a protected colonial market in Cuba and Puerto Rico, flourished until the loss of the latter colonies in 1900. Export-oriented wine industries existed in Andalusia (the region of Jerez, hence the English "sherry"), and in the province of Oporto ("The Port") in Portugal. In 1850 wines and brandy accounted for 28 percent of Spanish exports, but the dreaded phylloxera, a disease of the vine that had already struck France, spread into Spain in the last decades of the century with devastating effect. By 1913 wines accounted for less than 12 percent of Spanish exports.

Meanwhile a new source of foreign exchange, minerals and metals, developed to replace the lost earnings from wine. The famed mercury mines of Almadén had been producing since the sixteenth century; mercury, although profitable, did not loom large in the balance of payments. In the 1820s, however, the growing foreign demand for lead for plumbing resulted in the opening of the extremely rich lead deposits of southern Spain. As early as 1827 exports of pig lead accounted for more than 8 percent of total foreign earnings. From 1869 to 1898, when it was overtaken by the United States, Spain was the leading lead producer in the world. A new mining law in 1868 resulted in a great increase in the number of mineral concessions, for copper and iron as well as lead, mainly to foreign concerns. By 1900 exports of minerals and metals accounted for about one-third of total exports. Unfortunately for Spain, most exports went out as crude metal (lead and copper) or as ore (iron), with few backward linkages to the domestic economy.

Foreign capital also predominated in other modern sectors of the economy, especially in banking and railways. Prior to 1850 developments in both of these areas had been negligible; banking was dominated by the Bank of Spain, primarily an instrument of government finance, and only a few kilometers of railways had been built in the late 1840s. In the 1850s, in one of the frequent changes of government, the new regime gave special encouragement to foreign (mainly French) capitalists to create banks and railways. This they did, with government guarantees of interest on the capital invested in the railways during the period of construction. Unfortunately, when the main lines had been constructed and the guarantee of interest ceased, the railways had not developed sufficient traffic to meet operating costs, and most of the railways entered bankruptcy. The railways were built mostly with imported materials and equipment by foreign engineers and thus, like the mines, had few backward linkages. Not until near the end of the century did the railways become a paying proposition. Meanwhile most of the banks had been liquidated with greater or lesser profits for their foreign owners, leaving the field open for domestic entrepreneurs. Portugal obtained its first railway, a short line from Lisbon, in 1856, and its rail history was even sadder than that of Spain. Built with foreign (mainly French) capital, its railways endured fraud and corruption as well as bankruptcy and did little to aid the development of the economy.

Spain had some coal deposits (Portugal none), but they were not high quality and were poorly located for industrial exploitation. Nevertheless, in the last two decades of the nineteenth century a small iron and steel industry grew up along the north coast in the vicinity of Bilbao. Using the rich iron ores of the region and some imported coal and coke, the industry made slow headway against imports of iron, steel, hardware, and machinery, which it did not succeed in replacing. In the twentieth century the region became one of the wealthiest and economically most progressive in Spain. Nothing comparable occurred in Portugal.

Italy

Before 1860 Metternich's phrase for Italy, "a geographic expression," applied to the economy as well as its politics. An "Italian economy" did not exist. Left in the backwaters of economic change from the beginning of modern times, divided and dominated by foreign powers, Italy had long since lost its leadership in economic affairs. Wars and dynastic intrigues made it a battleground for foreign armies and looted it of both priceless artistic treatures and more utilitarian forms of wealth, while repeated monetary disturbances wiped out accumulated savings and shook the faith of investors.

The Congress of Vienna reimposed the bewildering mosaic of nominally independent principalities, but most, including the Papal States and the Kingdom of the Two Sicilies, were under the control or influence of the Habsburg Empire. Austria annexed Lombardy and Venetia directly; two of the economically most progressive provinces, and formerly the seats of famous industries and commerce, they were separated from the rest of Italy by Austria's high tariff barriers. The Kingdom of Sardinia, the only genuinely independent state, was a curious melange, an artificial nation composed of four major subdivisions with different climates, resources, institutions, and even languages. The island of Sardinia, from which the union took its name, languished in the backwaters of feudalism; its absentee landlords took no interest in improving their estates, with the result that its illiterate population lived under the most primitive conditions. Savoy, which gave the kingdom and later Italy its ruling dynasty, belonged culturally and economically to France. Genoa (and its hinterland Liguria), the commercial center, had for centuries before Napoleon maintained itself as an independent republic. Piedmont, surrounded on three sides by mountain peaks, formed a geographic continuation of the Lombard plain, but its altitude and climate set it apart from Lombardy as well. It contained roughly four-fifths of the kingdom's total population of about 5 million. Before 1850 it had little industry other than silk-throwing and a few small metallurgical establishments, but with the leadership of a few improving landowners its agriculture became the most advanced and prosperous of the peninsula.

Regional economic differentials, important in almost all countries, were especially marked in Italy. There the north–south gradient, still evident today, had existed since the Middle Ages. It may have been somewhat less noticeable in the nineteenth century, because of the general backwardness of the peninsula, but it remained. Agricultural productivity was higher in the north, especially in Piedmont and the Po valley, and some industry existed as well. And it was in the economically more progressive north that the movement for national unification began.

After the abortive revolutions and attempts at unification of the 1820s, 1830s, and 1848–49 had been suppressed by the Habsburgs, a remarkable individual came to the fore in the Kingdom of Sardinia. He was Count Camillo Benso di Cavour, a progressive landowner and agriculturist who had also promoted a railway, a newspaper, and a bank, and who in 1850 became minister of marine, commerce, and agriculture in his small country's newly established constitutional monarchy. The following year he added the portfolio of minister of finance, and in 1852 became prime minister. He stressed repeatedly that financial order and economic progress were the two "indispensable conditions" for Piedmont to assume, in the eyes of Europe, the leadership of the Italian pensinsula. To achieve these goals, he advocated foreign economic assistance, including foreign capital investment. Immediately upon taking office in 1850 he negotiated trade treaties with all the more important commercial and industrial nations of Europe. Between 1850 and 1855 exports increased by 50 percent while imports almost tripled; French investments financed the resulting heavy adverse trade balance. Throughout the remainder of the decade the French, with Cavour's encouragement, built railways, established banks and other joint-stock companies, and invested in the kingdom's growing public debt.

A part of the public debt had been contracted to liquidate the unsuccessful wars of 1848 and 1849, and more to prepare for the eventually successful war of 1859, in which the Kingdom of Sardinia, with the military and financial aid of France, defeated the Austrian Empire and prepared the way for the united Kingdom of Italy in 1861. The new nation, with a total population of approximately 22 million, had an average density of eighty-five inhabitants per square kilometer—already one of the highest in Europe. With the greater part of the labor force engaged in low-productivity agriculture, Italy had a long road to travel under the best of circumstances. Unification alleviated one of the major obstacles to economic development, the fragmentation of the market; but without the development of transportation and communications facilities, even this achievement would be illusory. The extension of Piedmont's progressive legislation and administrative system to the enlarged kingdom could not immediately alter the backward character of the institutions or the illiteracy and ignorance of the population of the remainder of the peninsula. No laws could remedy the poverty of natural resources, and only the wisest legislation and most judicious administration could overcome the scarcity of capital. Unfortunately for Italy, Cavour's exertions in those frantic years led to his premature death only three months after the proclamation of the kingdom, thus depriving the country of his wise and inspired leadership. His successors, though no less patriotic, lacked his experience, his finesse, and above all his subtle understanding of economic and financial questions. Italy remained dependent on foreign, especially French, investment and economic relations, but actions of the government repeatedly alienated foreign investors and finally, in 1887, drove Italy into a dramatic ten-year tariff war with France, with disastrous consequences for both economies.

Near the end of the 1890s, after the tariff war with France and with a new injection of foreign capital, this time from Germany, Italy experienced a small industrial growth spurt that lasted, with fluctuations, until after the beginning of World War I. Italy was not yet an industrial nation, but it had made a belated beginning.

Southeastern Europe

The five small countries that occupied the southeastern corner of the European continent—Albania, Bulgaria, Greece, Rumania, and Serbia—were, with the possible exception of Portugal, the poorest in Europe west of Russia. All had won independence from the Ottoman Empire at various dates after 1815, Albania as recently as 1913, and the heritage of Ottoman domination weighed heavily on their economies. At the beginning of the twentieth century all were predominantly rural and agrarian, with 70 or 80 percent of the labor force engaged in primary production and a similar proportion of total output consisting of agricultural products. Moreover, technology was primitive and productivity and per capita income correspondingly low. Although precise data are not available, rough figures suggest that, on average, the per capita income was less than that of neighboring Hungary, about half that of Bohemia, and about one-third that of Germany. Some slight variation existed within the group as well, with Rumania slightly better off than the others, and Albania the most backward.

Despite their poverty, high birth rates combined with moderately declining death rates engendered a population explosion from about the middle of the nineteenth century. In the half century before World War I the population grew at approximately 1.5 percent per year, among the highest rates for any European country or group of countries. The growing population pressure led to higher prices for farmland, land hunger, migration to urban areas and to the more developed countries to the West, and to some overseas migration, especially of Greeks to the United States.

No abundance of natural resources existed to ease the pressure of population. Much of the land was mountainous and unsuitable for cultivation, especially in Greece and to a lesser extent in Albania, Bulgaria, and Serbia. Rumania was best supplied with arable land, but with the primitive techniques of cultivation in use, even it was not highly productive. A few small, scattered coal deposits existed, but not enough to make any of the countries independent of imports, even considering the very small demand. Small deposits of nonferrous metals also existed, but these had scarcely begun to be exploited, by foreign capital, when World War I intervened. The most significant mineral resource was petroleum in Rumania. Several foreign firms, mainly German, began to tap this in the last decade of the nineteenth century.

In keeping with their agrarian character, the foreign trade of all the countries consisted of exports of agricultural products and imports of manufactured goods, mainly consumer goods. Cereals, mainly wheat, accounted for about 70 percent of the exports of Rumania and Bulgaria. Serbia, with less arable land, exported mainly live pigs and, shortly before the war, processed pork products, fresh plums, dried prunes, and its famous plum brandy, *slivovica*. Greece, with even less arable land and that not well suited to grain cultivation, exported mainly grapes and raisins as well as some wine and brandy.

In contrast with the slow diffusion of agricultural and industrial technology, the institutional technology of banks and foreign debts spread rapidly. By 1885 all four of the then-existing Balkan states had established central banks with exclusive powers of note issue. Joint-stock banks and other financial institutions developed rapidly, but with little connections to industrial finance. The new governments borrowed abroad, mainly in France and Germany, primarily to construct railways and other types of overhead capital, but also to purchase military equipment, to pay bloated bureaucracies, and increasingly to pay interest on previously acquired debt. In 1898 Greece became so obligated for foreign loans that it had to acquiesce to an International Financial Commission established by the great powers to oversee its finances. Eventually all other Balkan states except Rumania had to accept similar foreign control.

Much of the foreign borrowing was undertaken for railway construction, mainly for the account of the state. In 1870 the total length of railways in southeastern Europe amounted to less than 500 kilometers, mainly in Rumania and Bulgaria. By 1885 it came to 2000, in 1900 to more than 6000, and in 1912 to more than 8000 kilometers. Unfortunately, because of the lack of complementary industries, the railways had few backward linkages.

A small industrial sector did emerge in each of the countries, chiefly in consumer goods industries after about 1895, but nothing comparable to the industrial developments occurring in western Europe earlier in the nineteenth century. For practical purposes, one could say that modern industry had not yet penetrated southeastern Europe prior to World War I.

Imperial Russia

The Russian Empire at the beginning of the twentieth century was generally regarded as one of the great powers. Its territory and population, the largest by far of any European nation, merited that status. In gross economic terms as well, Russia loomed large: in total industrial production it ranked fifth in the world, after the United States, Germany, Great Britain, and France. It had large textile industries, especially cotton and linen, and heavy industries as well: coal, pig iron, and steel. It ranked second in the world (after the United States) in petroleum production, and for a few years at the end of the nineteenth century it held first place. Yet these large absolute amounts are misleading as a guide to Russia's economic strength. As Figure 10-4 shows, Russia's per capita production and consumption of coal were substantially below those of even Austria. Such was the case with almost every other category of production.

Russia was still a predominantly agrarian nation, with more than two-thirds of its labor force engaged in agriculture and producing more than half of the national income. Per capita income was no more than half that of France and Germany, and about one-third that of the United States and Great Britain. Productivity, especially in agriculture, was abysmally low, hampered as it was by a primitive technology and scarcity of capital. The institutional constraint of legalized serfdom, not removed until 1861, weighed heavily against the possibilities of growth in productivity even after the Emancipation (see Chapter 12, pp. 305, 308).

The beginnings of Russian industrialization have been traced back to the reign of Peter the Great and even earlier but, except for the Ural iron industry of the eighteenth century, these early industrial enterprises were hothouse undertakings connected with the needs of the Russian state, and were not economically viable. In the first half of the nineteenth century, especially from the 1830s onward, industrialization became more visible; it has been estimated that the number of industrial workers grew from less than 100,000 at the beginning of the century to more than half a million on the eve of Emancipation. Most of these workers were nominal serfs who made cash payments to their lords from their money wages, instead of the customary labor services. Paradoxically, there were also a number of serf entrepreneurs. The most dynamic, rapidly growing industry was cotton textiles, mainly in the Moscow region, with beet sugar refineries in the Ukraine a distant second. St. Petersburg boasted a number of large, modern cotton mills and also some metallurgical and machinery works, as did Russian Poland.

The Crimean War starkly revealed the backwardness of both Russian industry and Russian agriculture, and thus indirectly prepared the way for a number of reforms, the most notable of which was the emancipation of the serfs in 1861. Concurrently, the government encouraged a program of railway construction on the basis of imported capital and technology, and reorganized the banking system to permit the introduction of Western financial techniques. Signs of the effectiveness of the new policies became evident in the mid-1880s and in the "great spurt" of industrial production of the 1890s, when industrial output increased at an average rate of more than 8 percent, higher than even the best rates achieved in Western nations.

Much of the credit for this great spurt goes to the program of railway construction, especially that of the state-owned Trans-Siberian Railway, begun in 1891, and to the associated expansion of the mining and metallurgical industries (Fig. 10-5). The latter, in turn, owed much to foreign entrepreneurs and capital, who contributed decisively to the development of the great mining and metallurgical center of the southeastern Ukraine in the vicinity of the Donetz Basin.

The Donbas, as it is known, had large deposits of coal, but it was also remote from the main centers of population. Before the coming of the railway the coal was uneconomical to mine. Some 500 kilometers to the west, in the vicinity of Krivoi Rog, very rich iron ore deposits occurred, but for the same reason could not be economically exploited. In the 1880s French entrepreneurs persuaded the tsarist government to build a railway connecting the two areas and constructed blast furnaces at both sites, thus creating the first "cross-haul" metallurgical combine in the world. Production of both coal and pig iron soared; whereas in the 1870s domestic production of pig iron had satisfied only about 40 percent of demand, in the 1890s it accounted for three-quarters of a much larger consumption.

The government sought to encourage industrialization by several means. It borrowed abroad to finance the construction of the state-owned railways, and guaranteed the bonds of the railways belonging to companies. It placed orders for rails, locomotives, and other equipment for the state-owned railways with companies located in Russia (whether owned by Russians or foreigners) and instructed the private companies to do likewise. It placed high tariffs on imports of iron and steel products, but at the same time facilitated the introduction of the most recent equipment for the manufacture of iron and steel and engineering products. Producers in

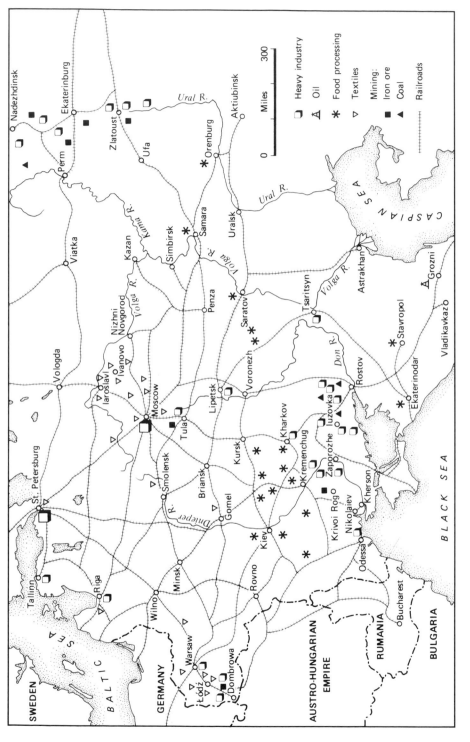

FIGURE 10–5. Industry and railways in Russia, about 1914. (From *Atlas of Russian and East European History*, by Arthur E. Adams. William O. McCagg, and Ian M. Matley. Copyright © 1966 by Frederick A. Praeger, Inc. Reprinted by permission of Praeger Publishers.)

Polish Silesia and St. Petersburg as well as the southeastern Ukraine benefited from these measures.

The boom of Russian industry in the 1890s was followed by a slump in the first years of the twentieth century, which in turn was followed by the disastrous (for Russia) Russo-Japanese War of 1904–5, and then by the revolution of 1905–6. Although the revolution was suppressed, it prompted a number of both political and economic reforms. The most important of the latter was the Stolypin agrarian reform (see Chapter 12, p. 305), which led to increased productivity in agriculture.

In the half century before World War I the Russian economy underwent substantial change in the direction of a more modern, technologically proficient system; but it was still far behind the more advanced Western economies, that of Germany in particular. Its economic weakness became acute during the war, contributing to Russian defeat and setting the stage for the revolutions of 1917.

Japan

The last and most surprising entry in the list of industrializing nations in the nineteenth century—and the only one entirely outside the Western tradition—was Japan. In the first half of the century Japan maintained its policy of exclusion of foreign, especially Western, influence more effectively than any other Oriental nation. From early in the seventeenth century the Tokugawa government had forbidden foreign trade (the Dutch were allowed to send one ship a year to a trading station that they maintained on a small island in Nagasaki's harbor, Japan's "window on the West") and had forbidden Japanese from traveling abroad. Society was structured into rigid social classes or castes, similar in some respects to the feudalism of medieval Europe. The level of technology was approximately that of Europe at the beginning of the seventeenth century. In spite of these constraints, however, the organization of the economy was remarkably sophisticated, with active markets and a credit system. The literacy level was substantially higher than those of southern and eastern Europe.

In 1853 and again in 1854 Commodore Matthew Perry, a U.S. naval commander, sailed into Tokyo Bay and, threatening to bombard the city, forced the Tokugawa shogun to open diplomatic and commercial relations with the United States. Soon other Western nations gained privileges similar to those granted to the United States. A key feature of these "unequal treaties" prevented the Japanese government from levying tariffs of more than 5 percent ad valorum; foreigners also gained rights of extraterritoriality (i.e., they were not subject to Japanese law). The weakness of the Tokugawa shogunate in the face of Western encroachments led to antiforeign riots and a movement to restore the emperor, who for centuries had performed only ceremonial functions, to a central position in the government. This movement, led by ambitious young *samurai* (members of the former warrior class), was fortuitously aided in 1867 by the accession of a vigorous, intelligent young emperor, Mutsuhito; the following year the emperor's party forced the shogun to abdicate and brought the emperor to Tokyo, the de facto capital. This event, marking the birth of modern Japan, is called the Meiji Restoration (Meiji meaning "enlightened government," which Mutsuhito chose to designate his reign). The Meiji era lasted from 1868 to the death of Mutsuhito in 1912.

Immediately upon coming to power the new government changed the tone of the antiforeign movement. Instead of attempting to expel the foreigners Japan cooperated with them but kept them at a polite distance. The old feudal system was abolished and replaced by a highly centralized bureaucratic administration modeled on the French system, with an army of the Prussian type and a navy like the British. Industrial and financial methods were imported from many countries, but especially the United States. Intelligent young men went abroad to study Western methods in politics and government, military science, industrial technology, trade, and finance, with the aim of adopting the most efficient methods. New schools were established in Japan on Western models, and foreign experts were brought in to train their Japanese counterparts. The government was careful, however, to set strict limits on their tenure, and to see that they left the country once their terms were over to prevent them from establishing positions of dominance.

One of the most vexing problems facing the new government was that of finance. Financial problems had been one of the causes of dissatisfaction with the old Tokugawa regime, and the new Meiji government inherited a mass of inconvertible paper money, which it was obliged to increase in the first years of the transition. In 1873 it enacted a land tax, assessed on the basis of the potential productivity of agricultural land regardless of the amount of actual produce. This had a doubly beneficial effect: on the one hand, it assured the government of a steady revenue (at the expense of the peasants, to be sure); second, it ensured that the land would be put to its best use, as those who were unable to maximize returns on it would lose it or be forced to sell to those who could.

Also in connection with its financial problems, the government set out to create a new banking system to replace the informal credit network of the Tokugawa era. In keeping with its policy of seeking the best of everything (a Prussian-style army, a British-style navy, etc.), it took as its model the National Banking System of the United States, created by the Union government in the closing years of the Civil War as a measure of war finance. Under this system banks could be established using government bonds as collateral for the issue of banknotes, which should be convertible into specie. (Not coincidentally, the Meiji government had just issued a large quantity of bonds to the former feudal lords and samurai as a replacement for their annual pensions.) Under this system, by 1876, 153 national banks had been established. Unfortunately, the following year the Satsuma Rebellion, a rising against the government by one of the largest western clans, broke out; although the government suppressed the rebellion, it did so at great cost and more issues of both inconvertible government money and national banknotes, resulting in rampant inflation.

A new finance minister, Count Matsukata, decided the banking system was at fault and, in addition to bringing about a drastic deflation of the currency in 1881, completely revamped the banking structure. He created a new central bank, the Bank of Japan, on the model of the latest fashion in central banks, the Banque Nationale de Belgique, which, although mainly privately owned, was under the close control of the government. It obtained a monopoly of note issue, the national banks losing their issue rights and being converted into ordinary commercial deposit banks on the English model. The Bank of Japan also acted as fiscal agent of the treasury.

From the time of the Meiji Restoration the government intended to introduce and domesticate virtually the full range of Western-style industries. To this end it built and operated shipyards, arsenals, foundries, machine shops, and experimental or model factories for the production of textiles, glass, chemicals, cement, sugar, beer, and a variety of other goods; it also imported Western technicians to instruct the native labor force and managerial hierarchy in the use of Western equipment. Such an undertaking was clearly a long-term proposition, however. In the meantime resources had to be found to pay for the imports of machinery and other equipment and the salaries of the foreign experts. As a predominantly agrarian economy at the time of the Restoration, and one with virtually no experience in foreign commerce, that was not an easy task.

Moreover, Japan had few natural resources. Smaller than the state of California, the island country is also quite mountainous, so that the proportion of arable land to the total was also smaller than that of California. Rice was the staple crop and also the staple of the diet, supplemented by fish and seafood from the teeming coastal waters. Japan did have some deposits of coal and copper ore, and before the 1920s these contributed to exports as well as to domestic consumption. For the most part, however, the agrarian sector had to bear the burden of providing the export revenues to finance the necessary imports.

Japan's two traditional textile industries based on domestic raw materials, silk and cotton, experienced very different fortunes. Soon after the opening of trade the cotton industry was wiped out completely by the machine-produced goods from the West, especially Great Britain. The silk industry, on the other hand, survived, and that part of it closest to the agrarian sector, the production of raw silk yarn from cocoons, actually flourished. Assisted by the introduction of modern equipment obtained from France, production of raw silk rose from little more than 2 million pounds in 1868 to more than 10 million in 1893, and about 30 million on the eve of World War I. The greater part of production was exported, and from the 1860s to the 1930s raw silk accounted for between one-fifth and one-third of export revenues. Some trade also developed in silk fabrics, which in 1900 accounted for almost 10 percent of export revenues; but high tariffs on fabrics in the countries that were the main markets for the raw silk, especially the United States, hampered the development of that industry.

The other major agrarian export was tea, which in the early years of the Meiji era was as important as silk; its relative importance, however, gradually declined with the growth of domestic population and income. The same was true to an even greater degree with rice; although small amounts were exported in the early years of the era, the growth of population was such that before the end of the century Japan depended partly on imports for its total consumption.

Although government initiative was responsible for introducing most elements of Western technology, it was not the government's intention to prohibit private enterprise. On the contrary, one of its slogans was "develop industry and promote enterprise." As soon as the mines, model factories, and other modern establishments (except for the arsenals and one steel mill, under military control) were operating satisfactorily, the government sold them (frequently at a loss in strict accounting terms) to private companies or corporations.

The cotton industry (mainly spinning, but with some mechanized weaving) made the most rapid progress. The technology was relatively simple, and it employed cheap, unskilled labor, mostly women and girls. It conquered the home market in the 1890s, and by 1900 exports of cotton yarn and cloth (mostly the former) accounted for 13 percent of total exports. The largest markets were China and Korea, which imported cheap, coarse yarn for hand weaving in peasant households.

The heavy industries—iron, steel, engineering, and chemicals—were slower to develop, and did so with large subsidies and tariff protection (the unequal treaties expired in 1898), but by 1914 Japan was largely self-sufficient in their products. World War I, of course, greatly increased the demand for them, and at the same time opened new markets. In fact, the war was a great boon for the Japanese economy as a whole. The deficit in the balance of trade of the last prewar years had been large, but the increased wartime demand, together with the diversion of European production to war uses, enabled Japanese producers to expand rapidly into foreign markets. By entering the war on the Allied side Japan was also able to take over German colonies in the Pacific and concessions in China. Exports, which amounted to 6 or 7 percent of gross national product in the 1880s and about 15 percent in the first decade of the twentieth century, jumped to 22 percent as early as 1915.

Overall, the economic transition of Japan from a backward, traditional society in the 1850s to a major industrial nation at the time of the First World War was a most remarkable feat. The growth rate of gross national product from the 1870s to the eve of the war averaged about 3 percent per year (estimates range from 2.4 to 3.6), as high as or higher than that of any European nation. Moreover, the growth rate was relatively stable; although it fluctuated somewhat, at no time did it fall below zero, as it frequently did in Europe and America during severe recessions or depressions. The rate of growth of mining and manufacturing output was still higher, about 5 percent for the period as a whole.

Japan's economic transition had political consequences as well. In 1894–95 Japan quickly defeated China in a short war and joined the ranks of the imperialist nations by annexing Chinese territory (notably Taiwan, which it renamed Formosa) and staking out a sphere of influence in China itself. Even more surprisingly, just ten years later Japan decisively defeated Russia on both land and sea. The rewards of this exploit were the southern half of the island of Sakhalin, the Russian leases on Port Arthur and the Liaotung peninsula of China, and Russian acknowledgment of Japanese predominance in Korea, which Japan annexed in 1910. Thus the Japanese proved they could play the white man's game.

11

The Growth of the World Economy

Although long-distance trade has existed from at least the beginnings of civilization, its importance grew enormously and rapidly in the nineteenth century. For the world as a whole the volume of foreign trade per capita in 1913 was more than twenty-five times greater than it had been in 1800. Throughout the century Europe accounted for 60 percent or more (as much as two-thirds) of the total of both imports and exports. The period of most rapid growth occurred between the early 1840s and 1873, when total trade increased at more than 6 percent annually—five times as fast as the population growth and three times as fast as the increase in production.

The international movement of people and capital—migration and foreign investment—also accelerated rapidly. By the beginning of the twentieth century it was possible to speak meaningfully of a world economy, in which virtually every inhabited portion participated at least minimally, though Europe was by far the most important. It was, in fact, the dynamic center that stimulated the whole.

At the beginning of the century two main types of obstacles, natural and artificial, hindered the flow of international commerce. The incidence of both fell significantly as the century progressed. The natural obstacle—the high cost of transportation, especially of land transportation—yielded to the railway and the improvements in navigation, culminating in the ocean-going steamship. The artificial obstacles—tariffs on imports and exports and also some outright prohibitions on the importation of certain commodities—likewise came down and even disappeared, although at the century's end a "return to protection" resulted in the imposition of higher import tariffs in a number of countries.

Britain Opts for Free Trade

Intellectual arguments for free international trade had been made even before Adam Smith's eloquent treatise, *The Wealth of Nations;* the latter elevated them to a new plane of respectability. Moreover, practical considerations forced governments to reconsider their prohibitions and high tariffs; smuggling was a lucrative occupation in the eighteenth century, and reduced both government revenue and legitimate entrepreneurial profits. The British government had begun to alter its protectionist stance in the late eighteenth century, but the outbreak of the French Revolution and the Napoleonic Wars deferred its efforts. Indeed, the British blockade and the

Continental System represented extreme forms of interference with international trade.

Adam Smith's case for free international trade derived from his analysis of the gains from specialization and the division of labor among nations as well as individuals. It rested on differences in the absolute costs of production, such as the cost of producing wine in Scotland as opposed to France. David Ricardo, in his *Principles of Political Economy* (1819), supposed (incorrectly) that Portugal had an absolute advantage in the production of both cloth and wine, compared with England, but that the relative cost of producing wine was cheaper; under those circumstances, he showed it would be to Portugal's advantage to specialize in the production of wine and to purchase its cloth from England. This was the principle of *comparative advantage,* the foundation of modern international trade theory.

Both Smith's and Ricardo's arguments for free trade rested on purely logical grounds. To have any practical effects on policy, these arguments had to persuade large groups of influential people that free trade would benefit them. One such group consisted of merchants involved in international trade. In 1820 a group of London merchants petitioned Parliament to permit free international trade. Although the petition had no immediate effect, it indicated a new trend of public opinion. Coincidently, at about the same time several relatively young men intent on modernizing and simplifying the archaic procedures of government rose to positions of influence in the governing Tory party. Among them was Robert Peel, the son of a wealthy textile manufacturer, who as home secretary reduced the number of capital offenses from more than 200 to about 100. (He also created the Metropolitan Police Force, the first of its kind, whose members were called "bobbies" or "peelers," at first in derision, but later with affection.) Another of the so-called Tory Liberals was William Huskisson, who, as president of the Board of Trade, greatly simplified and reduced the maze of restrictions and taxes that hampered the development of international commerce. The parliamentary reform of 1832 extended the franchise to the urban middle class, most of whom were favorably inclined toward freer trade.

The centerpiece and symbol of the protectionist system of the United Kingdom (which included Ireland from 1801) were the so-called Corn Laws, tariffs on imported bread grains. The Corn Laws had a long history, but they were strengthened appreciably at the end of the Napoleonic Wars at the behest of the landed interests, who were strongly represented in Parliament. The growth of population and increasing urbanization made self-sufficiency in foodstuffs virtually impossible, but Parliament stubbornly resisted attempts to alter the Corn Laws. After earlier unsuccessful attempts to repeal or modify them, Richard Cobden, a Manchester industrialist, formed the Anti-Corn Law League in 1839 and mounted a strong and effective campaign to influence public opinion. In 1841 the Whig government, then in power, proposed reductions in the tariffs on both wheat and sugar; when these measures went down in defeat, they called a new general election.

Earlier the Corn Laws and protectionism in general had not been party issues, since landowners formed the bulk of both the Whig and the Tory parties. In the electoral campaign the Whigs, seeking to capitalize on anti-Corn Law sentiment, proposed a reduction (not repeal) of the Corn Laws, whereas the Tories advocated the status quo. The Tories won, but the new prime minister, now Sir Robert Peel, had already decided on extensive revisions in the fiscal system, including the abolition of export taxes, elimination or reduction of many import taxes, but not the duties on grain, and imposition of an income tax to replace the lost revenue. Several of these measures had already taken effect, and the government would probably have proposed a reduction in the grain tariff, when in 1845 the disastrous potato blight struck Ireland (and Scotland to a lesser extent), reducing large numbers of the Irish population to starvation. Prodded by this catastrophe, Peel introduced a bill to repeal the Corn Laws which, with the support of most Whigs, passed in January 1846, over the opposition of the majority of his own party.

In the aftermath of Corn Law repeal the modern British political system— at least until 1914—began to take shape. Peel, ostracized by his own party, retired from politics. W. E. Gladstone, one of a small number of Tories who followed him in voting for repeal, joined the Whigs, becoming chancellor of the exchequer and eventually prime minister. The Whigs, subsequently known as Liberals, became the party of free trade and manufactures, whereas the Tories, also known as Conservatives, remained the party of the landed interest and, eventually, of imperialism.

Also in the aftermath, Parliament cleared the books of much of the old "mercantilist" legislation, such as the Navigation Acts, which were repealed in 1849. As the new party alignments settled down in the 1850s and the 1860s, with Gladstone as chancellor of the exchequer for much of the time, an uncompromising policy of free trade emerged. After 1860 only a few import duties remained, and those were exclusively for revenue on such non-British commodities as brandy, wine, tobacco, coffee, tea, and pepper. In fact, although most tariffs were eliminated altogether and the rates of duty on all others were reduced, the increase in total trade was such that customs revenue in 1860 was actually greater than that of 1842.

The Free Trade Era

The next major development in the movement toward free trade was a notable trade treaty, the Cobden-Chevalier, or Anglo-French, treaty of 1860. France had traditionally followed a policy of protection; this was especially true in the first half of the nineteenth century when the French government, at the behest of mill owners, strove to protect the cotton textile industry from British competition. Part of France's protectionist policy was a flat prohibition imposed on the import of all cotton and woolen textiles and very high tariffs on other commodities, including even raw materials and intermediate goods. Economists such as Frederic Bastiat pointed out the absurdity of such policies, but powerful vested interests in the French legislature were immune to rational argument.

The government of Napoleon III, which came to power in a coup d'état in 1851, desired to follow a policy of friendship with Great Britain. This was partly to gain political status and diplomatic respect. Even though the coup d'état had subsequently been ratified by a referendum, the government's legitimacy was still in question. After the Crimean War, in which Britain and France had been allies, Napolean III wished to cement these new ties of friendship. Moreover, although

France had traditionally followed a policy of protectionism, a strong current of thought favored economic liberalism. One of the leaders of this school was the economist Michel Chevalier, who had traveled widely both in Britain and the United States and had a cosmopolitan outlook. As professor of political economy at the Collège de France since 1840, he had been teaching the principles of economic liberalism and free trade. Appointed by Napoleon to the French Senate, he persuaded the emperor that a trade treaty with Britain would be desirable.

Another political circumstance in France made the treaty route attractive. Under the French constitution of 1851, which Napoleon himself had granted, the two-chamber parliament had to approve any domestic law, but the sovereign, the emperor, had the exclusive right to negotiate treaties with foreign powers, the provisions of which had the force of law in France. Napoleon tried in the 1850s to reduce the strongly protectionist stance of French policy, but because of opposition in the legislature he was unable to carry through a thorough reform of tariff policy. Chevalier was a friend of Richard Cobden, of Anti-Corn Law fame, and through Cobden he persuaded Gladstone, the British chancellor of the exchequer, of the desirability of a treaty. The thought in Britain at this time, after its move to free trade, was that the advantages of a free trade policy would be so obvious that other countries would adopt it spontaneously. Because of the strength of protectionist interests, however, this was not the case. Accordingly, a treaty negotiated by Cobden and Chevalier late in 1859 was signed in January 1860.

The treaty provided that Britain would remove all tariffs on imports of French goods with the exception of wine and brandy. These were considered luxury products for British consumers, so Britain retained a small tariff for revenue only. Moreover, because of Britain's long-standing economic ties with Portugal, which also produced wine, Britain was careful to protect the Portuguese preference in the British market. France, for its part, removed its prohibitions on the importation of British textiles and reduced tariffs on a wide range of British goods to a maximum of 30 percent; in fact, the average tariff was about 15 percent ad valorem. The French thus gave up extreme protectionism in favor of a moderate protectionism.

A major feature of the treaty was the inclusion of a most-favored-nation clause. This meant that if one party negotiated a treaty with a third country, the other party to the treaty would automatically benefit from any lower tariffs granted to the third country. In other words, both parties to the Anglo-French treaty would benefit from the treatment accorded to the "most favored nation." Britain, by this time on virtually complete free trade, had no bargaining power with which to engage in treaties with other countries, but the French still had high tariffs against imports of goods from other countries. In the early 1860s France negotiated treaties with Belgium, the Zollverein, Italy, Switzerland, the Scandinavian countries, and, in fact, with almost every country in Europe except Russia. The result of these new treaties was that when France granted a lower rate of duty, say, to imports of iron from the Zollverein, British iron producers automatically benefited from the lower rates.

Moreover, in addition to this network of treaties that France negotiated throughout Europe, the other European countries also negotiated treaties with one another, and all contained the most-favored-nation clause. As a result, whenever a new

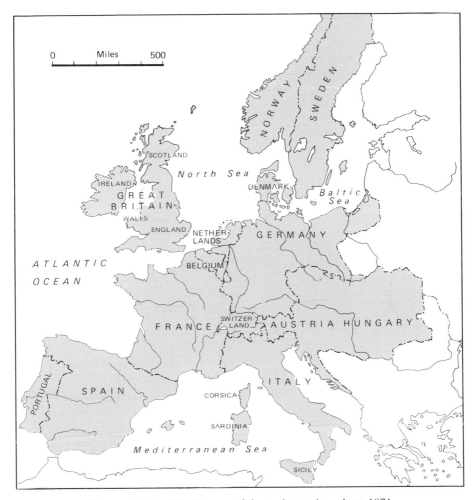

FIGURE 11–1.   The Europe of the trade treaties, about 1871.

treaty went into effect a general reduction of tariffs took place. For a decade or so, in the 1860s and the 1870s, Europe came as close as ever to complete free trade until after World War II (Fig. 11–1).

The consequences of this network of trade treaties were quite remarkable. International trade, which had already accelerated somewhat with the British reforms of the 1840s, increased at about 10 percent per year for several years (Fig. 11–2). Most of the increase took place in intra-European trade, but overseas nations also participated. (The American Civil War, which broke out the year after the Cobden-Chevalier treaty was signed, had a contrary effect. The northern blockade of the South cut off southern exports, causing a "cotton famine" in Europe, especially harmful in Lanchashire, and also restricted European exports of both consumer and capital goods to the South.) Another consequence of the treaties,

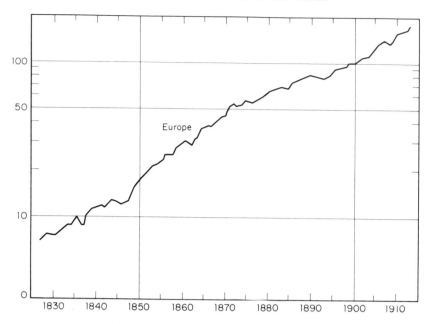

FIGURE 11–2. Index of the annual volume of exports of European countries (1899–1901 = 100). (From *Commerce extérieur et devéloppement économique de l'Europe au XIX siècle*, by Paul Bairoch, Paris, 1978.)

notably in France but also in several other countries, was a reorganization of industry forced by the greater competition; inefficient firms that had been protected by tariffs and prohibitions had to modernize and improve their technology or go out of business. The treaties thus promoted technical efficiency and increased productivity.

The "Great Depression" and the Return to Protection

Another consequence of the integration of the international economy brought about by freer trade was synchronization of price movements across national borders. In the preindustrial economy abrupt price fluctuations were generally local or regional, brought on by natural causes (droughts, floods, etc.) that affected the harvest. With increasing industrialization and international commerce, the fluctuations were more often related to the "state of trade" (fluctuations in demand), became cyclical in nature, and were transmitted from country to country through commercial channels. The cyclical nature of the movements became more pronounced as the century progressed.

Subsequent statisticians distinguished several varieties of "business cycles," as they came to be called: relatively mild short-term (2 or 3 years) "inventory cycles"; longer-term movements (9 or 10 years), frequently terminating in financial crises followed by depressions (1825–26, 1837–38, 1846–47, 1857, 1866, 1873, 1882,

1893, 1900–1901, 1907); and still longer-term (20 to 40 years) "secular trends." Complex interactions of both monetary and "real" factors caused the fluctuations, but authorities do not always agree on the relative importance of the two.

Fluctuations in production usually accompanied the price fluctuations. Again, authorities do not always agree on the direction of causation, but whereas a fall in prices might continue for a number of years, declines in output were usually relatively brief. The long-run trend was clearly upward, as is shown in Figure 9–1.

Prices in virtually all countries in Europe, and in the United States as well, reached a peak early in the century, near the end of the Napoleonic Wars. Causes were both real (wartime shortages) and monetary (the exigencies of war finance). Thereafter, until midcentury, in spite of short-term fluctuations, the secular trend was downward. Again, the causes were both real (technical innovations, improvements in efficiency) and monetary (repayment of war debts by governments). Prices bounded upward in the 1850s, primarily as a result of gold discoveries in California (1849) and Australia (1851), and then fluctuated for a couple of decades with no discernible trend.

In 1873, after a boom lasting several years, financial panics occurred in both Vienna and New York and quickly spread to most other industrial (or industrializing) nations. The ensuing fall in prices lasted until the mid- or late 1890s, and was known in Great Britain (until the greater catastrophe of the 1930s) as the "Great Depression." Eventually gold discoveries in South Africa, Alaska, Canada, and Siberia reversed the downward trend of prices and sent them gently upward again until World War I, which brought severe inflation.

The depression following the panics of 1873 was probably the most severe and widespread of the industrial era to that date. Industrialists incorrectly blamed it on intensified international competition as a result of the trade treaties, and more insistent pleas for renewed protection resulted. Concurrently, agriculturists—both landlords and small peasant farmers—joined in the demand for protection. Before 1870 they had not been troubled by overseas competition, because transportation costs on overseas shipments of bulky, low-value commodities such as wheat and rye had effectively protected them. In the 1870s dramatic reductions in transportation costs as a result of the extension of railways to the American Midwest and plains states, and subsequently in the Ukraine, Argentina, Australia, and Canada, together with equally dramatic reductions in ocean freight rates as a result of improvements in steam navigation, brought into production vast new areas of virgin prairies. In 1850 U.S. exports of wheat and flour, mainly to the West Indies, amounted to $8 million; in 1870 to $68 million, much of them bound for Europe; and in 1880 to $226 million. For the first time European farmers faced strenuous competition in their own markets.

At this juncture the situation of German agriculture was critical. Germany at the time was divided essentially into an industrializing west and an agrarian east. The Junkers of East Prussia, with their large estates, had long engaged in the export of grain through the Baltic to western Europe, including western Germany. This was the major exception to the fact that transport costs made the long-distance transportation of grain prior to the 1870s uneconomical. Thus, the Junker aristocrats had traditionally favored free trade because they were exporters. When they began to suffer from the fall in price of grain as a result of the large imports from America and Russia they demanded protection. The German population was growing rapidly. With industrialization cities were also growing rapidly. The Junkers wanted to preserve that large and growing market for themselves.

Otto von Bismarck, creator and chancellor of the new German Empire, formerly chancellor of Prussia, an astute politician, and himself a Junker landlord from East Prussia, saw his opportunity. The industrialists of western Germany had long been clamoring for protection; now that the East Prussian Junkers also demanded it, Bismarck "acceded" to their demands, denounced the Zollverein's trade treaties with France and other nations, and gave his approval to a new tariff law in 1879 that introduced protectionism for both industry and agriculture. This was the first major step in the "return to protectionism."

The protectionist interests in France, never reconciled to the Cobden-Chevalier treaty, gained political strength with defeat in the Franco-Prussian War and still more with the German tariff of 1879. In 1881 they succeeded in obtaining a new tariff law that explicitly reintroduced the principle of protectionism. Even so, the free traders retained considerable political clout, and in 1882 new trade treaties with seven continental countries maintained the basic principles of the Cobden-Chevalier treaty. Moreover, the tariff of 1881 did not attend to the protectionist demands of the agrarians. French agriculture, unlike that of East Prussia, was dominated by small peasant proprietors who, under the political system of the Third Republic, had the franchise and political power. After the elections of 1889 returned a protectionist majority to the Chamber of Deputies, they succeeded in passing the infamous Meline tariff of 1892. The tariff has been characterized as extremely protectionist, but a more apt term would be "refined protectionism." Though it did grant protection to some branches of agriculture, and retained the industrial protection of the tariff of 1881, it also contained several features favored by the free traders.

A tariff war with Italy from 1887 to 1898 caused a serious setback to French— and even more to Italian—commerce. Italy followed Germany in the return to protection and, largely for political reasons, chose particularly to discriminate against French imports. That was unwise, because France was Italy's largest foreign market. France retaliated with discriminatory tariffs of its own, and for more than a decade commerce between the two neighbors fell to less than half its normal figure. Germany and Russia also engaged in a brief tariff war in 1892–94.

A number of other countries also followed the German and French examples by raising tariffs. Austria-Hungary, which had a long history of protectionism, did enter into treaties with France and some other countries, but it retained a higher degree of protection than most and quickly returned to ultraprotectionism. Russia had never entered into the network of trade treaties set off by the Cobden-Chevalier treaty, and in 1891 it enacted a virtually prohibitive tariff. The United States, prior to the Civil War, had vacillated between very high and very low tariffs but, by and large, because of the influence of the southern planter aristocracy who depended on exports of cotton, had followed a low-tariff policy. After the Civil War, with the political influence of the South greatly diminished while that of the manufacturing interests of the Northeast and Midwest had increased, the nation became one of the most protectionist countries, and remained so to a large extent until after World War II.

There were a few holdouts for free trade during this return to protection, of which Great Britain was the most notable. Although political movements for "fair trade" and "empire preference" developed, they did not achieve any success before World War I. (The success of German merchants and industrialists in foreign and even in British markets did, however, inspire some retaliatory efforts. In 1887 Parliament passed the Merchandise Marks Act, which required foreign products to be labeled with the name of the country of origin. It was expected that the label "made in Germany" would discourage British consumers from buying those commodities; actually, just the reverse occurred.) The Netherlands specialized in processing such overseas imports as sugar, tobacco, and chocolate for reexport to Germany and other continental countries; thus they retained a largely free trade position, as did Belgium, which was heavily dependent on its export industries. Denmark, a predominantly agricultural nation, appeared to suffer from large-scale imports of cheap grain; but the Danes made a very quick adjustment from grain growing to livestock and dairy and poultry products, importing cheap grain as feed. Thus, Denmark also remained in the free trade bloc.

Much has been made in textbooks of the "return to protection"—probably too much. Although the rate of growth of international commerce slowed somewhat in the two decades after 1873, that rate was still positive, and it accelerated again in the two decades before World War I. In the immediate prewar decade it amounted to about 4.5 percent per year, almost as high as in the expansive middle decades of the century. The nations of the world, and especially those of Europe, depended more on international trade than ever before (Fig. 11–3). For the larger developed countries—Great Britain, France, and Germany—exports accounted for between 15 and 20 percent of total national income. For some of the smaller developed

Figure 11–3. Regional distribution of world trade, 1913.

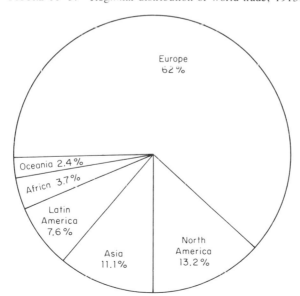

Europe
62%

Oceania 2.4%

Africa 3.7%

Latin
America
7.6%

Asia
11.1%

North
America
13.2%

countries, such as Belgium, Switzerland, the Netherlands, and the Scandinavian countries, the ratio was still higher. Even the lesser developed countries of Europe, those of the south and east, participated to a much greater degree than ever before in international commerce. Much the same can be said for other parts of the world. Although the United States, with its large, diversified economy, depended much less on the rest of world, it was nevertheless the world's third largest exporter in 1914. The self-governing dominions of the British Empire—Canada, Australia, New Zealand, and South Africa—and some of the British colonies were almost as dependent (or more so) as the mother country. Likewise, several nations of Latin America were involved in world markets for their exports of foodstuffs and raw materials, of which Europe took more than half.

In short, the world economy at the beginning of the twentieth century was more integrated and interdependent than it had ever been, or would be again until well after World War II. The peoples of the world, and those of Europe in particular, would discover, to their cost, how fortunate they had been in the agony and aftermath of world war.

The International Gold Standard

According to some authorities the high degree of integration achieved in the world economy in the late nineteenth century depended critically on general adherence to the international gold standard. According to others, that integration depended primarily on the central role of Great Britain, and of London, its financial as well as its political capital, in the world economy. Since Great Britain adhered to the gold standard for most of the century (although few other countries did), it is necessary to scrutinize the gold standard in some detail.

Throughout history various commodities (e.g., land, cattle, and wheat) have served as monetary standards, but gold and silver have always been the most prominent standards. The function of a monetary standard is to define the unit of account of a monetary system, the unit into which all other forms of money are convertible. Thus, in medieval England the "pound sterling" was legally defined as a pound weight of sterling silver; England at that time was on a silver standard, although the actual coins in use were only fractions of a pound. Technically—*de jure*—England remained on a silver standard until the French-Napoleonic Wars, although in the seventeenth and eighteenth centuries gold coins (the famous "guineas," which got their name from the region in Africa from which the gold originated) largely replaced silver coins in actual use. During the wars the Bank of England, with government sanction, "suspended payment"—that is, refused to pay out gold or silver in exchange for its banknotes—and, strictly speaking, the country had no monetary standard at all; it had fiat money, or "forced circulation."

After the wars the government decided to return to a metallic standard, but chose gold, the *de facto* standard of the eighteenth century, in preference to silver, although the pound was still called "sterling." The money of account (standard of value) was the gold sovereign or gold pound, defined as 113.0016 grains of fine (pure) gold. Under the terms of the Act of Parliament creating the gold standard three conditions had to be observed: (1) the Royal Mint was obliged to buy and sell unlimited quantities of gold at a fixed price; (2) the Bank of England—and, by extension, all other banks—was obliged to exchange its monetary liabilities (banknotes, deposits) into gold on demand; and (3) no restrictions could be imposed on the import or export of gold. This meant that gold served as the ultimate base or reserve of the entire monetary supply of the country. The amount of gold the Bank of England held in its coffers determined the amount of credit it could extend in the form of banknotes and deposits; these in turn (held as reserves by other banks of issue and deposit) determined the amount of credit they could extend. Thus, the movement of gold into and out of the country—a function of the balance of payments—caused fluctuations in the total money supply, which in turn caused fluctuations in the movement of prices. When the international gold flows were slight, or when inflows balanced outflows, as they usually did, prices tended to be stable; but large inflows, such as occurred after the gold strikes in California and Australia in 1849–51, could cause inflation, and sudden withdrawals, such as occurred periodically in the nineteenth century, brought on monetary panics.

For the first three quarters of the nineteenth century most other countries were on either silver or bimetallic (both gold and silver) standards; some had no metallic standard at all. But, because of the preeminent role of Britain in world commerce, almost all countries were affected by its economic fluctuations. Increasingly, as the world economy became more integrated, economic fluctuations tended to be transmitted internationally.

For a short time in the 1860s and 1870s France attempted to create an alternative to the international gold standard in the form of the Latin Monetary Union. Although France was nominally on a bimetallic standard, the gold discoveries in California and Australia caused a rise in the general price level and a decline in the price of gold relative to silver. France then changed to a *de facto* silver standard, and persuaded Belgium, Switzerland, and Italy to join it in 1865. (Italy's participation ended the following year during the war with Austria, when Italy adopted a *corso forzoso,* or forced circulation of paper money.) The objective was to maintain price stability. Each country defined its currency in terms of a fixed weight of silver (Belgium and Switzerland already used the franc, and Italy defined its new lira as equivalent to the franc). Subsequently Spain, Serbia, and Rumania joined the Union, defining their currencies as equal to the franc. But within a few years, as a result of the discovery of new silver deposits, the relative prices of gold and silver reversed themselves, and the nations of the Latin Monetary Union found themselves flooded with inflows of cheap silver. Rather than allow the inflation of prices that would result, they restricted their purchases of silver and eventually eliminated them altogether, returning to a pure gold standard.

Meanwhile, the first nation after Britain officially to adopt the gold standard was the new German Empire. In the wake of victory over France in the Franco-Prussian War, Bismarck, the German chancellor, extracted from the defeated nation an indemnity of 5 billion francs, unprecedented in history. On the basis of this windfall the government adopted a new money of account, the gold mark, and established the Reichsbank as its central bank and sole issuing agency. In view of Germany's increasing weight in international commerce, other nations joined the movement to the gold standard.

Prior to the Civil War the United States was technically on a bimetallic standard. During the war both North and South issued fiat paper money; the Confederate issues, of course, eventually became worthless, but the northern "greenbacks" continued to circulate, although at a discount to gold. In 1873 Congress passed a law declaring that the greenbacks would be redeemed in gold beginning in 1879. Concurrently, a long decline in prices that began in 1873 led to agitation by both farmers and silver producers against the "crime of '73," and demands for unlimited coinage of silver, but these were resisted. In effect, the United States was on a gold standard from 1879, although Congress did not legally adopt the gold standard until 1900.

Russia had been on a nominal silver standard throughout the nineteenth century but in fact, because of the precarious financial position of the government, had resorted to large issues of unredeemable paper money. In the 1890s, during the industrialization drive undertaken by the Minister of Finance, Count Witte, while the Russian government was borrowing large sums from France, Witte decided the country should go on the gold standard, which it did in 1897. In the same year Japan, having extracted a large indemnity from China after the 1895 war, used the proceeds to create a gold reserve in the Bank of Japan and officially adopt the gold standard. Thus, by the beginning of the twentieth century virtually all important trading nations had adopted the international gold standard. It endured for less than two decades.

International Migration and Investment

In addition to the freer movement of commodities symbolized by the free trade era, a great increase in the international movement of people and capital, the factors of production other than land, also occurred in the nineteenth century. Since international migration was dealt with in Chapter 8, a brief recapitulation is sufficient here.

Some international migration took place within Europe, but the most significant movement involved overseas migration. In the course of the century some 60 million persons left Europe for overseas destinations. The overwhelming majority went to countries with abundant land. The United States alone took 35 million and the newly settled areas of the British Empire another 10 million. About 12 or 15 million went to Latin America. The British Isles provided the largest number of emigrants; altogether about 18 million English, Welsh, Scots, and Irish settled abroad, mainly in the United States and the British dominions. German-speaking emigrants went to both the United States and Latin America. The latter also obtained many new citizens from Spain and Portugal. The late nineteenth and early twentieth centuries witnessed a large migration from Italy and eastern Europe. Italians went to the United States, but also to Latin America, especially Argentina. Emigrants from Austria-Hungary, Poland, and Russia went mainly to the United States. Some of the emigrants eventually returned to their native countries, but the vast majority remained overseas. Overall, this vast migration had beneficial effects; it relieved population pressures in the countries supplying the emigrants, thus lessening downward pressure on real wages; and it provided the resource-rich, labor-short countries to which they went with a supply of willing workers at wages higher than they could have obtained in their native lands. Finally, by means of human and cultural as well as economic ties, it promoted the integration of the international economy.

The export of capital, or foreign investment, further strengthened the integration of the international economy. Although foreign investment had occurred in the eighteenth century and before, it reached unprecedented magnitudes in the nineteenth and early twentieth centuries. It is useful to begin consideration of foreign investment in terms of *sources* (or resources), *motives,* and *mechanisms.*

In general, the resources available for investment abroad (just as for domestic investment) resulted from the tremendous increases in wealth and income generated by the application of new technologies. But unlike domestic investment, foreign investment requires special sources of funds generated by foreign trade and payments. Broadly speaking, there are two main categories of funds (gold or foreign exchange) that can be used for international investment: those arising from an export balance of commodity trade, and those arising from such "invisible" exports as shipping services, earnings from international banking and insurance, emigrant remittances, and the interest and dividends on previous foreign investments. These sources can operate in different combinations in different cases, as we will see.

The main motive for foreign investment is the expectation (not always realized) on the part of the investor of a higher rate of return abroad than at home.

The mechanisms for foreign investment consist of a host of institutional arrangements for transferring funds from one country to another: markets for foreign exchange, stock and bond markets, central banks, private and joint-stock investment banks, brokers, and many others. Most of these special institutional arrangements, although existing earlier, grew greatly during the nineteenth century.

Great Britain—or, more accurately, the private investors of Great Britain—was far and away the largest foreign investor before 1914. At the latter date British investments abroad amounted to about 4 billion pounds sterling (approximately $20 billion at current values), or 43 percent of the world total. This situation existed even though for most of the century Britain had a so-called unfavorable balance of trade, that is, it imported goods of a greater value than it exported. Thus, for Britain, the sources of its foreign investments consisted almost entirely of invisible exports. At the beginning of the century the earnings of Britain's merchant marine, the world's largest, accounted for most of its favorable balance of payments (not trade), and remained important to the end. Increasingly, however, earnings from international banking and insurance, and especially from previous foreign investments, contributed to the surplus. Indeed, after 1870 earnings from previous investments provided funds to cover all new investments, with a sizable surplus left to finance the deficit in the balance of commodity trade.

Before about 1850 British investors purchased government bonds of several European countries and invested in private enterprises there, especially the early French railways. They also bought the securities of American state governments, then engaged in large-scale programs of internal improvements (canals and railways), and those of Latin American countries as well. The revolutions of 1848 on the European continent deterred British investors from much further investment there. Instead, they turned their attention to American railways, mines, and ranches (American cowboys were financed in large part by British, especially Scottish, capital), to similar investments in Latin America, and, above all, to the British Empire. In 1914 the self-governing dominions accounted for 37 percent of British foreign investments and India for another 9 percent; the United States accounted for 21 percent and Latin America for 18 percent. Only 5 percent of Britain's total foreign investments were in Europe.

France (or the French) was the second largest foreign investor, with total investments in 1914 of more than 50 billion francs (about $10 billion). France actually began the century by borrowing abroad, mainly in Britain and Holland, to pay the large indemnity imposed by the Allies after the defeat of Napoleon. As noted, British capitalists also helped finance some of the early French railways. But France quickly established a large export surplus in commodity trade, which provided the bulk of the resources for foreign investment until the 1870s. After that the earnings on previous investments, as with the British, more than financed new investments.

In the first half of the century the French invested chiefly in their near neighbors: the securities of both revolutionary and reactionary governments in Spain, Portugal, and the several Italian states (the loans to the defeated parties were forfeited, of course); bonds of the new government of Belgium after the successful revolution of 1830; mines and other industrial enterprises in Belgium, both before and after 1830; and similar but smaller investments in Switzerland, Austria, and the German states, especially western Germany. Between 1851 and about 1880 French investors and engineers took it upon themselves to build the railway networks of much of southern and eastern Europe. They also invested in industrial enterprises in the same areas, and financed the perennial government deficits of the countries located there, as well as the Ottoman Empire and Egypt—to their subsequent sorrow, when both of those polities declared partial bankruptcy in 1875–76. After the Franco-Russian alliance of 1894 French investors, with the active encouragement of their government (and even before, without such encouragement), invested huge sums in both public and private Russian securities—also to their sorrow, when the Bolshevik government of V. I. Lenin repudiated all debts, public and private, incurred under the tsarist regime.

In 1914, at the outbreak of World War I, fully one-fourth of all French foreign investment was in Russia. About 12 percent each was in Mediterranean Europe (Iberia, Italy, Greece), the Near East (Ottoman Empire, Egypt, Suez), and Latin America, with smaller amounts in the United States, the Scandinavian countries, Austria-Hungary, the Balkans, and elsewhere. Unlike the British, the French put less than 10 percent of their investments into French colonies. Overall, the French contribution to the economic development of Europe was substantial, but as a result of wars, revolutions, and other natural and manmade disasters, especially the enormous disaster of World War I, the investors and their heirs paid dearly.

Germany presents the interesting case of a nation that made the transition from net debtor to net creditor in the course of the century. Disunited and poor at the beginning of the century, the German states had few foreign debts and even fewer foreign credits. In the middle decades of the century the western provinces benefited from an inflow of French, Belgian, and British capital; this capital helped develop powerful industries and a booming export surplus that provided the funds with which Germany repatriated the foreign capital and accumulated investments abroad. Most of these investments were in Germany's poorer neighbors to the east and southeast (including its ally, the Habsburg Monarchy), although Germans also had scattered investments in the United States, Latin America, and elsewhere (including miniscule amounts in the African and Pacific colonies). The German government, like the French, sometimes tried to use private foreign investment as a weapon of foreign policy; in 1887 it closed the Berlin stock exchange to Russian securities, and later it urged the Deutsche Bank to undertake the Anatolian (the so-called Berlin-to-Baghdad) Railway.

The smaller developed nations of western Europe—Belgium, the Netherlands, and Switzerland—all of which had benefited from foreign investment in their economies during the century, had likewise become net creditors by the end of the century. In 1914 their combined foreign investments amounted to about 6 billion dollars, almost as much as Germany's. Austria, the western half of the Habsburg Monarchy, invested in Hungary and also in the Balkans, although on balance the empire was a net debtor.

Of the recipients of foreign investment the United States was by far the largest (Fig. 11-4). As indicated, foreign, especially British, capital helped build railways, open up mineral resources, finance cattle ranches, and support numerous other endeavors. After the Civil War, however, and especially from the late 1890s, American investors began to purchase foreign securities, and even more important, American corporations began to invest directly abroad in a variety of industrial, commercial, and agricultural operations. Most of these investments were in the Western Hemisphere (Latin America and Canada), but some were in Europe, the Near and Middle East, and East Asia. In 1914, when total foreign investments in the United States amounted to slightly more than 7 billion dollars, American investors had invested almost half as much abroad. In the next four years of World War I, as a result of American loans to the Allies, the United States became the world's largest creditor nation.

Within Europe the largest single recipient of foreign investment was Russia. The Russian railway network, like the American, was built largely with foreign capital, which was channeled into both private securities (stocks and bonds) and government and government-guaranteed bonds. Foreigners, especially foreign banks, also invested heavily in Russian joint-stock banks, and in the great metal-lurgical enterprises of the Donbas, Krivoi Rog, and elsewhere. The largest borrower of all, however, was the Russian government, which used the money not only to build railways but also to finance its army and navy. The largest investors were the French, but Germans, British, Belgians, Dutch, and others also took part. After 1917, of course, the investors lost everything.

Most of the nations of Europe borrowed at one time or another during the nineteenth century. As indicated, Germany and some of the smaller developed nations made the transition from net debtor to net creditor. Of those that did not, the record for both productive uses of the funds and repayment was poorest in the

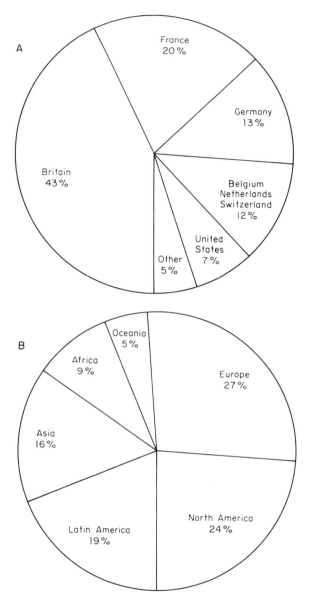

FIGURE 11–4.   Distribution of foreign investments in 1914: (A) by investing countries; (B) by recipients.

Mediterranean countries and those of southeastern Europe. Frequently the proceeds of both private investments and government loans were used wastefully and at times corruptly. As with domestic investment, a foreign investment, to contribute to economic development, must generate a stream of income sufficient to pay a positive rate of return and eventually to repay the original investment.

In brilliant contrast to the poor record of many investments in southern and eastern Europe (and the Ottoman Empire, Egypt, and North Africa), most investments in the Scandinavian countries not only paid for themselves but made positive contributions to the development of the economies in which they were made. Indeed, although the absolute amounts were relatively small, on a per capita basis foreign investments in Sweden, Denmark, and Norway were the largest in Europe. The amounts borrowed were invested wisely, and along with the high educational attainments of the populations of those countries, should be credited with the rapid development of their economies in the late nineteenth century.

Like the Scandinavian countries, Australia, New Zealand, and Canada had large foreign investments in relation to the size of their populations, which helps account for their high growth rates and high standards of living at the beginning of the twentieth century. By 1914 Canada had received the equivalent of $3.85 billion (1914 values), mostly from Great Britain, although U.S. citizens and firms had invested about $900 million there. Australia had $1.8 billion and New Zealand about $300 million—more than 95 percent in both cases from Great Britain. The greater part of the funds in all three cases was invested in public (government) securities, and went to finance social overhead capital (railways, port facilities, public utilities, etc.), although substantial sums also went into mining in Australia and Canada. This pattern of foreign investment permitted domestic investment to flow into directly productive activities in the most promising sectors of the economy. In view of the sparse populations and large land areas of all three countries, it is not surprising that they specialized in the production of goods requiring little labor in proportion to land: wool (with its by-product mutton) in Australia and New Zealand, and wheat in Canada. These products found ready markets in Europe, especially Britain, and accounted for the greater part of these countries' exports. Australia also exported some wheat and crude metals, and Canada exported metals, timber, and other forest products. With relatively high per capita incomes, all three countries developed domestic service industries and some manufacturing capacity, but remained dependent on Europe, mainly Britain, for most manufactured consumer and especially capital goods. (By the beginning of the twentieth century, however, the United States had replaced Britain as Canada's largest foreign market and supplier.)

Investments in Latin America and Asia, although substantial in total, were much smaller in relation to the populations of the recipient nations than the countries just considered. Moreover, they did not have the impressive amounts of human capital with which to work as did the others, and the institutional structures of their economies (except for Japan's) were not conducive to economic development. In these areas, and in Africa to an even greater degree, the principal result of foreign investment was development of sources of raw materials for European industries, without transformation of the internal structure of the economy. In 1914 foreign investments amounted to approximately $8.9 billion in Latin America, $7.1 billion in Asia, and slightly more than $4 billion in Africa. In every case Great Britain was the largest single source of funds, accounting for 42 percent in Latin America, 50 percent in Asia, and more than 60 percent in Africa.

A slightly more detailed examination of British investments in Latin America will provide a better understanding of the significance of foreign investment for the less developed countries in general, and for the world economy as a whole. Total British investments in the region rose from less than £25 million in 1825 to almost £1200 million in 1913. In the latter year Argentina was by far the largest recipient, with over 40 percent of the total, followed by Brazil with 22 percent and Mexico with 11 percent. Chile, Uruguay, Cuba, Colombia, and others received lesser amounts, but there was no country without some British investment. Of the total, almost 38 percent had been invested in government securities and a further 16 percent (i.e., more than 50 percent altogether) in railway bonds and similar securities. Most of these funds, as in Australia, New Zealand, and Canada, went to build railways and other infrastructure investments. Of the direct foreign investments (i.e., those in which the investor controlled the use of the funds) the largest part, again, went into railways, followed by public utilities (gas, electricity, waterworks, telephone and telegraph, tramways, etc.), financial institutions (banks and insurance companies), raw materials production (coffee, rubber, minerals, nitrates), miscellaneous industrial and commercial ventures, and shipping companies. In other words, except for the relatively small amounts invested directly in raw material production, most of the foreign investment provided infrastructure and superstructure to enable the dependent economies to participate in the international economy; production of commodities for both home consumption (mainly foodstuffs) and export (mainly raw materials, but also some foodstuffs) was left to the domestic population of landlords and peasants or landless laborers. Under these circumstances the countries of Latin America exchanged their primary products for European and American manufactures, and in doing so most of them became dependent on one or a very few staple commodities: wheat, meat, hides, and wool from Argentina, coffee and rubber from Brazil, nitrates and copper from Chile, tin from Bolivia, coffee from Colombia and Central America, and so on. They did not, as did the Scandinavian countries, for example, begin processing their own raw materials and so export a higher value-added product. Later critics of the system blamed the foreign investors, and their governments, for this state of affairs. In reality, most of the blame should be laid on the archaic social structures and political systems of the countries themselves.

The Revival of Western Imperialism

The vast continents of Asia and Africa participated only minimally in the commercial expansion of the nineteenth century until forced to do so by the military might of the West. Although parts of Asia, notably India and Indonesia, had been open to European influence and conquest since the beginning of the sixteenth century, much of the continent remained in isolation. The vast and ancient empire of China, as well as Japan, Korea, and the principalities of Southeast Asia, attempted to remain aloof from Western civilization, which they regarded as inferior to their own. They refused to accept Western diplomatic representatives, excluded or persecuted Christian missionaries, and allowed only a trickle of commerce with the West. Most of Africa, located within the tropics, had a climate oppressive to Europeans and a host of unfamiliar, frequently lethal diseases. It had few navigable rivers, making the interior largely inaccessible. The virtual absence of organized political states of the European variety and the low level of economic development made it unattractive to European merchants and entrepreneurs. Nevertheless, in spite of these negative features, a concatenation of events led inexorably to the involvement of both Asia and Africa in the evolving world economy before the end of the nineteenth century.

Africa (Fig. 11-5)

Cape Colony, at the southern tip of Africa, had been settled by the Dutch in the midseventeenth century as a victualing station for the East Indiamen on their way to and from Indonesia. The British captured it during the Napoleonic Wars, and afterwards encouraged British settlement. British policies, especially the abolition of slavery throughout the empire in 1834 and efforts to ensure more humane treatment for natives, angered the Boers or Afrikaaners (descendants of the Dutch colonists). To escape British interference the Boers began their Great Trek to the north in 1835, creating new settlements in the region between the Orange and Vaal rivers (which became the Orange Free State), north of the Vaal (the Transvaal, which became the South African Republic in 1856), and on the southeast coast (Natal). Despite the Boers' attempts to isolate themselves from the British, conflict continued throughout the century. In addition to conflicts between the British and the Boers, both groups clashed frequently with African tribes, encounters that ended sooner or later in defeat for the natives. Some tribes were almost exterminated, and those that survived were reduced to a state of servitude not far from slavery.

At first both Boer and British settlements were primarily agrarian, but in 1867 the discovery of diamonds led to a great influx of treasure seekers from all over the world. In 1886 gold was discovered in the Transvaal. These events completely altered the economic basis of the colonies and intensified political rivalries. They also helped the rise to power of one of the most influential persons in African history, Cecil Rhodes (1853–1902). Rhodes, an Englishman, came to Africa in 1870 at the age of seventeen and quickly made a fortune in the diamond mines. In 1887 he organized the British South African Company, and in 1889 secured a charter from the British government granting it extensive rights and governing power over a vast territory north of the Transvaal, later called Rhodesia.

Not content with mere profits, Rhodes took an active role in politics and became an ardent spokesman for imperialist expansion. He entered the Cape Colony legislature in 1880 and became prime minister of the colony ten years later. One of his major ambitions was to build a railway "from the Cape to Cairo"—all on British territory. President Kruger of the South African Republic would not join a South African union and also refused permission for the railway to cross the Transvaal. Rhodes thereupon hatched a plot to overthrow Kruger and annex his country. The

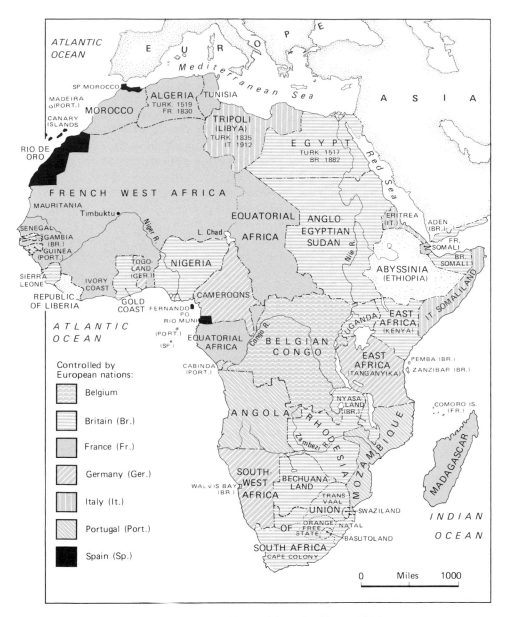

FIGURE 11–5. The partition of Africa to 1914.

plot failed; the British government in London denied any knowledge of the conspiracy and forced Rhodes to resign. It genuinely wished to avoid war with the Boers, but extremists on both sides pushed the issue to its fateful conclusion. In October 1899 the South African or Boer War began.

The British, who had only 25,000 soldiers in South Africa when the war began, suffered several early defeats, but eventually rallied with reinforcements and overran and annexed both the Transvaal and the Orange Free State. Soon afterward the British government changed its policy from repression to conciliation, restored self-government, and encouraged the movement for union with Cape Colony and Natal, which the British had earlier annexed. In 1910 the Union of South Africa joined Canada, Australia, and New Zealand as a fully self-governing dominion within the British Empire.

Before 1880 the only European possession in Africa, apart from British South Africa and a few coastal trading posts dating from the eighteenth century or earlier, was French Algeria. Charles X undertook to conquer Algeria in 1830 in an attempt to stir up popular support for his regime. The attempt came too late to save his throne and left a legacy of unfinished conquest to his successors. Not until 1879 did civil government replace the military authorities. By then the French had begun to expand from their settlements on the African west coast. By the end of the century they had conquered and annexed a huge, thinly populated territory (including most of the Sahara Desert), which they christened French West Africa. In 1881 border raids on Algeria by tribesmen from Tunisia furnished an excuse to invade Tunisia and establish a "protectorate." The French rounded out their North African empire in 1912 by establishing a protectorate over the larger part of Morocco (Spain claimed the small northern corner) after lengthy diplomatic negotiations, especially with Germany.

Meanwhile more momentous events took place at the eastern end of Islamic Africa. The opening of the Suez Canal by a French company in 1869 revolutionized world commerce. It also endangered the British "lifeline" to India—or so it seemed to the British. Great Britain had not participated in building the canal and, in fact, had actually opposed it. But once the canal opened it became a cardinal tenet of British foreign policy to seek control over both it and its approaches in order to prevent them from falling into the hands of an unfriendly foreign power. This purpose was fortuitously favored by the financial difficulties of the khedive (king) of Egypt. The khedive and his predecessors, in an effort to build Egypt up to the status of a great power, had incurred enormous debts to European investors (mainly French and British) for such purposes as an abortive attempt at industrialization, the construction of the Suez Canal, and an attempted conquest of Sudan. This financial stringency enabled Benjamin Disraeli, the British prime minister, to purchase on behalf of the British government the khedive's shares in the canal company at the end of 1875. In an effort to bring some order into the tattered finances of the country (it suspended payments on its debt in 1876) the British and French governments appointed financial advisers who soon constituted the effective government. Egyptian resentment of foreign domination resulted in widespread riots and the loss of European lives and property. To restore order and protect the canal the British bombarded Alexandria in 1882 and landed an expeditionary force.

The British prime minister, once again the Liberal Gladstone, assured the Egyptians and the other great powers (who had been invited to participate in the occupation but declined) that the occupation would be temporary. Once in, however, the British found to their chagrin that they could not easily or gracefully withdraw. Besides continued nationalist agitation, the British inherited from the government of the khedive the latter's unfinished conquest of Sudan. In pursuing this objective, which seemed to be justified by the importance of the Upper Nile to the Egyptian economy, the British ran head-on into the French, who were expanding eastward from their West African possessions. At Fashoda in 1898 rival French and British forces faced one another with sabers drawn, but hasty negotiations in London and Paris prevented actual hostilities. At length the French withdrew, opening the way for British rule in what became known as the Anglo-Egyptian Sudan.

One by one the Turkish sultan's nominal vassal states along the North African coast has been plucked away until only Tripoli remained, a long stretch of barren coastline backed by an even more barren hinterland. Italy, its nearest European neighbor, was a latecomer as both a nation and an imperialist. It had managed to pick up only a few narrow strips on the East African coast and had been humiliatingly repulsed in an attempted conquest of Ethiopia in 1896. It watched with bitter, impotent envy while other nations picked imperial plums. In 1911, having carefully made agreements with the other great powers to gain a free hand, Italy picked a quarrel with Turkey, delivered an impossible ultimatum, and promptly occupied Tripoli. The war was something of a farce, since neither side was vigorous enough to overcome the other. The threat of a new outbreak in the Balkans, however, persuaded the Turks to make peace in 1912. They ceded Tripoli to Italy, and the Italians renamed it Libya.

Central Africa was the last area in the "dark continent" to be opened to European penetration. Its inaccessability, inhospitable climate, and exotic flora and fauna were responsible for its sobriquet and formidable reputation. Prior to the nineteenth century the only European claims in the region were those of Portugal: Angola on the west coast and Mozambique on the east. Explorers such as the Scottish missionary David Livingstone and the Anglo-American journalist H. M. Stanley aroused some interest during the 1860s and 1870s. In 1876 King Leopold of Belgium organized the International Association for the Exploration and Civilization of Central Africa, and hired Stanley to establish settlements in the Congo. Agitation for colonial enterprise in Germany resulted in the formation of the German African Society in 1878 and the German Colonial Society in 1882. A reluctant Bismarck, for domestic political reasons, allowed himself to be converted to the cause of colonialism. The discovery of diamonds in South Africa stimulated exploration in the hope of similar discoveries in central Africa. Finally, the French occupation of Tunis in 1881 and the British occupation of Egypt in 1882 set off a scramble for claims and concessions.

The sudden rush for territories created frictions that might have led to war. To head off this possibility, and incidentally to balk British and Portuguese claims, Bismarck and Jules Ferry, the French premier, called an international conference on African affairs to meet in Berlin in 1884. Fourteen nations including the United States sent representatives. The conferees agreed on a number of pious resolutions, including one calling for the suppression of the slave trade and slavery, which still flourished in Africa. More important, they recognized the Congo Free State headed by Leopold of the Belgians, an outgrowth of his International Association, and laid the ground rules for further annexations. The most important rule provided that a nation must effectively occupy a territory to have its claim recognized.

In this fashion the dark continent was carved up and made to see the light. Before the outbreak of World War I only Ethiopia and Liberia, established by emancipated American slaves in the 1830s, retained their independence. Both were nominally Christian. Annexation was one thing, however, effective settlement and development quite another. The African subjects would have to wait a long time before receiving the fruits, if any, of European tutelage.

Asia (Fig. 11-6)

Internal decay had seriously weakened the Manchu dynasty, which had ruled over China since the middle of the seventeenth century. That gave Westerners the opportunity to force their way into the empire from which they had for so long been excluded. British commercial interests provided the initial occasion for intervention. Chinese tea and silks found a ready market in Europe, but British traders could offer little in exchange until they discovered that the Chinese had a marked taste for opium. The Chinese government forbade its importation, but the trade flourished by means of smugglers and corrupt customs officials. When one honest official at Canton seized and burned a large shipment of opium in 1839, the British traders demanded retaliation. Lord Palmerston, the foreign secretary, informed them that the government could not intervene for the purpose of permitting British subjects to violate the laws of the country with which they traded, but the military and diplomatic representatives on the spot disregarded these instructions and took punitive action against the Chinese. Thus began the Opium War (1839–42), which ended with the dictated Treaty of Nanking. Under it China gave Britain the island of Hong Kong, agreed to open five more ports to trade under consular supervision, established a uniform 5 percent import tariff, and paid a substantial indemnity. The opium trade continued.

The ease with which the British prevailed over the Chinese encouraged other nations to seek equally favorable treaties, which were accordingly granted. Such a show of weakness by the Chinese government provoked demonstrations that were both antigovernment and antiforeign and led to the Taiping Rebellion (1850–64). Government forces eventually defeated the rebels, but in the meantime the general lawlessness gave Western powers another excuse for intervention. In 1857–58 a joint Anglo-French force occupied several principal cities and forced more concessions, in which the United States and Russia also participated.

China's history for the remainder of the nineteenth century followed a depressingly similar pattern. Concessions to foreigners led to fresh outbreaks of antiforeign violence and lawlessness, leading in turn to further foreign reprisals and concessions. In the end, China avoided complete division by the great powers only because of great power rivalry. Instead of outright partition, Britain, France, Germany, Russia, the United States, and Japan contented themselves with special treaty

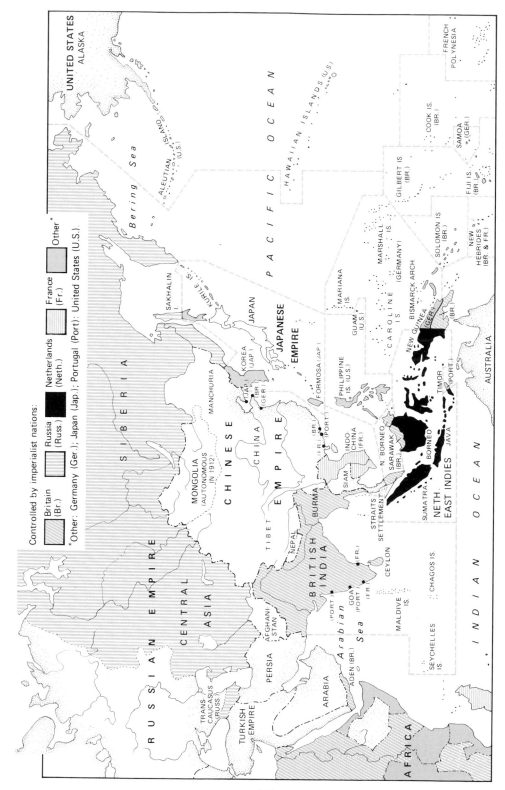

Controlled by imperialist nations:

Britain (Br.)

Russia (Russ.)

Netherlands (Neth.)

France (Fr.)

Other*

* Other: Germany (Ger.); Japan (Jap.); Portugal (Port); United States (U.S.).

FIGURE 11–6. Imperialism in Asia and the Pacific to 1914.

ports, spheres of influence, and long-term leases of Chinese territory. At the initiative of the American secretary of state, John Hay, the great powers agreed in 1899 to follow an "open door" policy in China by not discriminating against the commerce of other nations in their own spheres of influence.

Continued humiliations resulted in a final desperate outburst of antiforeign violence known as the Boxer Rebellion (1900–1901). "Boxers" was the popular name given to the members of the secret Society of Harmonious Fists, whose aim was to drive all foreigners from China. In uprisings in several parts of the country they attacked Chinese converts to Christianity and murdered hundreds of missionaries, railroad workers, businessmen, and other foreigners, including the German minister to Peking. The first attempts of British and other military forces to occupy Peking were repulsed. A second and larger joint expedition took the capital, meted out severe reprisals, and exacted further indemnities and concessions. Thereafter the Chinese Empire was in a state of almost visible decay. It succumbed in 1912 to a revolution led by Dr. Sun Yat-sen, a Western-educated physician, whose program was "nationalism, democracy, and socialism." The Western powers did not attempt to interfere in the revolution, but neither were they worried. The new Republic of China remained weak and divided, its hopes of reform and regeneration long postponed.

Korea in the nineteenth century was a semiautonomous kingdom under nominal Chinese rule, although the Japanese had long had claims there. The bitter rivalry between China and Japan for predominance, as well as the general poverty of the country, discouraged Western diplomats and traders. Although Korea had been the principal cause of the war between China and Japan in 1894, the treaty of 1895 ending it did not result in Japanese annexation. Japan remained content with China's recognition of Korea's "independence." After the defeat of Russia in 1905 and a series of rebellions against Japanese-imposed puppets, Japan formally annexed Korea in 1910.

Indochina is the name frequently given to the vast peninsula of Southeast Asia because the culture of the area is essentially a mixture of classical Indian and Chinese civilizations. During the nineteenth century the British, operating from India, established control over Burma and the Malay states and eventually incorporated them into the empire. French missionaries had been active in the eastern half of the peninsula since the seventeenth century, but in the first half of the nineteenth they were subjected increasingly to persecution, giving the French government an excuse for intervention. In 1858 a French expedition occupied the city of Saigon in Cochin China, and four years later France annexed Cochin China itself. Once established on the peninsula the French found themselves involved in conflict with the natives, which obliged them to extend their "protection" over ever-larger areas. In the 1880s they organized Cochin China, Cambodia, Annam, and Tonkin into the Union of French Indochina, to which they added Laos in 1893.

Thailand (or Siam, as it was called by Europeans), between Burma on the west and French Indochina on the east, had the good fortune to remain an independent kingdom. It owed its independence to a series of able and enlightened kings, and to its position as a buffer between French and British spheres of influence. Although it was opened to Western influence by gunboat treaties, like most of the rest of Asia, its rulers reacted with conciliatory gestures and at the same time tried to learn from the West and modernize their kingdom. Few other non-Western nations were so fortunate.

Explanations of Imperialism

Asia and Africa were not the only areas subject to imperial exploitation, nor were the nations of Europe the only ones to engage in it. Once Japan had adopted Western technology it pursued imperialistic policies much like those of Europe. The United States, despite strong domestic criticism, embarked on a policy of colonialism before the end of the century. Some British dominions were far more aggressively imperialistic than the mother country itself. The expansion of South Africa, for example, took place mostly through South African initiative and frequently against the wishes and explicit instructions of the government in London. British annexation of southeastern New Guinea in 1884, after the Dutch had claimed the western half and the Germans the northeast, was a direct result of the agitation of the Queensland government in Australia.

A distinction is sometimes made between imperialism and colonialism. Thus, neither Russia nor Austria-Hungary had overseas colonies, but both were clearly empires in the sense that they ruled over alien peoples without their consent. The imperial powers did not establish colonies in China, yet China was clearly subject to imperial control. The countries of Latin America experienced no new attempts at conquest by outside powers, but it was frequently alleged that they constituted part of the informal empires of Britain and the United States as a result of economic dependence and financial control.

The causes of imperialism were many and complex. No single theory explains all cases. One of the most popular explanations of modern imperialism involves economic necessity. In fact, modern imperialism has been called "economic imperialism," as if earlier forms of imperialism had no economic content. One such explanation goes as follows: (1) competition in the capitalist world becomes more intense, resulting in the formation of large-scale enterprises and the elimination of small ones; (2) capital accumulates in the large enterprises more and more rapidly, and since the purchasing power of the masses is insufficient to buy all the products of large-scale industry, the rate of profit declines; and (3) as capital accumulates and the output of capitalist industries goes unsold, the capitalists resort to imperialism to gain political control over areas in which they can invest their surplus capital and sell their surplus products.

Such is the essence of the Marxist theory of imperialism, or actually the Leninist theory, for Marx did not foresee the rapid development of imperialism even though he lived until 1883. Building on the foundation of Marxist theory and in some cases modifying it, Lenin published his theory in 1915 in his widely read pamphlet, *Imperialism, the Highest Stage of Capitalism.*

Lenin was by no means the first person to advance an economic interpretation of imperialism. He borrowed heavily from John A. Hobson, the liberal British critic of imperialism, who in turn adopted, in revised form, many of the arguments of the advocates of imperialism. One such was Captain A. T. Mahan, an American naval officer and author whose dictum was "trade follows the flag." Jules Ferry, a French politician who twice became prime minister, was chiefly responsible for the largest colonial acquisitions of France. Interestingly, Ferry did not use economic arguments in defending his actions before the French Assembly; instead, he stressed French prestige and military necessity. Only after he had been permanently retired from office did he write books in which he emphasized the economic gains France would supposedly realize from its colonial empire.

The advocates of imperialism argued that, in addition to offering new markets and outlets for surplus capital, colonies would provide new sources of raw materials and serve as outlets for the rapidly growing populations of the industrial nations. The argument that colonies would serve as outlets for surplus population is easily seen to be fallacious. Most colonies were located in climates that Europeans found oppressive. Most emigrants preferred to go to independent countries, such as the United States and Argentina, or to the self-governing territories of the British Empire. It is true that colonies did in some cases furnish new sources of raw materials, but access to raw materials (or to any purchasable commodity) did not require political control. In fact, North and South America and the self-governing dominions of Australasia were the largest overseas suppliers of raw materials for European industry.

The justification of colonies as markets for surplus manufactures was also fallacious. The colonies were neither needed for this purpose nor used for it after they were acquired. Before 1914 little more than 10 percent of French exports went to French colonies, in spite of favored status. The colonies were too sparsely populated and too poor to serve as major markets. Moreover, as in the case of raw materials, political control was not required. British India was indeed a large market, for in spite of its poverty it purchased large quantities of European products—but not from Britain alone. The Germans sold far more to India than in all their own colonies together. France sold more to India than to Algeria. In spite of protective tariffs, the industrial, imperialist nations of Europe continued to trade predominantly with one another. The largest external market for German industry was Britain, and one of the largest markets for British industry was Germany. France was a major supplier and a major customer for both Britain and Germany. The United States was also a large customer and supplier of European countries.

Perhaps the most important argument for imperialism as an economic phenomenon concerned the investment of surplus capital, at least according to Marxist theory. Here again the facts do not substantiate the logic. Britain had the largest empire and the largest foreign investments, but over half of Britain's foreign investments were made in independent countries and the self-governing territories. Less than 10 percent of French investments before 1914 went to French colonies; the French invested mainly in other European countries—Russia alone, itself an imperialist nation, took more than a quarter of French foreign investments. German investments in German colonies were negligible. Some of the imperialist nations were actually net debtors; besides Russia, these included Italy, Spain, Portugal, Japan, and the United States.

If the economic interpetation of imperialism fails to explain the burst of imperialism in the late nineteenth century, how can it be explained? Major responsibility must be assigned to sheer political opportunism, combined with growing aggressive nationalism. Disraeli's conversion to imperialism (he had been an anti-imperialist earlier in his career) was motivated principally by the need to find new issues with which to oppose the Liberal Gladstone. Bismarck encouraged French imperialism as a means of deflecting French ideas for revenge on Germany, but at first he had rejected it for Germany itself; when at last he allowed himself to be persuaded, he did so to strengthen his own political position and deflect attention from social questions in Germany.

Power politics and military expediency also played important roles. Britain's imperial policy throughout the century was dictated primarily by the supposed necessity of protecting Indian frontiers and the "lifeline." This explains the British conquest of Burma and Malaya, Baluchistan and Kashmir, as well as British involvement in the Near and Middle East. The occupation of Egypt, undertaken reluctantly by Gladstone with the promise of early withdrawal, was deemed necessary to protect the Suez Canal. Other nations emulated the successful British, either in the hope of gaining similar advantages or simply for national prestige.

The intellectual climate of the late nineteenth century, strongly colored by Social Darwinism, likewise favored European expansion. Although Herbert Spencer, the foremost popularizer of Social Darwinism, was an outspoken anti-imperialist, others applied his arguments for "survival of the fittest" to the imperial struggle. Theodore Roosevelt spoke grandly of "manifest destiny," and Kipling's phrase, "the lesser breeds without the law," reflected the typical European and American attitude toward the nonwhite races. The historical roots of European racism and ethnocentrism, however, reach far deeper than Darwinian biology. Christian missionary activity itself was an expression of old beliefs in European, or Western, moral and cultural superiority. Throughout their history—at least, until the midtwentieth century—Europeans and Christians have been expansionist and evangelical. In the final analysis, modern imperialism must be regarded as a psychological and cultural phenomenon as well as a political or economic one.

12

Strategic Sectors

In our brief sketch of the patterns of development within individual countries in Chapters 9 and 10, we either completely ignored or merely mentioned a number of features of the development process that should be discussed more fully in a detailed treatment of the history of industrialization. For even a summary treatment such as this one, however, three areas of activity must be discussed in some detail for the process of industrialization to be intelligible—agriculture, finance and banking, and the role of the state in economic affairs.

Agriculture

It has already been pointed out that one of the major structural changes in the economy that occurred in the nineteenth century was a decline in the relative size of the agricultural sector. That does not imply, however, that agriculture ceased to be important; quite the contrary. The prerequisite for a decline in the relative size was an increase in agricultural productivity, with the size of the decline of the former being proportional to the increase of the latter. In other words, the ability of a society to raise its standards of consumption above a bare subsistence level and to shift a significant portion of its labor force into other, potentially more productive activities depends on a prior increase in agricultural productivity. (This statement ignores the possibility of importing food supplies, which most industrializing countries, especially Great Britain, did to some extent in the nineteenth century; but they also had highly productive agrarian sectors.)

An increase in agricultural productivity can contribute to overall economic development in five potential ways:

1. The agricultural sector can supply a surplus population (labor force) to engage in nonagricultural occupations.
2. The agricultural sector can supply foodstuffs and raw materials for the support of the nonagricultural population.
3. The agricultural population can serve as a market for the output of manufacturing industries and service trades.
4. By means of either voluntary investment or taxation, the agricultural sector can furnish capital for investment outside agriculture.
5. By means of agricultural exports, the agricultural sector can furnish foreign

exchange to enable the other sectors to obtain necessary inputs of either capital goods or raw materials that are not available domestically.

It is not necessary for the agricultural sector to perform all five of these functions for a society to develop economically, but it is difficult to imagine a situation in which economic development could occur without the assistance of agriculture in at least two or three of them. And for that to happen, agricultural productivity must increase.

At the beginning of the nineteenth century British agriculture was already the most productive in Europe. This fact was intimately related to the early emergence of British industrialism. Although the agricultural population continued to grow in absolute numbers until the 1850s, it had long provided a surplus for nonagricultural activities, a surplus that became marked in the latter half of the eighteenth and the first half of the nineteenth centuries. (Generally, it was the sons and daughters of farmers, not the farmers themselves, who left the countryside for urban occupations.)

Likewise, British agriculture met the majority of the nation's needs for food and some raw materials, such as wool, and barley and hops for the brewing industry. In the first half of the eighteenth century it even provided a surplus of grain for export; although that disappeared after 1760, British farmers continued to supply the greater part of the national consumption of food, even after the repeal of the Corn Laws. Indeed, the period from the midforties to the midseventies was the era of "high farming," when British agriculture, like British industry, was at its relative peak. Technical improvements—light iron ploughs, steam threshers, mechanical harvesters, and the widespread use of commercial fertilizers—increased productivity even more than the earlier introduction of convertible husbandry and its associated techniques. After about 1873, with the growing inundation of cheap American grain, British farmers cut back on their plantings of wheat, but many of them switched to the production of higher value-added meat and dairy products, frequently using imported grain as feed.

The prosperous agrarian sector also provided a ready market for British industry. Indeed, before the midnineteenth century the nation's rural population constituted a larger market for most industries than did foreign nations. Although there were few instances of agricultural income being invested in industry (except the coal industry, in which wealthy landowners frequently financed the development of mines on their estates), landed wealth did contribute substantially to the creation of social overhead capital: canals and turnpikes in the eighteenth century, and railways in the nineteenth. All in all, therefore, British agriculture played a major role in the rise of British industry.

The role of agriculture on the Continent differed from that in Great Britain, and from one region to another as well. In general, as suggested in Chapters 9 and 10, there was a rather close correlation between agricultural productivity and successful industrialization, with a gradient running from the northwest to the south and east. Agrarian reform was frequently a prerequisite for substantial improvement in productivity. There are many types of agrarian reform, however, and not all produced the intended result.

Basically, agrarian reform involves a change in the system of land tenure. The enclosure movement in England, which resulted in the creation of relatively large compact farms in place of the open field system, could be regarded as a kind of agrarian reform. The French Revolution, which abolished the Old Regime and confirmed France's independent peasant proprietors in the possession of their small farms, was a different type of land reform. French-type reforms were imposed in some of the territories occupied by the French, notably Belgium and the left bank of the Rhine. The Prussian reforms of 1807 and afterwards (see Chapter 9), on the other hand, although they emancipated the serfs, obliged the latter to cede much of their land to their former lords, resulting in the creation of even larger estates. Sweden and Denmark abolished serfdom in the latter part of the eighteenth century and instituted enclosure movements which, by the middle of the nineteenth century, had created a class of substantial peasant proprietors.

Elsewhere, agrarian reform had a less happy outcome. In the Habsburg Monarchy Joseph II attempted to alleviate the burdens on the peasantry in the 1780s, with indifferent results; full emancipation had to await the revolution of 1848. In Spain and Italy half-hearted attempts at agrarian reform ran afoul of the governments' needs for revenue and were effectively jettisoned. The Balkan states inherited their systems of land tenure from the period of Turkish rule, but made no serious attempts to alter them. Small peasant proprietorships—with population growth and no rule of primogeniture to prevent subdivision of properties—characterized Serbia and Bulgaria. Greece and Rumania, on the other hand, although they also had small peasant proprietors, had large estates cultivated by tenant farmers as well. Neither system was conducive to high agricultural productivity.

Imperial Russia had the distinction of undergoing two very different types of agrarian reform within two successive generations. The emancipation of the serfs, carried out reluctantly in 1861 following defeat in the Crimean War, did not fundamentally alter the structure of Russian agriculture. The former serfs, although freed from their lords, now belonged compulsorily to the peasant commune, the *mir;* to leave, they had to obtain a special passport, but even if they left they were still obligated to pay their share of taxes and redemption payments. Techniques remained unchanged, and the cultivable strips were periodically redistributed among families to compensate for changes in family size. Under the circumstances it is small wonder that productivity remained low and peasant unrest increased. In the wake of the revolution of 1905–6, the government abolished further redemption payments and decreed the so-called Stolypin Reform (named for the minister who devised it), which provided for private ownership of land and the consolidation of strips into compact farms. As a result of this "wager on the strong," productivity in Russian agriculture began to rise, but the entire country was soon overwhelmed by war and revolution.

The performance of French agriculture is, at first sight, as contradictory and paradoxical as that of French industry. Although the classic home of the small peasant proprietor, frequently accused of being subsistence-oriented and technically backward, France also had many progressive farmers. In 1882, when *morcellement* (subdivision of property) was at its peak, there were some 4.5 million parcels of ten hectares (twenty-five acres) or less, but these accounted for only 27 percent of the land. They were located mainly in the less fertile south and west. On the other hand, over 45 percent of the land was in holdings of forty hectares or more, mainly in the more fertile north and east. These prosperous farms produced a marketable surplus to feed the growing urban population at steadily rising levels of nutrition. Moreover, in spite of the French peasants' legendary attachment to the soil, more than 5 million persons left agriculture for other employment (as in Britain, mostly the sons and daughters of farmers rather than the farmers themselves). There is also some evidence that savings originating in agriculture were applied to industrial investment, or at least to overhead capital. Finally, the wine industry, which is, after all, a part of agriculture, served as a major source of export earnings.

In Belgium, the Netherlands, and Switzerland agriculture had long been market-oriented. Productivity in all three countries was among the highest on the Continent. In Switzerland agriculture employed an average of 500,000, rising to a maximum of about 650,000 in 1850, and then falling to about 450,000 in 1915; but the *relative* decline was from over 60 percent of the labor force at the beginning of the nineteenth century to about 25 percent in 1915. Similar stories could be told for Belgium and the Netherlands.

Great variety characterized the performance of agriculture in the various German states, and later in the new German Empire. In the southwest, Baden and Württemberg had large numbers of small peasant proprietors like those in France; these were not necessarily inefficient. In the north and east, in Mecklenburg and the Prussian provinces of Pomerania and East and West Prussia, great estates worked by hired labor were the rule; these were not necessarily highly efficient. Traditionally, these large estates had been exporters of grain to western Europe since at least the fifteenth century (see Chapter 5). They continued this role in the nineteenth century, until the large-scale imports of American and Russian grain forced prices down and brought about the return to protection, as recounted in the previous chapter. By that time the German population had grown so large that a surplus for export no longer existed, even if prices had been competitive. Indeed, by the 1890s Germany imported about 10 percent of its cereal consumption.

The emancipation of the serfs in Prussia following the Edict of 1807 did not immediately bring about any great changes. As long as the peasants remained on their holdings they continued to perform their customary obligations and enjoy their customary rights. But with the gradual growth of population and the more rapid increase in the demand for labor in the Rhineland from the midcentury on, the population was substantially redistributed from east to west. The absolute numbers in the agricultural labor force continued to grow until 1914, reaching 10 million in 1908; but as a proportion of the total labor force it fell from 56 percent at midcentury to under 35 percent by 1914.

Agriculture contributed importantly to the economic development of both Denmark and Sweden, though not that of Norway. If, however, one looks at the entire primary sector, which includes forestry and fishing as well as agriculture, the picture is quite different. In all countries the primary sector provided the bulk of the food supply as well as the increased labor force for the other sectors (and, in the cases of Norway and Sweden especially, an increase in the labor force for American agriculture through migration). The primary sector also provided a market for domestic industry and, at least in Sweden where the state built the railways, contributed to capital formation by means of taxation. The most spectacular way in which the primary sectors of the Scandinavian countries contributed to their economic development, however, was by means of exports. As noted in Chapter 10, timber and timber products constituted the greater part of Swedish exports before 1900, and oats were also important in the middle decades of the century. After the decline of the oat trade Sweden exported some meat and dairy products. Timber was also an important component of Norway's exports, but the fisheries were even more so; in 1860 they accounted for 45 percent of commodity exports, and still over 30 percent just before World War I. As already mentioned, virtually all of Denmark's exports consisted of high value-added agricultural products.

Finland, which was ruled by the tsar of Russia as a grand duchy, is sometimes included with the Scandinavian countries. Unlike them, however, it did not experience any substantial structural change in the nineteenth century. It remained predominantly agrarian, with low-productivity agriculture and low average incomes. Its major export was timber, with some woodpulp late in the century.

The Habsburg Monarchy, like Germany, was marked by regional variations. At the beginning of the nineteenth century about three-fourths of the labor force in the Austrian half of the empire (including Bohemia and Moravia) was engaged in agriculture, a still higher proportion in the Hungarian half. By 1870, when the figure for Austria had fallen to around 60 percent, Hungary had just reached the position that Austria held at the beginning of the century. On the eve of the First World War the proportion in Austria proper and Bohemia had fallen below 40 percent, but it was still above 60 in Hungary.

The growth of agricultural output, both total and per worker, seems to have been reasonably satisfactory throughout the century in both halves of the empire. The peasant population constituted an adequate if not a dynamic market for textiles and other consumer goods. The Hungarian half of the monarchy "exported" agricultural produce, especially wheat and flour, to the Austrian half in exchange for manufactures and also capital investments. The failure of the empire as a whole to develop substantial agricultural exports can be attributed mainly to two factors: the difficulties of transportation, and the fact that the domestic market absorbed the bulk of production. Austro-Hungarian agriculture, like Austro-Hungarian industry, faithfully reflected the empire's location between west and east.

As previously indicated, Spain, Portugal, and Italy—to which we may now add Greece, similar in many respects—underwent no meaningful agrarian reform in the nineteenth century. With well over half the population involved in agriculture, even in the early years of the twentieth century, productivity and incomes remained among the lowest in Europe. Such populations could not serve as thriving markets for industry, let alone provide capital to the latter. Although all four countries exported some fruits and wine, for which their climates suited them, they all remained partly dependent on imports for their supplies of bread grain.

The small countries of southeastern Europe, to an even greater extent than those of the Mediterranean, remained mired in backward and unproductive agriculture, which provided neither markets for industry nor much of a surplus of food, raw materials, or labor for urban markets. It did, however, supply a small surplus for export, as described in Chapter 10.

Imperial Russia also remained overwhelmingly rural and agrarian on the eve of the Great War. Nevertheless, agriculture played a somewhat different role in Russia than in southeastern Europe or the Mediterranean. Backward as it was, Russian agriculture managed to feed the Russian people and provide a surplus for export, which proved to be critical for Russia's drive to industrialize in the late nineteenth and early twentieth centuries. It was formerly believed that the export surplus came at the expense of the peasants through heavy taxation ("hunger exports"); but the results of recent research suggest that agricultural productivity and rural living standards were rising, at least after 1885. If this is true, then Russia was on a path of economic development similar to that taken by the nations of western Europe and the United States.

Agriculture played a dynamic role in the process of American industrialization and in the rise of the United States to the position of the world's leading economic power. From the colonial period agriculture supplied in abundance not only the food and raw materials for the nonfarm population, but also the greater part of American exports. The southern colonies sent tobacco, rice, and indigo to Europe in exchange for the manufactured goods needed by the growing colonial economy. New England and the middle colonies exchanged fish, flour, and other foodstuffs with the West Indies for sugar, molasses, and the Spanish silver dollars that ultimately became the basis of the American monetary system. In the first half of the nineteenth century cotton became "king" of exports, with more than 80 percent of total production going abroad, mainly to Lancashire. After the Civil War, with the opening of the trans-Mississippi West by railways and the decline of ocean freight rates, corn and wheat became major export commodities. In that era between 20 and 25 percent of total agricultural production went to foreign markets, compared with only 4 to 5 percent of manufacturing output.

American agriculture was market-oriented from its beginnings; although there were instances of domestic production of household goods and "homespun" clothing, for example, American farmers early relied on dispersed rural craftsmen and small-scale industries for their tools and other manufactured goods. It has been estimated that in the 1830s rural households (farm and village) accounted for more than three-fourths of total consumer expenditure. Although the proportion fell with the growth of the urban population, the absolute amount continued to rise. Before the end of the century such mail-order firms as Sears Roebuck and Montgomery Ward found it profitable to supply the rural population with standardized, mass-produced consumer goods.

The rapid natural increase of the rural population, in particular, also provided a labor force for nonagricultural occupations. This source of labor was supplemented, especially from the 1880s, by emigrants from Europe (most of them also of rural origin), but most of the nonagricultural labor force was still native-born. The rural origins of many notable business leaders (e.g., Henry Ford), politicians, and statesmen (e.g., Abraham Lincoln) are symptomatic in that respect.

Data are not adequate to determine, with any certainty, whether or not American agriculture contributed significantly to the formation of nonagricultural capital; probably it did not. (On the other hand, the large amount of capital embodied by agricultural structures and equipment was almost wholly of agricultural origin.) In any case, it is clear that the agricultural sector of the American economy contributed very positively to the industrial transformation of the United States.

The United States neither experienced nor needed an agrarian reform of the European style, but it did experience an extraordinary stimulus to the agricultural economy in the disposition of the public domain. After the Revolutionary War the federal government acquired title to most of the trans-Appalachian West, and after the Louisiana Purchase and subsequent territorial acquisitions, to most of the trans-Mississippi West. From the beginning the government followed a policy of sales to private individuals (and some companies) in fee simple—in other words, a free market in land. At first, however, the minimum parcel was so large (640 acres) it discouraged individuals of modest means, especially when sales were for cash or short-term credit. The policy gradually evolved, however, toward smaller parcels and lower prices, a tendency that culminated in the Homestead Act of 1862, which granted settlers 160 acres free, provided they lived on and cultivated the land for five years.

In perhaps no other country did agriculture play so vital a role in the process of industrialization as in Japan. At the time of the Meiji Restoration the population was approximately 30 million—a high-density population by Western standards. By the time of World War I the population had grown to more than 50 million, giving it a very high density indeed. In spite of this, and of the scarcity of arable land, Japanese agriculture sufficed to feed the population for most of the prewar period (some rice was imported from the colonies after 1900) and to furnish the greater part of Japanese exports, as indicated in Chapter 10. By means of the land tax of 1873, agriculture also financed the greater part of government expenditure (94 percent in the 1870s, still almost half in 1900) and thus, indirectly, a part of capital formation. In spite of their poverty Japanese peasants constituted the largest market for Japanese industry. Finally, they also furnished labor for that industry; the labor force in agriculture fell from 73 percent in 1870 to 63 in 1914, while the proportion in industry rose from less than 10 to almost 20.

Finance and Banking

The process of industrialization in the nineteenth century was accompanied by a proliferation in the number and variety of banks and other financial institutions necessary to provide the financial services required by the greatly enlarged and increasingly complex economic mechanism. Although all banking systems have certain common features, determined by the functions they perform, systems also differ in structure by nationality, since structure is determined primarily by legislation and by historical evolution unique to each nation. From a broad spectrum of possible forms of interaction between the financial sector and the other sectors of the economy that require its services, one can isolate three type-cases: (1) that in which the financial sector plays a positive, growth-inducing role; (2) that in which the financial sector is essentially neutral or merely permissive; and (3) that in which inadequate finance restricts or hinders industrial and commercial development.

The origins of Britain's banking systems have already been sketched in Chapter 7. (It must be stressed that the English and Scottish banking systems were distinct until the second half of the nineteenth century; the Irish system was also distinct, whereas that of Wales was an adjunct of the English system.) At the beginning of the nineteenth century the Bank of England—in reality, the Bank of London—was still secure in its monopoly of joint-stock banking; the numerous small "country banks" in the provinces were all obliged to use the partnership form of organization, which made them susceptible to financial panics and crises. After an especially severe crisis at the end of 1825 Parliament amended the law to permit other banks to adopt the joint-stock form as long as they did not issue banknotes; and a few years later Parliament passed the Bank Act of 1844, which shaped the structure of British banking until World War I and afterward.

Under the Bank Act of 1844 the Bank of England traded its monopoly of joint-stock banking for a monopoly of note issue. It remained primarily a government bank (although privately owned), providing financial services to the government; increasingly, however. it also became a bankers' bank, and by the end of the century it had consciously accepted the functions of a central bank. Along with the Bank of England, the British banking system (the Bank Act of 1845 extended the provisions of the act of 1844 to Scotland, and effectively amalgamated the two systems) featured a number of joint-stock commercial banks that accepted deposits from the public and lent to business enterprises, generally at short term. The number of these banks, both in London and the country, grew rapidly to the 1870s; thereafter, by merger and amalgamation, the number shrank until by 1914 only about forty remained, five of them headquartered in London but with branches throughout the nation, controlling almost two-thirds of the total assets of the system.

Another feature of the British banking system, the private merchant bankers of London, was much less visible than the other two features. Maintaining a low profile, these private firms, such as N. M. Rothschild & Sons, Baring Brothers, and J. S. Morgan & Co. (Morgan was an American, the father of J. Pierpont Morgan, Sr.), engaged primarily in financing international trade and dealing in foreign exchange, but they also participated in underwriting issues of foreign securities, discussed in Chapter 11, which they listed on the London Stock Exchange. That institution specialized almost entirely in foreign investments, leaving to the provincial stock exchanges the function of raising capital for domestic enterprise.

In addition to the institutions just discussed, Britain had a number of other specialized financial institutions: savings banks, building and loan societies, and so on. Although the resources at their command were not insignificant, they did not play a prominent role in the process of industrialization. Overall, the British banking system responded rather passively to the demands placed on it, neither hastening nor retarding the process of economic development.

The French banking system, like that of England, was dominated by a politically inspired bank doing most of its business with the government, the Bank of France. Created by Napoleon in 1800, it quickly acquired a monopoly of note issue and other special privileges. For a time under Napoleon, and at his insistence, it operated a few branches in provincial cities, but gave them up as unprofitable after Napoleon's fall. Like the Bank of England, it became in effect the Bank of Paris, allowing a few banks of issue on its model to operate in major provincial cities. It successfully blocked all other requests for joint-stock banks presented to the government before 1848, and in the revolutionary upheaval of that year took over the departmental banks of issue as its own branches.

Before 1848 France had no other joint-stock banks and no counterparts to the English country banks. It was, in effect, underbanked, because the provincial notaries who provided some brokerage functions could not possibly fill the role of the missing banks. In an effort to do so, several entrepreneurs formed *commandite* banks in Paris in the 1830s and 1840s. Even these could not fulfill the demand for banking services, and in any case they all succumbed in the financial crisis that accompanied the revolution of 1848.

France did have one other important type of financial institution in the first half of the nineteenth century, however. This was *la haute banque parisienne*, private merchant bankers similar to those of London, whose leading member was De Rothschild *frères*, founded by James (Jacques) de Rothschild, brother of London's Nathan. (They and their three brothers, sons of an eighteenth-century German court Jew, Meyer Amschel Rothschild, established branches of the family bank in Frankfurt, Vienna, and Naples as well as in London and Paris during the Napoleonic era.) As in London, the principal activities of these private banks (they referred to themselves as "merchants") were the finance of international trade and dealings in foreign exchange and bullion, but following the Napoleonic Wars they began to underwrite government loans and other securities, such as those of canal and railway companies.

Following the coup d'état of 1851 and the proclamation of the Second Empire the next year, Napoleon III sought to lessen the dependence of the government on Rothschild and other members of *la haute banque* by creating new financial institutions. He found eager collaborators in the persons of the brothers Emile and Isaac Pereire, former employees of Rothschild who had decided to go their own way. With the blessings of the emperor, in 1852 they founded both the Société Générale de Crédit Foncier, a mortgage bank, and the Société Générale de Crédit Mobilier, an investment bank that specialized in railway finance. Subsequently the government allowed the formation of other joint-stock banks, some of which followed the examples of the Crédit Mobilier (whose operations were patterned, in part, on those of the Société Générale de Belgique; see Chapter 9), and others the example of the English joint-stock commercial banks. The French banks, both private and joint stock, also led the way in promoting French foreign investment. Overall, the French banking system in the first half of the nineteenth century, hobbled by government conservatism and the restrictive policies of the Bank of France, failed to live up to its full potential in promoting the development of the economy; in the second half of the century it was somewhat more expansive, but less so than those of Belgium and Germany.

The origins of the Belgian banking system were briefly sketched in Chapter 9. The Société Générale de Belgique and the Banque de Belgique performed wonders in promoting the industrialization of their small country, but the very latitude of their powers, together with their intense rivalry, led them into difficulty. In 1850 the government created the Banque Nationale de Belgique as a central bank with a monopoly of note issue, freeing the others and those subsequently authorized to get on with ordinary commercial and investment banking functions. Overall, the Belgian banking system gets very high marks for its role in promoting the development of its economy.

The Dutch had fallen far from the commanding position in European finance and commerce that they had occupied in the seventeenth century, but they still had reserves of financial power. When the kingdom of the United Netherlands took the place of the defunct Dutch Republic in 1814, the Nederlandsche Bank took the place of the Bank of Amsterdam, which had been liquidated during the French occupation. In addition, the Dutch financial system included several old established private bankers, led by Hope and Company, whose business consisted mainly of underwriting government loans, and the *kassiers,* money changers and bill brokers.

In the 1850s, after the successes of the Société Générale de Belgique and, more recently, the Crédit Mobilier, Dutch businessmen became convinced they could foster the industrialization of their country with similar institutions. They made four separate proposals for *mobilier* banks in 1856, but the government, relying on advice from the Nederlandsche Bank, rejected them all. In 1863, when four new proposals came in, two from Amsterdam and two from Rotterdam, the government relented and authorized all four. They met with varied fortunes. One of them, an affiliate of the French Crédit Mobilier, quickly overextended itself and had to be liquidated in 1868. The others fared somewhat better and participated in the Dutch industrial build-up that occurred in the last decades of the century.

Switzerland, which developed as a major world financial center in the twentieth century, was much less important before 1914. To be sure, Geneva was one of Europe's key financial centers in the Renaissance, and Swiss private bankers were still important in the eighteenth century. Nevertheless, the bases of Switzerland's subsequent prominence were laid in the nineteenth century. In the 1850s, 1860s, and 1870s numerous new banks were founded on the model of the French Crédit Mobilier, including several of those subsequently famous: the Schweizerische Kreditanstalt (1856), the Eidgenossischen Bank in Bern (Banque Federale Suisse, 1864), and the Schweizerische Bankgesellschaft (Swiss Bank Corporation, 1872, which had roots in a local bank in Winterthur dating from 1862). Two other major banks, the Schweizerische Bankverein and the Schweizerische Volksbank, came about through mergers of banks dating from 1856 and 1869, respectively.

In the first half of the nineteenth century it could not be said that a German banking system existed. The several sovereign states, with their separate monetary and coinage systems, prevented the emergence of a unified financial system. Prussia, Saxony, and Bavaria had monopolistic note-issuing banks (the earliest of them, that of Bavaria, founded in 1835), but these were closely regulated by their governments and catered mainly to government finance. Numerous private banks existed, especially in important commercial centers like Hamburg, Frankfurt, Cologne, Dusseldorf, and Leipzig, and in the Prussian capital, Berlin; but they were concerned primarily with the finance of local and international commerce or, in some instances, with the placement of private fortunes. From the 1840s onward, however, a number of them began to engage in promotional finance, founding and underwriting new industrial enterprises and, especially, railways. This was the harbinger of a new era in German banking.

The distinguishing feature of the German financial system, as it developed in the second half of the century, was the joint-stock "universal" or "mixed" bank, engaged in both short-term commercial credit and long-term investment or promotional banking. Called *Kreditbanken* (misleadingly, as all banking operations involve the use of credit), they took over and extended the promotional operations of the private bankers just mentioned. (In fact, in several instances the *Kreditbanken* were mere extensions of the private bankers themselves.)

The first of these new institutions was the Schaaffhausen'scher Bankverein of Cologne, founded in the revolutionary year 1848. It was something of an anomaly as well as a novelty, however, inasmuch as it was erected on the bankrupt remains of a private bank, Abraham Schaaffhausen and Company; the panic-stricken government in Berlin had been persuaded to depart from its usual practice of forbidding joint-stock charters to banks in order to stem the financial crisis. It required several years to get its affairs in order, and only later did it function as a true *Kreditbank*. Meanwhile the Prussian government reverted to its former policy, and authorized no more joint-stock banks until 1870.

The first conscious example of the new type of bank was the Bank für Handel und Industrie zu Darmstadt, popularly known as the Darmstädter, established in the capital of the Grand Duchy of Hesse-Darmstadt in 1853. Its promoters, private bankers in Cologne, had intended to establish it in that city, but encountered the refusal of the government. They next tried the important financial center of Frankfurt, but the Senate of that free city, dominated by its own powerful private bankers, also rebuffed them. The government of the grand duke, whose capital city was located only a few miles south of Frankfurt, proved more cooperative. The new bank was modeled on, and received both financial and technical assistance from, the French Crédit Mobilier, founded the previous year. From the beginning it operated throughout Germany.

Faced with the refusal of the Prussian government to authorize joint-stock charters for banks, ambitious promoters used the device of the *Kommanditgesellschaft* (similar to the French *société en commandite*), which did not require government authorization. A number of these were established in the 1850s and 1860s, of which the most notable were the Diskonto-Gesellschaft of Berlin and the Berliner Handelsgesellschaft. Meanwhile, some of the smaller German states did not have the Prussian government's antipathy to joint-stock banks, and allowed them to be formed. Finally, in 1869 the North German Confederation, which was the euphemistic name given to the greatly enlarged Prussia after the Austro-Prussian War, adopted a law modeled on those of Britain and France that permitted free incorporation.

With that law, and in the euphoria induced by the Prussian victory over France in 1870, more than one hundred new *Kreditbanken* were created before the crisis of June 1873. The depression that followed winnowed out most of them, the weaker and more speculative ones; then a process of concentration and amalgamation, similar to that in Britain, led to the domination of the financial scene by about a dozen huge banks, with networks of branches and affiliates throughout Germany and abroad. The most famous of these were the "D-banks"—the Deutsche Bank, Diskonto-Gesellschaft, Dresdner, and Darmstädter—all with capitals of more than 100 million marks and headquartered in Berlin. They not only catered to the needs of German industry (it was said that they accompanied enterprises "from the cradle to the grave") but also facilitated the extension of Germany's foreign commerce by providing credit to exporters and foreign merchants.

One further major institutional innovation, the Reichsbank, created in 1875, capped Germany's financial structure. It, too, was in part a consequence of Prussia's victory over France, and the huge indemnity it brought. In name, it was merely a transformation of the Prussian State Bank, but its resources and powers were greatly enlarged. It had a monopoly of the note issue and acted as a central bank. As such, it was able to support the *Kreditbanken* in difficult times, and thus made it possible for them to take greater risks than they normally would have.

The development of German banking in the second half of the nineteenth century was one of the most striking accompaniments—indeed, as some would have it, a cause—of the equally rapid process of industrialization. Too much stress has probably been put on the role of the banks; naturally, many other elements contributed to the success of German industry, and that very success in turn contributed to the success and prosperity of the banking system. The fact remains, however, that the banks did play a prominent role in industrial development; altogether, the German banking system at the beginning of the twentieth century was probably the most potent in the world.

Austria (or the Habsburg Monarchy) acquired its modern banking system at about the same time as Germany did. True, it had created the Austrian National Bank in 1817, but that was a privileged enterprise, like the Banks of England and France, for dealing with the disordered state of public finances after the Napoleonic Wars. It also had some private banks, of which the House of Rothschild was by far the most prominent. (All five Rothschild brothers had been made barons of the Austrian Empire in the 1820s in partial recompense for their role in restoring Austrian state finances.) But the first modern joint-stock bank was the Austrian Creditanstalt, created in December 1855. Its creation was a direct result of the rivalry of the Pereire brothers and the Rothschilds. The Pereires made a bid for it at the same time that they successfully purchased the Austrian State Railways for the Crédit Mobilier, but the Rothschilds, who had been the Habsburgs' "court Jews" since the time of Napoleon, wrested it away from them. It remains today, after numerous transformation, one of the most powerful financial institutions in central Europe.

In addition to the Creditanstalt, a number of other important joint-stock banks were created in Vienna, Prague, and Budapest, as well as smaller banks in provincial towns, but largely for reasons of natural endowment and institutional constraints, they did not exhibit the dynamism of the German banking system.

Although the economy of Sweden was relatively backward in the first half of the nineteenth century, it had a long banking tradition. The Sveriges Riksbank (the forerunner of the National Bank of Sweden), founded in 1656, was in fact the first bank to issue true banknotes. Some private note-issuing banks also dated from the first half of the nineteenth century. Nevertheless, the modern history of banking in Sweden, like that of so many other European countries, dates from the 1850s and 1860s, and derived its inspiration from the example of the Crédit Mobilier. The Stockholms Enskilda Bank, founded in 1856, was the first of the new type in Sweden; it was followed by the Skandinaviska Banken in 1864 and the Stockholms Handelsbank (subsequently the Svenska Handelsbank) in 1871. All three, as well as some smaller provincial banks, engaged in mixed banking (commercial and investment) operations with considerable success. It could be debated whether the successful transformation of the Swedish economy contributed to the prosperity of the banks, or vice versa; but it is clear that the two progressed together.

In the first half of the nineteenth century Denmark had a central bank, the Nationalbank, which was privately owned but government-controlled, and a number of small savings banks. Like Sweden, its modern banking history dates from the 1850s, and like Sweden, it was dominated by three large joint-stock banks, all based in Copenhagen: the Privatbank (1857), the Landsmanbanken (1871), and the Handelsbanken (1873). Norway and Finland were less advanced financially than Denmark and Sweden, but in all four countries the general levels of literacy made the population better able to take advantage of banking facilities.

The Latin nations of the Mediterranean also obtained modern financial institutions in the 1850s and 1860s, but mainly at French initiative and employing French capital. Spain had a bank of issue, the Banco de San Carlos (later renamed the Banco de España), dating from 1782 (and founded by a Frenchman), but like other banks of its type it was concerned primarily with government finance. The important commercial and industrial city of Barcelona also had a bank of issue dating from the 1840s, but it did not engage in promotional activities. The Pereires attempted to establish a Spanish affiliate in 1853, at the time of the Darmstädter promotion, but failed to secure authorization from the reactionary Spanish government of the moment. In 1855, however, after a change of government brought in a ''moderate'' faction, they persuaded the finance minister to introduce a bill in the Cortes authorizing the government to charter banking enterprises on the model of the Crédit Mobilier. Early the following year they organized the Sociedad General de Crédito Mobiliario Español.

The law authorizing the Crédito Mobiliario Español permitted the government to charter similar institutions without further authorization from the Cortes. Other French entrepreneurs lost no time in presenting themselves; almost simultaneously four institutions modeled on the Crédit Mobilier sprouted on Spanish soil—three of them relying on French capital, including one sponsored by the Rothschilds. All of them participated in the feverish outburst of railway promotion and construction that ensued, and several, especially the Crédito Mobiliario, engaged in other industrial and financial enterprises, including Spain's first modern insurance company. Indeed, what little economic development Spain achieved in the nineteenth century was largely a result of the activities of these French-inspired credit companies.

Shortly after obtaining the charter for the Crédito Mobiliario Español the Pereires contracted with the Portuguese government for a similar company in Lisbon. The upper chamber of the Portuguese parliament refused to ratify the agreement, however. Later that year another French financial adventurer, who had assisted the government in raising a loan, obtained a charter for a Portuguese Crédit Mobilier, but it was short-lived. The promoter went bankrupt in the crisis of 1857, and the company went down with him. Subsequently French entrepreneurs contributed to the formation of two mortgage banks on the lines of the Crédit Foncier, but no other promoters considered Portugal a suitable area for investment banking.

The Pereires also wished to establish a filial in the rapidly developing state of Piedmont. Cavour, the guiding genius of that development, welcomed their interest as a counterweight to the influence that Rothschild exercised over all of the little kingdom's financial relations; but in the end he decided against alienating that financial power, and granted to the latter's Cassa del Commercio e delle Industrie the sole charter for a joint-stock investment bank in Piedmont. It participated in a number of Rothschild enterprises in Italy, Switzerland, and Austria as well as in Piedmont proper, but bad management and "irresponsibility on the part of important financiers connected with it" (in the words of a financial journal) resulted in large losses. The Rothschilds withdrew in 1860, and the bank stagnated until 1863 when the Pereires bought a controlling interest, increased its capital, and renamed it the Società Generale de Credito Mobiliare Italiano. In subsequent years it became identified with virtually every new enterprise in Italy, including railways, ironworks, and steel mills. It had close connections in high government circles and ranked after the Banca Nazionale as the most important bank in Italy. In the midst of the crisis of 1893, however, the revelation of serious scandals in its internal organization and in its relations with the government forced it into liquidation.

Most other Italian banks founded in the 1860s also used French capital, but only one other, the Banca di Credito Italiano, owed as much to French initiative as the Credito Mobiliare. It, too, fell prey to the crisis of 1893. In the following year, to fill the void, two new large banks were established, this time on German initiative and with German capital: the Banca Commerciale Italiana in Milan and the Credito Italiano in Genoa. Although the German capital withdrew around 1900 (replaced in part by French capital), these two institutions played major roles in Italy's industrial spurt in the years before World War I.

French bank promoters carried on the search for concessions in southeastern Europe in the 1850s, but the time was not yet ripe there. Both Serbia and Rumania rejected offers to establish mobilier-type banks then, but in 1863 the Crédit Mobilier obtained an agreement with the Rumanian government, only to have ratification blocked by the parliament. Two years later, after a coup d'état, the ruling Prince Cuza conceded a charter to French and British capitalists for the Banca Romaniei. Finally, in 1881, the Rumanians obtained a Societate de Credit Mobiliar under French sponsorship.

The Crimean War dramatically revealed the economic backwardness of Russia vis-à-vis the West, and led the government of the tsar to a campaign of railway construction and the emancipation of the serfs. It also led it to an overhaul of the financial and banking systems. The major financial institution was the State Bank, founded in 1860. It was wholly owned by the government and under the immediate supervision of the finance ministry. In the beginning it did not issue banknotes—fiat paper money was issued directly by the state printing office—but when Russia went on the gold standard in 1897 it obtained a monopoly of note issue. The State Bank controlled the state savings banks and created the Peasants' Land Bank (1882), the Land Bank for the Nobility (1885), and the Zemstvo and Urban Bank (1912). It owned shares in the Loan and Discount Bank of Persia (1890) and in the Russo-Chinese Bank (1895), created to facilitate Russian penetration of those countries.

The banking system also contained a variety of small institutions—cooperative, communal, mortgage, and other types of banks—but the most important, after the State Bank, were the joint-stock commercial banks. The first of these was the St. Petersburg Private Commercial Bank, founded in 1864. By 1914 they numbered fifty, mostly headquartered in St. Petersburg and Moscow, with nationwide networks of branches totaling more than 800 bank offices. The twelve largest banks, eight of them based in St. Petersburg, controlled about 80 percent of total assets. Another distinctive feature of these banks was the extent of foreign influence. Many of them had been founded or were managed by French, German, British, and other bankers. Foreign banks, especially French ones, owned much of their stock. In 1916 foreign banks owned 45 percent of the capital of the ten largest banks; more than 50 percent of that belonged to the French. The Russian joint-stock banks, in cooperation with their foreign partners, contributed greatly to Russian industrialization after 1885, which was also carried out in large part by foreign entrepreneurs and technicians (see Chapter 10).

European financiers also contributed their expertise to their neighbors in the Near and Middle East. The first joint-stock bank to be established in the area (and the first British joint-stock bank in any foreign country), the Bank of Egypt, commenced operation in 1855. It aroused the opposition of the numerous French private bankers in Alexandria, who protested to their consul, but in vain. In time the French set up joint-stock banks of their own.

A similar development occurred in the venerable, and decrepit, Ottoman Empire. In 1856 a group of British capitalists organized the Ottoman Bank in Constantinople as an ordinary commercial bank. Some years later it solicited a charter as the exclusive bank of issue, but French-educated reform ministers at the time desired a connection with the French financial market. In 1863 they forced the Ottoman Bank to amalgamate with a French group led by the Crédit Mobilier in a new institution, the Banque Impériale Ottomane. It was a most unusual institution, combining the functions of a central bank with a monopoly of note issue with those of ordinary commercial and investment banking. In addition, the bank was charged with withdrawing the paper money and money of bad alloy, collecting and transmitting taxes in areas served by its branches, and servicing the public debt. Profits on the first *seven months* of operation amounted to almost 20 percent of paid-in capital. The bank prospered throughout the prewar decades, and even came to terms with the nationalistic Mustapha Kemal (Ataturk) after World War I.

Persia (modern Iran) had a similar institution, the Imperial Bank of Persia, founded by British interests in 1889. The promoters had intended to use the bank to finance railway construction, but the Russian government, fearful of British penetration of its southern flank, exerted diplomatic pressure on the shah to prevent the railways from being built. The bank, thus founded "by mistake" and operated by nonprofessionals in the financial area, contributed little to Persian economic development.

This was not the case with many other British overseas banks. Beginning in the 1850s, a number of banks with British charters and British capital were established overseas, especially in India and Latin America. They were not branches of British domestic banks, but were generally founded by British merchants operating abroad. One of the most famous was the Hongkong and Shanghai Bank, which played a prominent role in Chinese finance and is today a major multinational corporation. The main function of these banks was to finance international trade, but they also took part in issuing securities of foreign governments and corporations. In time they faced competition from both local banks and branches of other European banks.

Banking in the United States had a checkered career in the nineteenth century. In the early years of the republic the struggle between the Hamiltonians, who favored a strong role for the federal government, and the Jeffersonians, who preferred to leave policy to the individual states, was clearly reflected in the history of banking. The Hamiltonians succeeded at first, getting the congressionally chartered first Bank of the United States (1791–1811), but when its charter expired the proponents of states' rights and of state-chartered banks, which were already numerous and were jealous of the larger institution, prevented its renewal. A second Bank of the United States (1816–36) met the same fate at the hands of the Jacksonian Democrats. Thereafter, a variety of institutional experiments were made until the Civil War. Some states allowed "free banking" (anyone could start a bank), others operated state-owned banks, and still others tried to prohibit banks altogether. In spite of this seeming confusion, the economy obtained the banking services it needed and continued to grow rapidly.

During the Civil War, and partly as a measure of war finance, Congress created the National Banking System, which allowed federally chartered banks to compete with the state-chartered banks. The competition was unfair, because Congress also imposed a discriminatory tax on the note issues of the state banks, which forced many of them to convert to national banks. In time, however, they discovered that it was possible to carry on a banking business by means of deposits subject to checks, and state banks staged a strong comeback in the closing decades of the century.

Both the state and the national banking systems suffered from excessively restrictive rules and regulations. For example, branch banking was generally prohibited. Banks could not deal in international finance, which meant that the country's large volume of imports and exports was financed from Europe and by the relatively small number of private merchant banks, such as J. P. Morgan & Co., which were not hampered by the restrictions that applied to corporate banks. Some believed that the absence of a central bank also made the country more susceptible to the periodic financial panics and depressions that occurred. To remedy this defect, Congress in 1913 created the Federal Reserve System which, among other things, relieved the national banks of their note-issuing function, but also freed them to engage in international finance.

In sum, the experience of the United States with both rapid economic growth and a changing, somewhat chaotic banking system seems to show that, though banks are necessary for economic growth in complex industrial societies, a rational banking system is not.

The Role of the State

Few topics in the economic history of the nineteenth century are more widely misunderstood than the role of the state, or government, in the economy. On the one hand is the myth of laissez faire—namely, that apart from enacting and (more or less) enforcing the criminal laws, the state strictly abstained from any interference in the economy. On the other hand is the Marxist notion that governments acted as the "executive committees" of the ruling class, the bourgeoisie. The historical reality, however variegated and complex, was far different from either of these simplistic formulations.

The government can play a variety of possible roles with respect to the economy. The most fundamental function of government in the economic sphere, one that cannot be avoided or abdicated, is the creation of the legal environment for economic endeavor. This can range from a pure "hands-off" policy to one of total state control. The cardinal sin in this area is neither intervention nor nonintervention, but ambiguity. The "rules of the game" should be clear, unequivocal, and enforceable. They include, as a minimum, the definition of rights (property and other) and responsibilities (contractual, legal, etc.). Theft is a crime in both free enterprise and socialist societies.

The second broad category of ways in which government participates in the economy includes promotional activities short of directly productive ones. These include tariffs, tax exemptions, rebates, and subsidies, as well as measures such as establishing tourist or immigration bureaus. Not all activities in this category are necessarily conducive to growth; for example, a protective tariff may perpetuate an inefficient industry.

Similar in some respects to promotional activities, but usually with a different goal in mind, are the regulatory functions of government. These range from measures to protect the health and safety of specific groups of workers to detailed controls of prices, wages, and output. The purpose of such regulations *may* be to foster growth—for example, by prohibiting or regulating private monopolies—but more often the purpose is not growth-related; it is intended rather to eliminate inequity or exploitation. In the latter case the unintended side effects of the regulation may be to retard growth.

Finally, governments can engage in directly productive activities. These range from such benign devices as providing educational facilities to total government ownership and control of all productive assets, as in the former Soviet Union. Such government participation may be essentially entrepreneurial or innovational, such as New York State's Erie Canal and Japan's model factories, and hence favorable to private enterprise; or it may compete with or supplant private enterprise, as with government ownership of public utilities or telegraph facilities.

With these manifold possibilities before us, let us look at the actual historical record to see what role governments actually played in the nineteenth century. Since some of the functions of government, especially the first one, have already been dealt with in Chapter 8, and commercial policy was discussed in Chapter 11, this brief survey focuses on the other functions.

Great Britain is generally regarded as the home of laissez faire, or minimal government. How large or small was that government? Throughout the century, after the Napoleonic Wars, the proportion of central government expenditure of the United Kingdom to gross national product was generally less than 10 percent—in peacetime, about 6 to 8 percent. (To measure the actual size of government, one would have to add local government expenditures, probably no more than 2 to 3 percent.) Is that a large or small proportion? Compared with twentieth-century proportions, ranging from 30 to 50 percent or more, it is certainly small. On the other hand, one-tenth. . . ?

In spite of its reputation as the home of minimal government, the size of the government in the United Kingdom (or Great Britain) was probably typical for that of Europe as a whole; if anything, it was slightly larger, in relative terms, than that of most continental nations. In both the German Empire and the United States the ratio of central government expenditures to national income was generally less than 5 percent, but of course these were both federal nations; state and local government expenditures combined exceeded those of the central government. Ironically, the nations in which the ratio generally exceeded that of Great Britain were the poorer countries of southern and eastern Europe, such as Spain, Italy, and Russia. The ratio in the Balkan nations in the early 1900s ranged from 20 to 30 percent.

So much for the size of government, insofar as it can be measured in pecuniary terms. What of the activities of government, both those that promoted and those that retarded economic development? Again, let us begin with Great Britain. Most people take it for granted that one of the functions of government is to deliver the mail (although in recent years the gross inefficiency of the postal service, and the rise of alternative private courier services, is bringing that assumption into question). Before the nineteenth century private courier services coexisted with bumbling, inefficient government postal services, which were maintained more for the purposes of censorship, espionage, and revenue than for service. Modern postal service began in 1840 when Sir Rowland Hill, the postmaster-general of the United Kingdom, introduced the prepaid, flat-rate penny post. Within a few years most Western nations adopted similar systems. When the electric telegraph became operational a few years later it seemed logical to add it to the government's postal monopoly. The same policy was followed later in the century after the invention of the telephone. Most continental countries followed Britain's example, but in the United States both the telegraph and telephone were left to private enterprise.

A most unusual example of a private enterprise was the East India Company. Although founded in the seventeenth century as a strictly commercial enterprise, by the beginning of the nineteenth century it had become the ruler of India, "a state within a state." In the aftermath of the Sepoy Rebellion of 1857, in which the native militia revolted against their officers, public opinion became conscious of the anomaly and forced the dissolution of the company, with its governmental functions taken over by the India Office. That seemed right and normal, but a few years later, in 1875, the Tory prime minister, Benjamin Disraeli, made the government a stockholder in one of the largest private enterprises of the day by buying the shares of the khedive of Egypt in the Suez Canal Company, chartered in France. That action was justified on the grounds of national defense. Similar reasons were advanced for the purchase of the Anglo-Persian Oil Company in 1914, on the eve of World War I. The reasoning may have been valid, given the conditions and assumptions of the times, but the actions were scarcely in keeping with Britain's reputation (then and later) as a minimalist state.

In one area Britain really did live up to its reputation. Nowhere did Britain lag further behind other Western nations than in public support of education. Until 1870 the only schools available were those operated by private or religious foundations, most of which charged fees, except for the parish schools in Scotland. As a result, fully half the population received no formal education at all. Only the well-to-do received more than the rudiments. That factor more than any other served to preserve Britain's archaic class structure in an age of otherwise rapid social change, and contributed to the relative decline of Britain's industrial leadership. The Education Bill of 1870 provided state support for existing private and church-connected schools that met certain minimum standards. Not until 1891, however, did education become, even in principle, both free and universal up to the age of twelve. As late as the 1920s only one in eight of the eligible population attended a secondary school.

In higher education England also lagged behind the Continent and the United States. Until state scholarships were instituted in the twentieth century, Oxford and Cambridge were open only to the sons of the wealthy, mainly the aristocracy. By contrast, Scotland, with a much smaller population, had four ancient and flourishing universities open to all qualified applicants. London's University College, in existence from 1825, became the University of London in 1898 with the addition of more colleges. In 1880 Manchester became the first provincial city to acquire a new university. By the beginning of the twentieth century several others were established, but even after World War I only four persons per thousand in the appropriate age group were enrolled in a university.

Most continental countries had long traditions of state paternalism or *étatisme*, in the French term. In several of them the state owned forests, mines, and even industrial enterprises. The latter produced military and naval equipment, but not only that; the French had their *manufactures royales* for the production of procelain, crystal, tapestries, and so on, as did other governments. In the eighteenth century, as the superiority of British technology in certain industries became apparent, governments sponsored efforts to obtain access to that technology, by espionage or otherwise. Both France and Prussia, for example, undertook to produce coke-smelted pig iron in state-owned furnaces; neither experiment was commercially successful, however, and it remained for private entrepreneurs to reintroduce the process long after the Napoleonic Wars.

This example suggests how the states were obliged to modify their traditions of paternalism in the course of industrialization. A more vivid example comes from the Ruhr mining industry. In Prussia, as in France and several other countries, mining, even in privately owned mines, had to be carried out under the supervision of engineers of the royal mining corps. This was called the *Direktionsprinzip* (direction principle). That sufficed in the Ruhr as long as mining was confined to the relatively shallow deposits in the Ruhr valley proper; but when the riches of the "hidden" coalfield north of the Ruhr were discovered in the late 1830s and 1840s, the conservatism of the royal mining corps became a hindrance. The new mines required greater capitalization for the deeper shafts, steam pumps, and other mining equipment. The mining companies, several of them operated by French, Belgian, and British entrepreneurs, began a long drawn-out struggle with the Prussian authorities, which finally ended in 1865 with the replacement of the *Direktionsprinzip* by the *Inspektionsprinzip* (inspection principle), under which the government engineers merely inspected the mines for safety.

The rapidly developing technology of transportation—specifically, that of the railways—obliged all governments to become involved. The British, true to their minimalist tradition, did the least, leaving promotion, construction, and most operating details to private initiative; but even in Britain Parliament had to pass the enabling legislation to allow the companies to buy the land for rights of way, and the Railway Act of 1844 laid down a number of rules and regulations, including a maximum fare for third-class passengers. (The act also provided that the government might purchase the railways when their charters expired, but that provision was not acted on until after World War II.)

Elsewhere governments took a much greater interest in railways. As we have seen, the new Belgian state in the 1830s undertook to build and operate a basic railway network for its own account. After it was completed it allowed private companies to build branch lines, but when these encountered financial difficulties in the 1870s the government bailed them out. France endured a prolonged debate on the question of state versus private ownership; in the end the proponents of private ownership won out, but with numerous provisos that allowed the government a large role. And when one important company went bankrupt the government took it over to continue the service. As noted in Chapter 9, the German states followed different policies at the beginning of the railway era, some building for the account of the state, others leaving the job to private enterprise. Subsequently, after the creation of the empire, Bismarck established the Imperial Railway Office, whose function was to buy up private companies and use the railways consciously as instruments of economic policy, for example, by giving favorable rates on goods destined for export. (Earlier, however, in 1865, Bismarck sold the Prussian government's shares in the Cologne-Minden Railway to raise money for the war with Austria when the Prussian Diet refused to grant taxes for the purpose.)

The Austro-Hungarian Empire's policy on railways fluctuated, as did the Russian's, first favoring state ownership and operation, then private companies, and finally veering back to the state. In other countries, if they did not begin with state-owned networks, as Sweden did in 1855, they sooner or later came around to the principle of state ownership. Even where they did not, as in France before World War I, the state exercised considerable regulatory powers. In the United States the federal government left railway policy to the states before the Civil War, but shortly thereafter gave large land grants to private companies to stimulate the construction of the transcontinentals. In 1887, in response to complaints from farmers and others, Congress created the Interstate Commerce Commission to regulate the railways.

These few examples do not exhaust the instances in which the state took an active part in the economy—far from it. But they do illustrate the varied and sometimes contradictory roles of government. If, in retrospect, the nineteenth century appears to be one in which government was less pervasive than in previous centuries, or the one that followed, this does not mean the government played no role at all.

13

Overview of the World Economy in the Twentieth Century

Spurred by the accelerating pace of technological change, buffeted by the two most destructive wars in history, the world economy in the twentieth century took on new and unprecedented dimensions. Nowhere were these dimensions more evident than in the behavior of population.

Population

The population of Europe more than doubled in the nineteenth century, but that of the world outside the areas of European settlement increased by little more than 20 percent. In the twentieth century, on the other hand, population growth in Europe decelerated while that of the rest of the world accelerated at unprecedented rates. Most of that growth has occurred since the Second World War, as is clear from Table 13-1.

As a first approximation we can say that the cause of the tremendous increase in numbers has been the decline in crude death rates, especially in non-Western countries. Western nations underwent a ''demographic transition'' (from a regime of high birth and death rates to one much lower) in the late nineteenth and early twentieth centuries. Most non-Western nations are currently undergoing a similar transition. As a result of the diffusion of Western technology of public health and sanitation, medical care, and agricultural production, death rates in Third World countries have declined dramatically while birth rates have responded much more slowly. This is shown for selected countries in Table 13-2.

A major contributory factor in the decline in the overall death rate was the decline in infant mortality (deaths under the age of one year). This is shown in Table 13-3.

A major consequence of the decline in death rates was a sharp increase in the average life span. This is frequently measured by the concept ''life expectancy at birth,'' the average number of years that persons born in a given year will live. At the beginning of the twentieth century this figure was generally below 50 even in the advanced countries. For example, in the United States in 1900 the average for both sexes of the white population was 47.3 and only 33.0 for the nonwhite population. In Sweden, which had exceptionally long-lived people, the averages in the decade

TABLE 13-1.   World Population
by Continents (millions)

Region	1900	1950	1989
Africa	120	222	628
Asia	937	1366	3052
East Asia[a]		671	1317
Other		695	1735
Europe[b]	401	392	497
Russia/USSR	126	180	286
North America	81	166	274
Latin America[c]	63	165	439
Oceania	6	13	26
World total	1608	2504	5201

[a]Japan, Korea, China, Southeast Asia.
[b]Excluding Russia/USSR and European Turkey.
[c]Central America, Mexico, and the Caribbean with Latin America.

Sources: 1950 and 1989, United Nations *Demographic Yearbook,* 1989; 1900, W. S. and E. S. Woytinsky, *World Population and Production: Trends and Outlook* (New York, 1953); the 1900 figure for Russia is from the 1897 census. Both dates and population figures are approximations.

1881–90 were 48.5 for males and 51.5 for females. On the other hand, in the non-Western world life expectancy at the beginning of the century was quite low, scarcely above the levels attained in the ancient Roman Empire. For example, in India the life expectancy at birth as late as 1931 was only 26.8. By midcentury the life expectancy in the advanced Western nations had risen to the 60s or even higher, while in the rest of the world it was usually still below the Western figures for the beginning of the century. Since World War II, however, large gains have been recorded almost everywhere, as indicated in Table 13-4.

A close correlation exists between these statistics, especially those of life expectancy, and various measures of welfare such as per capita income, nutritional levels, and standards of health care. Thus, in countries with high average incomes the population is better fed and has better medical care, as a rule, than in countries with markedly lower incomes; consequently, death rates are lower and life expectancy correspondingly greater.

The process of urbanization, so marked in Europe in the nineteenth century, continued in the twentieth century, spreading to other areas of the world. (A possible wave of the future, however, can be observed in Great Britain, the first industrial-urban nation: there the proportion of the urban population has declined slightly in recent decades as prosperous business and professional people have forsaken city residences in favor of small villages from which they commute to the city.) In advanced industrial nations cities are usually centers of affluence as well as of culture, since productivity and incomes are normally higher in urban than in rural

TABLE 13-2. Crude Birth Rate (CBR) and Crude Death Rate (CDR), Selected Countries (per thousand total population)

Country	Ca. 1900–1910		1950		1987[a]	
	CBR	CDR	CBR	CDR	CBR	CDR
Australia	26.5	11.2	23.3	9.6	15.0	7.2
Austria	34.7	23.3	15.6	12.4	11.4	11.2
England and Wales	27.2	15.4	15.8	11.7	13.6	11.3
France	20.6	19.4	20.7	12.8	13.8	9.5
Israel			34.5	6.9	22.7	6.7
Japan	32.2	20.7	28.2	10.9	11.0	6.2
Rumania	39.8	25.8	19.6	20.0	16.0	10.7
Russia/USSR	46.5[b]	29.7[b]	26.7[c]	9.7[c]	19.8	9.9
Spain	34.4	25.2	20.1	10.9	11.3	7.9
Sweden	24.8	14.9	16.4	13.7	12.5	11.1
United States	24.3	15.7	23.5	9.6	15.6	8.7
Argentina			25.4	9.0	21.3	8.0
Brazil			41.4	12.1	28.6	7.9
Mexico			44.7	15.9	29.0	5.3
Bangladesh			49.1	27.3	42.2	15.5
India			40.5	24.2	32.0	11.9
Indonesia			45.0	26.4	27.4	11.2
Egypt			44.2	19.0	41.1	9.2
Nigeria			51.8	29.0	49.8	15.6
Tunisia			30.1	9.5	27.5	7.4
Zaire			48.2	26.3	45.6	13.9

[a] Or nearest year.
[b] Ca. 1900–1910, European Russia only.
[c] 1945.

Sources: Ca. 1900–1910, W. S. and E. S. Woytinsky, *World Population and Productions: Trends and Outlook* (New York, 1953); 1950, 1987, United Nations, *Demographic Yearbook,* various issues. Both dates and rates are approximations.

occupations. This is not necessarily true of Third World nations, however. In them a large proportion of urban inhabitants consists of unemployed or underemployed migrants from the countryside living in miserable shantytowns on the fringes of the city centers (Fig. 13-1). Mexico City, for example, grew from about 2 million inhabitants in the 1940s to more than 15 million in the 1980s—mostly as a result of the influx of illiterate, unskilled, and unemployed peasants. Such mushroom growth has characterized most of the large cities of Latin America, Asia, and even Africa, placing unbearable pressures on the urban infrastructure. Table 13-5 depicts the growth of the urban population in selected countries since World War II. Too much reliance should not be placed on the absolute numbers, because of different definitions of "urban" in different countries, as well as the usual hazards of population statistics; but the general trend is undoubtedly correct.

The growth of cities came about primarily as a result of internal migration, as the surplus population from rural areas and small towns sought the greater opportunities and freedom of city life and lights. International migration, such a prominent feature of nineteenth-century population history (at least for Europe and regions of European settlement overseas), also continued, although under somewhat different circumstances. Most nineteenth-century migration had been motivated by economic pressures at home and opportunities abroad. These factors remained important in the twentieth century, but political oppression (or its threat) in the wake of wars and revolutions also played a major role.

TABLE 13-3.   Infant Mortality, Selected Countries
(per thousand live births)

Country	Ca. 1901–1910	1950	1987
Australia	87	23.8	8.7
Austria	209	55.6	—
England and Wales	128	27.9	9.0
France	132	46.2	7.8
Israel		41.8	10.0
Japan	156	52.7	4.8
Russia/USSR	251	75.2	13.5
Spain	164	62.5	10.9
United States	97	28.1	10.4
Argentina		64.8	26.2
Brazil		107.3	45.3
Mexico		57.7	29.2
India		185	130
Indonesia		95.2	125
Egypt			75
Nigeria		87.2	124
Tunisia		193	125
Zaire			104

Sources: Ca. 1900–1900, W. S. and E. S. Woytinsky, *World Population and Production: Trends and Outlook* (New York, 1953); 1950, 1987, United Nations, *Demographic Yearbook,* various issues. Both dates and rates are approximations.

TABLE 13-4.   Life Expectancy at Birth in Selected
Countries (years)

Country	Ca. 1950		Ca. 1989	
	M	F	M	F
Australia	67	73	73	79
Czechoslovakia	61	66	67	75
United Kingdom	66	71	72	78
Israel	70	74	74	77
Japan	56	60	76	81
Russia/USSR	61	67	65	74
Spain	60	64	73	79
Sweden	70	73	74	80
United States	66	71	71	78
Argentina	63	69	65	73
Brazil	51	57	62	68
Mexico	48	51	62	66
Bangladesh	38	37	57	56
China	45	47	68	71
India	42	41	53	52
Indonesia	37	38	55	57
Egypt	41	44	59	62
Nigeria	33	36	49	52
Tunisia	43	45	65	66
Zaire	37	40	51	54

Source: United Nations, *Demographic Yearbook,* various issues.

The nineteenth-century type of international migration reached its culmination in the years immediately preceding World War I, with an annual average of more than a million people leaving Europe for overseas destinations, principally the United States. World War I temporarily halted that flow in part, and the U.S. adoption after the war of restrictive immigration legislation further curtailed it. Whereas immigration to the United States in the decade preceding the war averaged about 1 million per year, immigration in the 1920s was less than half that figure. The depression of the 1930s severely constrained the opportunities in America, and World War II further reduced the tide of immigration, which averaged less than 50,000 per year from 1930 to 1945. After the war many refugees from the wartime devastation and new political repressions swelled the numbers of immigrants, which rose from about 100,000 in the late 1940s to more than half a million in the 1980s.

FIGURE 13–1. Third World shantytown. This settlement on the outskirts of Lima, Peru, is typical of the living conditions of much of the Third World's urban population. Lacking sanitary facilities and frequently even running water, such settlements are breeding grounds for epidemic disease. (From *A House of My Own: Social Organization in the Squatter Settlements of Lima, Peru*, by Susan Lobo, Copyright 1982 by University of Arizona Press, Tucson. Reprinted with permission.)

TABLE 13-5.   Urban Population
(percent of total)

Country	Ca. 1950	Ca. 1985
Australia	79	85
Czechoslovakia	51	75
Israel	78	89
Japan	38	77
Russia/USSR	48	65
Spain	55	90
United States	64	75
Argentina	74	85
Brazil	36	72
Mexico	43	66
Bangladesh	—	13
India	—	25
Indonesia	15	22
Egypt	38	45
Nigeria	10	16
Tunisia	37	53
Zaire	22	39

Source: United Nations, *Demographic Yearbook.*

The character of American immigration has also changed in recent decades. Formerly the immigrants were overwhelmingly European; today many more come from Asia and Latin America. Many of the latter (probably a majority of those from Mexico and some Central American countries) are illegal entrants, "wetbacks" seeking work, or political refugees from Central America and the Caribbean.

The character of European immigration and emigration has also changed in the twentieth century. In the nineteenth century Europe provided the bulk of international migrants, but today western Europe has become a haven for political refugees and, temporarily at least, a land of opportunity for the impoverished masses of Mediterranean Europe, North Africa, and parts of the Middle East.

The process began in the wake of the Russian Revolution of 1917, when many subjects of the tsar (former aristocrats and others) chose residence in the West, especially France, rather than remain in their homeland under the Soviet regime. The process accelerated massively after World War II with the redrawing of the boundaries of eastern Europe. Millions of German-speaking people were expelled or fled, but many of other nationalities also took advantage of the postwar chaos to flee from what they regarded as oppressive political regimes—a process that was repeated on a smaller scale in the wake of the abortive Hungarian Revolution of 1956 and the invasion of Czechoslovakia in 1968.

West Germany bore the brunt of the flood of refugees, which seemed a heavy burden at first; but with the economic revival of continental western Europe in the 1950s and 1960s, with its strong demand for labor, the burden proved to be a blessing. Indeed, the demand for labor outran the supply of refugees, with the result that several countries—notably France, Switzerland, and Belgium as well as West Germany—invited ''guest workers'' from Portugal, Spain, Italy, Greece, Yugoslavia, Turkey, and North Africa to supplement their own native labor forces. In most instances these migrations were temporary, or intended as such, but they also led to some permanent immigration.

Another novel current of migration involved European Jews and, eventually, Jews from other parts of the world. Following World War I the British, who had been given a League of Nations mandate over Palestine, allowed a limited number of Zionists to settle there. During and after World War II, with the revelation of Nazi atrocities toward the Jews, thousands of survivors of the Holocaust sought refuge there. At first the British resisted, and deported many who had immigrated illegally; but after the proclamation of the state of Israel in 1948 the floodgates opened and millions of Jews entered, not only from Europe but also America, Asia, and Africa. In recent years many Jews from the former Soviet Union left for Israel or other destinations.

Resources

The unprecedented growth of population in the twentieth century, as well as the growing affluence of at least a part of the world, resulted in an unprecedented demand on the world's resources. Although occasional temporary shortages of some commodities occurred, especially in wartime, and fears were expressed in the last quarter of the century concerning the exhaustion of certain critical resources, the world economy responded reasonably well to the demands made on it. That it did so was a result, in large part, of the increasing interaction of science and technology with the economy. Agronomists discovered new ways to increase the yield of crops, engineers discovered new ways to increase the yield of minerals, scientists discovered new uses for existing resources, and, indeed, created new resources from old in the form of synthetic products.

The most important development in terms of resources in the twentieth century has been a change in the nature and sources of primary energy. In the nineteenth century coal became the most important source of energy in industrializing nations, largely replacing wood, charcoal, wind, and water power. In the twentieth century coal has been largely, though not wholly, displaced by new energy sources, especially petroleum and natural gas. Although petroleum was first commercially produced in the nineteenth century, it was then used mainly for illumination and secondarily as a lubricant. The development of internal combustion engines at the end of the nineteenth century greatly extended its possibilities, and it also competed with coal and water power in the production of electricity and with the former for space heating. In the second half of the twentieth century it acquired new importance as a raw material for the production of synthetics and plastics.

At the beginning of the twentieth century the dominance of coal was unquestioned. In 1928 it still accounted for 75 percent of world energy production, petroleum for about 17 percent, and water power for about 8 percent. (These figures omit the contributions of work animals, wood fuel, manure, etc., but these were negligible in the industrialized economies.) Around 1950 coal still accounted for almost half of total energy, with petroleum and natural gas up to 30 percent, but by the 1980s those proportions had been more than reversed.

In the light of its major importance and manifold uses, petroleum has acquired great geopolitical significance. Petroleum deposits are widely scattered throughout the world, but most production comes from a relatively small number of geographical areas. Ironically, Europe, although abundantly endowed with coal, has the smallest petroleum reserves of any major land mass. The United States, Russia, and possibly China, on the other hand, have plentiful resources of both coal and petroleum. Petroleum production was first developed on a large scale in the United States. As of 1950 more than 60 percent of total cumulative world production since the beginnings of commercial exploitation had taken place in the United States, and in that year the country still produced more than 50 percent of the world's output. Since that time, however, although it is still a major producer, the United States had become a net importer of petroleum. The countries of the Middle East surrounding the Persian Gulf are now, collectively, the largest source of supply for the world market. Russia is also a major producer. Table 13-6 shows the percentages of world primary energy production by type of fuel and geographic location.

TABLE 13-6.   World Primary Energy Production, 1988 (percentage of total)

	Crude Oil	Natural Gas	Coal	Hydroelectric	Nuclear	Total
North America	7.9	7.5	6.8	1.7	2.0	25.8
Central and South America	2.6	0.7	0.2	1.0	—	4.5
Western Europe	2.5	2.3	2.7	1.6	2.2	11.0
Eastern Europe and USSR	7.7	8.1	8.0	0.8	0.9	25.6
Middle East	9.5	1.4	—	—	—	10.9
Africa	3.5	0.8	1.3	0.1	—	5.8
Far East and Oceania	3.9	0.9	9.0	1.2	0.8	16.3
World total	37.6	22.3	27.9	6.4	5.7	100.0

Source: U.S. Department of Energy, International Energy Annual, 1988 (Washington, D.C., 1989).

Technology

Technological change, the major driving force behind nineteenth-century industrialization, continued that role undiminished in the twentieth century. Indeed, it appears that the pace of change accelerated, although our measures for gauging such changes are rather crude and unreliable. It is indubitable, nevertheless, that new technology affects, in profound and countless ways, the daily life of virtually every human being, even those to whom the technology itself is foreign. In earlier ages the mark of success of human societies was their ability to adapt to their environments. In the twentieth century the mark of success is the ability to manipulate the environment and adapt it to the needs of society. The fundamental means of manipulation and adaptation is technology—specifically, technology based on modern science. A major reason for the more rapid pace of social change in the twentieth century is the marked acceleration of scientific and technological progress.

The recent history of transportation and communications provides a graphic example of the acceleration of technological change (Fig. 13-2). At the beginning of the nineteenth century the speed of travel had not changed appreciably since the Hellenistic era. By the beginning of the twentieth century people could travel at velocities of up to eighty miles per hour by means of steam locomotives. The development of automobiles, airplanes, and space rockets dwarfed even that achievement in speed and also in range and flexibility.

Until the invention of the electric telegraph, communication over appreciable distances was limited to the speed of human messengers. The telephone, radio, and television added immeasurably to the convenience, flexibility, and reliability of long-distance communication. In 1931 President Hoover made the first trans-Atlantic telephone call, talking to his advisers in Europe. Subsequently, almost instantaneous communication with most inhabited portions of the world became commonplace, and it is even possible to "communicate" with (or to receive communications from) other planets in the solar system by means of space capsules and electronic systems. Each of these achievements has depended increasingly on the application of basic science.

The scientific basis of modern industry has resulted in hundreds of new products and materials. Already in the nineteenth century chemists had created numerous synthetic dyes and pharmaceuticals. Beginning with the invention of rayon in 1898, dozens of artificial or synthetic textile fibers have been created. In the twentieth century plastic materials made from petroleum and other hydrocarbons have replaced wood, metals, earthenware, and paper in thousands of uses ranging from lightweight containers to high-speed drilling machines. The increasing use of electrical and mechanical power, the invention of hundreds of new labor-saving devices, and the development of automatic instruments of control have brought about changes in the conditions of life and work more far-reaching than the so-called industrial revolution in Great Britain. In an extreme instance, a single worker can oversee the operation of a huge petroleum refinery.

The ability of science and technology to grow rapidly depends on a host of accessory developments, some of them stemming from the progress of science

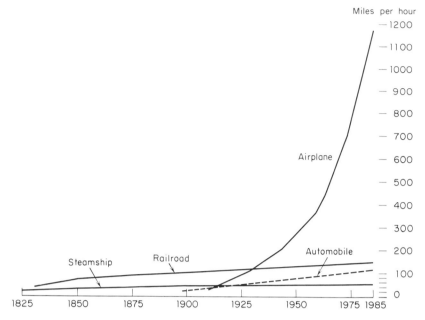

FIGURE 13–2. Range of practical speeds for continental and intercontinental travel.

itself. A major example is the electronic computer, which performs thousands of complicated calculations in a fraction of a second. The first mechanical calculating machine beyond the simple abacus was invented in the 1830s and was powered by steam. By the beginning of the twentieth century a few rudimentary mechanical devices were in use, chiefly for commercial purposes, but the age of the electronic computer did not dawn until World War II. Its progress since then has rivaled the speed with which it operates. Without it many other scientific advances, such as the exploration of space, would have been impossible.

This example brings to mind the role of scientific research and in particular the ways in which it is financed. Although many new developments in chemistry and biology have been stimulated by their commercial applications in agriculture, industry, and medicine, most basic research requires expenditures so vast, and with so little prospect of immediate return, that governments have been obliged to finance it either directly or indirectly. Moreover, the requirements of war and national rivalry have led governments to devote huge resources to scientific research and development for military purposes. Military crash programs resulted in the development of radar and other electronic communication devices, in the successful harnessing of atomic energy, and in the development of space rockets and artificial satellites. Such achievements are scarcely imaginable without the financial resources of governments.

Another requirement for scientific and technical advance is a sizable pool of educated manpower—or brainpower. By the beginning of the twentieth century virtually all Western countries had high literacy rates, in sharp contrast with the low literacy rates of most of the rest of the world. The widening technical gap between the developed and underdeveloped parts of the world is reflected in differences in educational levels as well as in differences in income.

Mere literacy, however, as important as it is for the initiation and sustenance of economic development, is not sufficient for the high-technology world of the late twentieth century. The ability of individuals to participate fully and effectively in the new scientific-technological matrix of civilization, whether as scientists and technicians or in its commercial and bureaucratic superstructures, increasingly requires advanced study at the college or university level and beyond. That is another reason for the widening gap between rich and poor nations.

The application of scientific technology has greatly increased the productivity of human labor. Output per worker, or per worker-hour, is the most meaningful measure of economic efficiency. In agriculture, still the major source of supply for the majority of the world's foodstuffs and raw materials, productivity has increased greatly in Western nations by means of scientific techniques of fertilization, seed selection and stock breeding, pest control, and by the use of mechanical power. Unfortunately, these techniques are not yet widely used in Third World countries. At midcentury output per worker in agriculture in the United States was more than ten times as great as in most Asian countries, and about twenty-five times as high as in most African countries. In the 1960s, as a result of the "green revolution" (new techniques especially adapted for tropical climates), agricultural productivity increased substantially in some Asian nations—India, for example, became self-sufficient in food production—but a large gap in productivity remained between rich and poor countries.

The rise in power production has been even more remarkable. World power production increased more than fourfold between 1900 and 1950, and has more than quadrupled since then. Most of the increase took place in areas of European settlement and in forms that had been in their infancy at the beginning of the century. For example, the generation of electricity (which is a *form*, not a *source* of energy) has increased more than a hundredfold. Electric energy is far cleaner, more efficient, and more flexible than most other forms of energy. It can be transmitted hundreds of miles at a fraction of the cost of transporting coal or petroleum. It can be used in massive concentrations to smelt metals or in tiny motors to operate delicate instruments, as well as to provide illumination, heat, and air conditioning. Its application to domestic appliances has helped to revolutionize patterns of family life, the status of women, and the employment of domestic servants. From a total production of less than 1 trillion kilowatt-hours in 1950 (up from virtually zero in 1880), world production rose to more than 12 trillion kilowatt-hours in 1990, an average annual rate of 6.8 percent. Table 13-7 indicates the pattern of production by source and by geographic region.

Several features of the table are worth noting. First is the preponderance of North America (principally the United States) and Europe in *all* types of electricity both in 1950 and the 1980s, in contrast to the tiny shares of Africa and South America. (Oceania also has a tiny share, but a far smaller population than the other two areas.) Second, the relative share of North America and Europe declined between 1950 and 1989, even though the absolute amounts increased enormously. (The data on which the table is based show that both produced more hydroelectricity in 1989 than all types in 1950.) The relative shares of all other regions increased between those dates, but the largest gainer of all was Asia, reflecting the emergence of Japan as a major economic power. Aggregation in the table hides other equally interesting facts. For example, in 1950 France obtained electricity from steam and water power in roughly equal proportions; in 1984 it obtained almost 60 percent of its electricity from nuclear power—the largest proportion of nuclear-generated electricity in any country—only 23 percent from thermal stations, and 17 percent from water power. Sweden and Switzerland also produced significant proportions of electricity with nuclear energy: for Sweden, 40 percent (against 55 percent hydro-

TABLE 13-7.   Production of Electricity, 1950 and 1989
(percentage of world total)

		Total	Thermal	Hydro	Nuclear & Other
World	1950	100	64.2	35.6	0.1
	1989	100	64.8	18.3	16.8
Africa	1950	1.5	1.4	0.1	—
	1989	2.7	2.3	0.4	a
North America	1950	47.3	30.9	16.3	—
	1989	32.0	21.1	5.2	5.5
South America	1950	1.8	0.8	1.0	—
	1989	3.8	0.8	2.9	a
Asia	1950	6.5	1.9	4.5	—
	1989	20.8	14.9	3.5	2.4
Europe	1950	31.6	19.7	11.6	0.1
	1989	24.1	13.1	4.1	6.9
Oceania	1950	1.2	0.8	0.4	a
	1989	1.6	1.2	0.3	—
USSR	1950	9.5	8.2	1.2	—
	1989	15.1	11.3	1.9	1.9

a Less than 0.1 of 1 percent.

Source: United Nations, *Energy Statistics Yearbook 1989* (New York, 1991).

electric) and for Switzerland, 37 percent (versus 61 percent hydroelectric). In 1950 both relied on water power for more than 95 percent of their electricity. Norway, which does not have a nuclear power program, still relies on water power for 99.9 percent of its electricity.

Petroleum and natural gas, which accounted for only a tiny fraction of total energy at the beginning of the century, surpassed coal as a source of energy around 1960, and in the 1980s accounted for more than 60 percent of total world production. The internal combustion engine, the most important consumer of petroleum, was an invention of the nineteenth century, but it produced a revolution only when applied to two of the most characteristic technological devices of the twentieth century, the automobile and the airplane. A few automobiles were built in the last years of the nineteenth century, but not until Henry Ford introduced the principle of mass production with a moving assembly line in 1913 did the automobile become more than a rich man's toy (Fig. 13-3). Ford's technique was soon imitated by other

FIGURE 13–3.　Ford's first assembly line. The moving assembly line, pioneered by Henry Ford in 1913, made the automobile an object of mass production and consumption, and symbolized both a new method of production and a new style of life. (From the collections of Henry Ford Museum and Greenfield Village.)

manufacturers in the United States and Europe, and the automobile industry became one of the largest employers of any manufacturing industry, as well as providing unprecedented opportunities for personal mobility.

The automobile came to symbolize twentieth-century economic development in much the same way that the steam locomotive symbolized that of the nineteenth century. In addition to being a large employer in its own right, the automobile industry stimulated the demand for several other industries. Just as locomotives and trains needed railways and iron, so automobiles needed roads and cement. In the United States and other leading auto producers, the industry was the largest consumer of steel, rubber (both natural and, later, synthetic), and glass. The rapid emergence of Japan as a major economic power in the second half of the twentieth century owed much to its success as an exporter of automobiles. The automobile also had a profound impact on social mores and customs, from courtship to commuting.

The technique of assembly-line production was adopted by other industries, including the aircraft industry in World War II. The air age began with the fifteen-second flight of the Wright brothers on a beach in North Carolina in 1903. Airplanes were discovered to have military applications in World War I, at first for observation, and later for bombardment. After the war they were used to carry mail and eventually fare-paying passengers. The commercial aviation industry developed rapidly in the 1930s, along with the technology, and on the eve of World War II trans-Atlantic service became available. Up until that time all successful aircraft had been powered by gasoline-fired piston engines driving propellers. During the war the Germans began experimenting with jet-propelled planes and also with rockets. Although their experiments did not prevent them from losing the war, they did set the stage for further developments in both aviation and the exploration of space—the latter carried out mainly by Russians and Americans, who scrambled in 1945 to obtain the services of the German rocket scientists. By 1960 jet-propelled aircraft had rendered propellers obsolete for commercial passenger traffic and, in the United States at least, had also made the railway obsolescent for passenger traffic.

The most spectacular application of science to technology has occurred in the exploration of space. As recently as the 1940s manned flight in space was chiefly a subject for science fiction. While comic strips portrayed scantily clad men and women of the twenty-fifth century flying through space with rockets strapped to their shoulders, learned men made calculations that purported to prove that no vehicle could ever attain the velocity required to leave the earth's gravitational field. During World War II scientists gained much valuable experience with jet engines and military rockets, but few people expected it would be possible for humans to survive in outer space even if they could reach it. New developments such as more powerful rocket engines, electronic signaling and control devices, and computers for rapid calculation of trajectories converged to make space flight a real possibility. On October 4, 1957, scientists in the Soviet Union put a capsule into orbit around the earth. The Space Age had begun.

Further progress occurred rapidly, largely stimulated by national rivalry. A second orbiting Russian rocket followed a month later, and early in 1958 the United States placed a capsule in orbit. Within a few years both nations rocketed astronauts into space and successfully retrieved them. Unmanned satellites were put into more or less permanent orbits to relay scientific information back to earth by means of radio and television, and other rockets were sent to the moon, Venus, Mars, and outer space with similar aims. In December 1968 the United States put a space crew in orbit around the moon, but outdid even that feat the following year. On July 20, 1969, astronauts Neil Armstrong and Edwin Aldrin, supported by astronaut Michael Collins and a crew of thousands of scientists and technicians on earth, became the first men to set foot on the moon. Truly, the human race had created a new age. One measure of the difference between that new age and all previous eras of human achievement lay in the manner in which the event was publicized. When Columbus discovered the New World (which he mistook for the Indies) the event was witnessed only by the actual participants, and it was months, even years, before the news reached a larger public. The first step of a man on the moon, by contrast, was witnessed by hundreds of millions all over the earth by means of television relays—the largest audience, in fact, ever to witness a single event to that time.

Institutions

Altered by technological change and by changes in the use of natural resources, strained by the growth of world population, and alternately bruised and soothed by political changes outside the scope of the economy proper, the institutional structure of the world economy at the end of the twentieth century differed greatly from what it had been at the beginning of that century. The number of such institutional changes is almost countless, and they vary from insignificant to earth-shaking. They can be categorized, however, under a few headings: changes in international relations, in national institutions, and within nations, such as the role of government, the nature and size of business firms, and the role of education. Some of these institutional changes are treated in somewhat greater detail in subsequent chapters. Only the broad outlines are sketched here.

International Relations

The pre-1914 world economy was dominated, literally and figuratively, by Europe (especially western Europe) and the United States. In political terms, the overseas empires of West European nations—primarily the British, French, and Germans, but also the Dutch, Belgians, Danes, and Italians—together with the vast land empire of imperial Russia gave them control of more than three-quarters of the earth's surface and almost as large a fraction of world population. Economically, Europe and the United States (without their empires) accounted for well over half of

TABLE 13-8.   World Trade (Imports and
Exports), 1913 (percentage of total)

Area	Percentage
World	100
Europe	58.4
United States and Canada	14.1
Asia	12.1
Latin America	8.3
Africa	4.4
Oceania	2.7

Source: W. S. and E. S. Woytinsky, World Commerce and
Governments (New York, 1955), p. 45.

total production and trade. Table 13-8 shows the percentage distribution for world trade; comparable figures for production or income are not available, but it is certain that the European-American predominance in production was even greater than in trade.

World War I and its concomitant, the Russian revolutions of 1917, brought fundamental changes in this structure. Tsarist Russia disappeared, its place taken by the Soviet Union, with a novel form of economic organization. The Habsburg Empire of east-central Europe also disappeared, replaced by several new or enlarged national states, economically deprived and struggling. Germany lost its overseas empire as well as a substantial part of its own territory and population. The remaining European empires exploited their colonies with increased nationalist fervor. Japan, which had a small prewar empire, enlarged it and became an important economic power. Europe proper suffered a decline in its share of world trade and production, mainly though not exclusively to the United States, the British dominions, and Japan. Finally, and partly at least as a result of the war, the 1920s and 1930s witnessed the rise of Fascist dictatorships in Italy, Germany, and several other European nations, also with novel forms of economic organization.

World War II brought in its wake a fundamental reorganization of international relations, with important economic consequences. Europe lost its hegemony in both politics and economics. Instead, a rivalry between the two new superpowers, the United States and the Soviet Union, replaced the age-long bickering of the traditional European great powers. As a consequence of this rivalry, Europe was divided more clearly and decisively than ever before between East and West: an eastern bloc under Soviet domination, and a western group of mainly democratic nations, most of which were tied politically and economically to the United States.

In the immediate aftermath of the war the old imperial nations attempted to retain or reimpose their authority in their former overseas possessions, but the new political and economic realities soon disabused them of their illusions. The Arab nations of the Middle

East and North Africa rather quickly threw off the controls that the French and British had exercised between the wars, except Algeria, which engaged in a long and bloodly struggle before the French reluctantly granted its independence in 1962. The various colonies of Southeast Asia, occupied by Japan during the war, also quickly gained their independence, although France again sought, unsuccessfully, to retain its control in Vietnam. Britain, rather than face a war on the Indian subcontinent, agreed in 1947 to the creation of two new nations there, the predominantly Hindu Union of India and the Islamic state of Pakistan. (East Pakistan, separated from the dominant western part of the country by the breadth of India, subsequently declared its independence as Bangladesh.) Britain also granted independence to Ceylon, subsequently renamed Sri Lanka.

Japan, devastated by American bombardment, including the only two atomic bombs ever exploded during hostilities, underwent nearly five years of occupation by American military forces, during which it radically reshaped virtually all of its major institutions (with the notable exception of the imperial dynasty) under the supervision of American authorities, emerging as a truly democratic nation. The outbreak of the Korean War, coincident with the restoration of Japanese sovereignty, provided a powerful economic stimulus for Japan, which it used to good effect. Within a few decades Japan had become the second largest economy in the world.

China, which had more or less successfully resisted Western incursions for more than two centuries, underwent two radical changes—revolutions—in the twentieth century, as well as decades of civil and international war. In 1911 a group of young, Western-oriented reformers overthrew the venerable Qing (Ch'ing) dynasty and attempted to create a modern democratic republic. They never really gained control of the country, however, and in the 1930s the Japanese invasions, first of Manchuria and then of China proper, prevented any sustained economic development. Immediately after the end of World War II the Chinese Communist party began its assault on the government, which it eventually drove out in 1949. For a few years the Chinese Communists allied themselves with the Soviet Union and attempted to model their economy along Soviet lines. After breaking with the Soviet Union in 1960 they tried various other experiments and expedients without great success. Eventually, in the 1970s, they reestablished diplomatic and economic relations with the United States and other Western nations, and began a new era of economic development with a curious blend of public and private enterprise.

Decolonization and the creation of new nations, together with the attempts of other Third World nations (e.g., those of Latin America) to modernize and achieve sustained economic development, introduced a new element into international economic relationships. A North-South dimension (developed versus underdeveloped—euphemistically referred to as ''developing'') was added to the East-West confrontation. Partly to facilitate constructive dialogue and prevent outright hostilities, a number of new international organizations were established.

A few international organizations date from the nineteenth century—for example, the International Red Cross, established in Geneva in 1864, and the Universal Postal Union, created in 1874 with headquarters in Berne, Switzerland—but the twentieth century has been especially prolific in creating them. Literally hundreds of organizations exist, mostly of little or no economic significance, but some affect the performance of the world economy in major ways.

The League of Nations, created by the Treaty of Versailles in 1919, was intended by Woodrow Wilson, whose brainchild it was, to guarantee world peace, and thus prosperity. The failure of the U.S. Senate to ratify the treaty, and of the United States to join the League, along with weaknesses in its structure, condemned it to failure. Its economic achievements were more lasting than its political ones, but scarcely impressive. Its Economic Section collected and published useful statistics and technical reports, and introduced standardized accounting methods, but it proved powerless to deal with the really important economic issues of the interwar period. One of the League's subagencies, the International Labor Organization (ILO), survived the League and continues as a subagency of the United Nations. It investigates working and living conditions of workers, publishes its findings, and makes recommendations regarding them; but its recommendations are not binding.

The League's successor, the United Nations, has had a slightly better record as a peacekeeper and has spawned several specialized agencies dealing with economic and related affairs. Two of those actually preceded the creation of the United Nations and have played a major role in the world economy: the International Monetary Fund (IMF) and the International Bank for Reconstruction and Development (World Bank), both authorized at a conference in July 1944 at Bretton Woods, New Hampshire, in anticipation of the Allied victory in World War II. More will be said about them in Chapter 15.

Several other international and supranational organizations, notably the Organization for European Economic Cooperation (OEEC) and the European Economic Community (EEC, or Common Market), are also discussed in Chapter 15.

The Role of Government

Another major institutional change affecting all nations in the twentieth century is the greatly enlarged role of government in the economy. In the heyday of economic nationalism—the seventeenth century—absolute monarchs attempted to bend the economy to their wills, but their resources were too limited and their instruments too puny for them to have much effect. In the nineteenth century, on the other hand, under the influence of the classical economists, governments in general deliberately limited their participation in the economy. The growth of government in the twentieth century is related in part to the financial necessities of the two world wars and to other considerations of national defense—but only in part.

In the Soviet Union, and in other Soviet-style economies, the government assumed total responsibility for the economy through a system of comprehensive economic planning and control. (The major changes in the Soviet system in the early 1990s, and the collapse of the Soviet empire, are discussed in Chapter 16.) During the two world wars most of the belligerent nations also adopted very far-reaching controls and government participation, but with some exceptions (to be noted later), economically productive activities in peacetime in the advanced industrial democracies are assumed to be the province of private individuals and corporations. This does not, however, mean a reversion to nineteenth-century notions (or myths) of laissez faire. In the interwar period all governments attempted, generally with little success, to pursue policies of economic recovery and stabilization. After World War II they tried even more deliberately, with greater sophistication, and generally with greater success. Most adopted some form of economic planning, although not as comprehensive or compulsory as that of the Soviet Union. For this the label of "mixed economies" has been applied to the nations of western Europe.

The exceptions mentioned previously are of two types: directly productive activities carried out by or on behalf of the government, and transfer payments, or redistribution of income by means of taxation and expenditure. Even in the nineteenth century municipal governments, for example, operated waterworks, gasworks, and other public utilities, and in a few instances national governments built or subsequently nationalized railways (see Chapters 8 and 12). In the twentieth century state-owned industries became much more common, sometimes as a result of the failure of private enterprise (e.g., passenger railways in the United States), at other times because of the ideological commitment of the ruling political party. More will be said about these cases in subsequent chapters.

The other main reason for the growth of government—transfer payments—also has roots in the late nineteenth century, but it did not achieve large dimensions until afer World War II. In the 1880s Bismarck, the German chancellor, introduced compulsory sickness and accident insurance for workers and a very limited pension system for the overaged and disabled, largely for paternalistic reasons. These innovations were gradually copied and extended in other countries, mostly after World War I; the United States, for example, did not adopt comprehensive social insurance (including unemployment compensation) until the New Deal reforms of the 1930s. After World War II, as a result of strong political pressures, most democratic governments greatly extended their systems of social security and other transfer payments. For this reason they have become known in some quarters as "welfare states."

A few figures will give meaning to the phrase "growth of government." In the nineteenth century, in peacetime, government expenditure as a percentage of national income was generally less than 10 percent, sometimes much less. For example, in the United States for the years 1900 to 1916, federal government expenditures amounted to only 2.5 percent of national income (to be sure, the sum of state and local government expenditures in the United States in those days amounted to more than the federal budget). But even in Great Britain on the eve of World War I, when the country was engaged in an armaments race with Germany, total government expenditures amounted to only 8 percent of national income. During the war, on the other hand, while government expenditures rose to 28 percent of national income in the United States, it exceeded 50 percent in most of the European belligerents. After the war government spending came down, but not by much and not for long. For example, in the United Kingdom government expenditures averaged about 20 percent throughout the 1920s and 1930s, much of it accounted for by interest payments on the war debt and much of the rest by the "dole," Britain's system of unemployment compensation. The U.S. federal budget, after falling briefly to less than 5 percent of national income in the late 1920s, averaged about 12 percent during the New Deal years and then rocketed to more than 50 percent during World War II. Once again, after the war expenditures, as a proportion of much larger national incomes, came down modestly, but not for long. In the 1950s in both western Europe and the United States government expenditures amounted to between 20 and 30 percent of national income, depending on the extent of public enterprise, but they have since risen to between 30 and 40 percent, or more.

The Forms of Enterprise

The joint-stock limited-liability company, or modern corporation, was already well established in the leading industrial countries at the beginning of the twentieth century, but for the most part it was used only in large-scale, capital-intensive industries. In other activities, such as wholesaling and retailing, artisanal production, the service trades, and especially agriculture, the unincorporated family firm predominated. The long-term trend, however, favored the spread of the corporate form of enterprise to ever-wider spheres of activity. Large corporate multiunit enterprises, "chain stores," came to dominate retailing in industries as diverse as fresh produce and high-technology electronics; they integrated backward to the production stage, in many cases eliminating the wholesale function altogether. In other cases producers of sewing machines, farm machinery, and automobiles, for example, integrated forward, relying on franchised dealers to handle the retailing function. A related development was the appearance of the corporate conglomerate, huge corporations engaged in the production and sale of dozens or even hundreds of products, ranging from heavy producer goods to such consumer goods as cosmetics and fashions. This development was facilitated by the use of holding companies, corporations whose only business was to own (and manage) other corporations. Although the corporate form of enterprise was brought into existence initially by the technologically determined requirements of large-scale production, and accordingly favored ever-larger units of organization, it could also be adapted to smaller scales of operation; by the latter half of the century even independent professional practitioners, in law and medicine, for example, were incorporated for fiscal purposes.

These trends in the use of the corporate form of organization were pioneered primarily in the United States in the latter part of the nineteenth century, but they spread rapidly in Europe and elsewhere in the twentieth century. One reason for this was that it enabled enterprises to compete successfully with another primarily American phenomenon, the multinational firm. Multinational firms were not entirely novel, nor were they exclusively American—the Medici bank in the fifteenth century, based in Florence, had branches in other countries—but until the twentieth century they were relatively rare. Now, they are quite common. An outstanding example is the Nestlé Company (food products), headquartered in the small city of Vevey, Switzerland, but with production and sales facilities on every continent and in virtually every country of the world. In recent years its sales turnover has exceeded the Swiss government budget!

Organized Labor

At the beginning of the twentieth century the right of workers to organize and bargain collectively was recognized in most Western nations, and in a few (e.g., Great Britain and Germany), organized labor wielded considerable power in the labor market. Even in those countries, however, organized labor was a minority, no more than a fifth or a fourth of the total labor force. The interwar years witnessed a growth in union membership in the industrial nations and a spread of labor organization to other, less developed nations. In the United States, for example, union membership in the 1920s amounted to only about one-tenth of the nonagricultural labor force, mainly skilled workers; by 1940, largely as a result of New Deal legislation favorable to organized labor and a campaign by the latter to organize unskilled and semiskilled factory workers, the proportion had grown to more than one-quarter. It reached a peak of almost 36 percent in 1945, as a result of the marked expansion of the war industries, then dropped off slightly. Since the mid-1950s, with the growth of service occupations and high-technology industries, union membership as a percentage of the labor force has again declined to less than one-quarter.

Trends in union membership in western Europe, although not identical with those in the United States, have been similar. A major difference, however, is that in Europe trade unions are much more closely identified with specific political parties than in the United States. In Great Britain, for example, the Labour party is supported mainly, although not exclusively, by union members and other, unorganized, workers. In a stunning upset in the general election of 1945, immediately after the war, it won a clear victory over the wartime prime minister, Winston Churchill, and, with a socialist program proceeded to nationalize several key industries. Although it lost the election of 1952, it alternated in power with the Conservative party for the next thirty years; in the wake of a decisive defeat in 1979, however, it split in two, with the less doctrinaire members forming a new Social Democratic party.

The Social Democratic party in pre-World War I Germany was a worker-supported party, the largest in Germany, although it never succeeded in forming a government before the war. Under the Weimar Republic it participated in most of the coalition governments of that fragile democracy, but with the advent of the Nazi dictatorship of Adolph Hitler in 1933 it was forcibly dissolved, along with all other political parties except the Nazis.

The Nazis abolished not only the political parties but the trade unions as well. All workers were compelled to become members of the Labor Front, an organization run by members of the Nazi party to ensure labor discipline. Similar developments took place in Italy, the Soviet Union, and other totalitarian countries. At the time of the 1917 revolution the members of the Russian trade unions (which existed illicitly under the tsarist regime) expected they would be called on to play a leading role in the reform and reorganization of the Russian economy and society. They were grievously disappointed when the government used the unions, not as defenders of workers' rights, but as instruments to instill labor and party discipline.

14

International Economic
Disintegration

Fundamental economic change normally occurs over a long period of time. The consequences of changes in population, resources, technology, and even institutions may spread out over a period of years, decades, and even centuries. Political changes, on the other hand, can occur quite abruptly, in a period of days or weeks, sometimes bringing in their wake abrupt economic changes as well. Such was the case with World War I. The intricate but fragile system of the international division of labor that had grown gradually in the century prior to August 1914, and that had brought unprecedented levels of well-being and even affluence to the populations of Europe and some overseas outposts of Western civilization, suddenly disintegrated with the outbreak of war. After more than four years of the most destructive war the world had yet witnessed, world political leaders sought a "return to normalcy," but, like Humpty Dumpty in the nursery rhyme, the world economy could not easily be put together again.

The Economic Consequences of World War I

Before it became known in history books as the World War (later the First World War) the war of 1914–18 was known to millions of Europeans who experienced it as the Great War. In retrospect it seems a tragic prelude to the war of 1939–45, but for the generation who lived before 1939 its emotional and psychological as well as physical impact clearly justified the name. For concentrated destructiveness it surpassed anything in human history until the mass air raids and atomic bombings of World War II. Military casualties numbered about 10 million killed and twice that number seriously wounded; direct civilian casualties also amounted to about 10 million, and another 20 million died from war-caused famine and disease. Estimates of the direct money costs of the war (i.e., for military operations) range from 180 to 230 billion dollars (1914 purchasing power), and the indirect money costs as a result of property damage to more than 150 billion dollars. Most of the damage—destruction of housing, industrial plants and equipment, mines, livestock and farm equipment, transportation and communications facilities—occurred in northern France, Belgium, a small area of northeastern Italy, and the battlefields of eastern Europe.

Ocean shipping also suffered greatly, primarily as a result of submarine warfare. Such estimates are subject to a large margin of error, and are probably too low. Not included are losses of production occasioned by shortages of manpower and raw materials for industry, excessive depreciation and depletion of industrial plant and equipment without adequate maintenance and replacement, and overcropping and lack of fertilizer and draft animals in agriculture. In central and eastern Europe, cut off from economic relations with the rest of the world and further disrupted by the marching and countermarching of armies, the fall in agricultural output reduced large areas to the point of mass starvation.

Even more damaging to the economy, in the long run, than the physical destruction, the disruption and dislocation of normal economic relations did not cease with the war itself but continued to take its toll in the interwar period. Prior to 1914 the world economy had functioned freely and, on the whole, efficiently. In spite of some restrictions in the form of protective tariffs, private monopolies, and international cartels, the bulk of economic activity, both domestic and international, was regulated by free markets. During the war the governments of every belligerent nation and those of some nonbelligerents imposed direct controls on prices, production, and the allocation of labor. These controls artificially stimulated some sectors of the economy, and by the same token artificially restricted others. Although most of the controls were removed at war's end, the prewar relationships did not reestablish themselves either quickly or easily.

An even more serious problem resulted from the disruption of foreign trade and the forms of economic warfare to which the belligerents—Britain and Germany, in particular—resorted. Before the war Britain, Germany, France, and the United States, as the world's leading industrial and commercial nations, were also one another's best customers and leading suppliers. Commercial intercourse between Germany and the others halted immediately, of course, although the United States, in its neutral phase, attempted to maintain relations. It was hampered in this by the retaliatory actions of both Britain and Germany. Britain, with its mastery of the seas, immediately imposed a blockade on German ports, as it had done against Napoleon a century earlier. The blockade was, on the whole, quite effective. Not only were German ships prohibited from the seas, but the British fleet also harassed neutral shipping and sometimes confiscated their cargoes. This caused some friction with the United States, but that was offset by the countermeasures of the Germans. Unable to attack the British navy frontally, especially after the battle of Jutland, the Germans resorted to submarines—a novel instrument of war—in an effort to halt the flow of overseas supplies to Britain. The submarines avoided the British navy as much as possible, but attacked unarmed vessels, neutral as well as British, passenger as well as merchant ships. The sinking of the British liner *Lusitania* off the coast of Ireland in 1915, with the loss of more than a thousand lives (including about 100 Americans) brought a strong protest from the United States. For a time the German High Command moderated its policy, but in January 1917, desperate to bring Britain to its knees, it unleashed unrestricted submarine warfare. That was a major cause of America's entry into the war which, in turn, ensured the eventual Allied victory.

Closely related to the disruption of international trade and the imposition of government controls, the loss of foreign markets had even longer-lasting effects. Germany, of course, was completely cut off from overseas markets and, without the ingenuity of its scientists and engineers (for example, the inventors of the Haber-Bosch process for the fixation of atmospheric nitrogen, an essential ingredient for both fertilizer and gunpowder), would have been forced to capitulate much sooner than it did. But even Britain, with its control of the seas and large merchant marine, had to divert resources from normal uses to war production. By 1918 its industrial exports had fallen to about half their prewar level. Consequently, overseas nations undertook to manufacture for themselves or to buy from other overseas nations goods they had formerly purchased in Europe. Several Latin American and Asian countries established manufacturing industries, which they protected after the war with high tariffs. The United States and Japan, which had already developed important manufacturing industries before the war, expanded into overseas markets formerly regarded as the exclusive preserve of European manufacturers. The United States also greatly increased its exports to the Allied and neutral countries of Europe.

The war also upset the equilibrium in world agriculture. By greatly increasing the demand for foodstuffs and raw materials at the same time that some areas went out of production or were cut off from markets, the war stimulated production in both established areas, such as the United States, and relatively virgin areas such as Latin America. This led to overproduction and falling prices in the 1920s. Wheat, sugar, coffee, and rubber were especially vulnerable. American farmers increased their acreage in wheat during the war, and also bought new land at war-inflated prices. When prices fell many were unable to pay off their mortgages and went into bankruptcy. Malaya, the source of much of the world's natural rubber, and Brazil, which accounted for 60 to 70 percent of the world's coffee, both tried to raise prices by holding supplies off the market; but as they did so new producers came in and drove prices down again. Producers of cane sugar in the Caribbean, South America, Africa, and Asia suffered from the protected and subsidized producers of beet sugar in Europe and the United States.

In addition to losing foreign markets, the belligerent nations of Europe suffered a further loss of income from shipping and other services. The German merchant marine, completely bottled up during the war, had to be handed over to the Allies in payment of reparations after the war (see p. 350). Germany's submarine warfare took a heavy toll on the British merchant fleet, while the United States, with a government-subsidized program of wartime shipbuilding became a major competitor in international shipping for the first time since the American Civil War. London and other European financial centers likewise lost some of their income from banking, insurance, and other financial and commercial services that were transferred to New York and elsewhere (e.g., Switzerland) during the war.

Another major loss from the war was that of income from foreign investments (and in many cases the investments themselves). Before the war Britain, France, and Germany were the most important foreign investors. Since Britain and France imported more than they exported, the income from foreign investments helped pay for the import surplus. Both were obliged to sell some of their foreign investments to finance the purchase of urgently needed war materials. Other investments declined in value as a result of inflation and related currency difficulties. Still others suffered default or outright repudiation, notably the large French investments in

Russia, which the new Soviet government refused to recognize. Overall, the value of British foreign investments declined by about 15 percent (in contrast to a continuously rising value before the war), and those of France by more than 50 percent. Germany's investments in belligerent countries were confiscated during the war, and subsequently all were liquidated for reparations payments. The United States, on the other hand, converted itself from a net debtor into a net creditor as a result of its booming export surplus and its large loans to the Allies.

A final dislocation in both national and international economies resulted from inflation. The pressures of wartime finance forced all belligerents (and some non-belligerents) except the United States off the gold standard, which had served in the prewar period to stabilize, or at least to synchronize, price movements (see Chapter 11). All the belligerents resorted to large-scale borrowing and the printing of paper money to finance the war. This caused prices to rise, though they did not all rise in the same proportion. At the end of the war prices in the United States averaged about 2.5 times higher than they were in 1914; in Britain they were about 3 times higher, in France about 5.5 times, in Germany more than 15 times, and in Bulgaria more than 20 times higher than in 1914. The great disparity in prices, and consequently in the values of currency, made the resumption of international trade difficult, and also caused severe social and political repercussions.

Economic Consequences of the Peace

The Peace of Paris, as the postwar settlement came to be known, instead of attempting to solve the serious economic problems caused by the war, actually exacerbated them. The peacemakers did not intend this to happen (except in the treatment of Germany); they simply failed to take account of economic realities. Two major categories of economic difficulty resulted from the peace treaties: the growth of economic nationalism and monetary and financial problems. For neither of these difficulties were the peace treaties solely to blame, yet in both the treaties added to the problems instead of ameliorating them.

The actual treaties were named for the suburbs of Paris in which they were signed. The most important was the Treaty of Versailles, with Germany. It restored Alsace-Lorraine to France and permitted the French to occupy the coal-rich Saar valley for fifteen years. It gave most of West Prussia and a part of mineral-rich Upper Silesia to newly recreated Poland. With other minor border adjustments, it deprived Germany of 13 percent of its prewar territory and 10 percent of its 1910 population. These losses included almost 15 percent of its arable land, about three quarters of its iron ore, most of its zinc ore, and a quarter of its coal resources. Of course, its colonies in Africa and the Pacific had already been occupied by the Allies (including Japan), who were confirmed in their possession.

In addition, Germany had to surrender its navy, large quantities of arms and ammunition, most of its merchant fleet, 5000 locomotives, 150,000 railroad cars, 5000 motor trucks, and various other commodities. It also had to accept restrictions on its armed forces, Allied occupation of the Rhineland for fifteen years, and several other damaging or merely humiliating conditions. Most humiliating of all was the famous "war guilt" clause, Article 231 of the Treaty of Versailles, which declared that Germany accept "the responsibility of Germany and her allies for causing all the loss and damage . . . as a consequence of the war. . . ." The statement was intended to justify Allied claims to monetary "reparations," but the Allies themselves were badly divided on both the nature and amount of the reparations, to the extent that they could not agree in time for the signing of the treaty and had to appoint a Reparations Commission with instructions to report by May 1, 1921. John Maynard Keynes, an economic adviser to the British delegation at the peace conference, was so distressed that he resigned his position and wrote a best-selling book, *The Economic Consequences of the Peace*, in which he predicted dire consequences, not only for Germany but for all Europe, unless the reparations clauses were revised. Although Keynes's reasoning has been disputed, the subsequent course of events seemed to bear out his prediction.

The break-up of the Austro-Hungarian Empire in the last weeks of the war resulted in two new states, Austria and Hungary, each much smaller than the old areas of the same names. Czechoslovakia, created from former Austrian and Hungarian provinces, and Poland, recreated from former Austrian, German, and (mostly) Russian lands, also became new nation-states. Serbia obtained the South Slav provinces of Austria-Hungary and united with Montenegro to become Yugoslavia. Rumania, allied with the Western powers, obtained much territory from Hungary, whereas Bulgaria, a vanquished enemy, lost land to Greece, Rumania, and Yugoslavia. Italy gained Trieste, the Trentino, and the German-speaking South Tyrol from Austria. The hoary Ottoman Empire lost virtually all of its territory in Europe except for the immediate hinterland of Istanbul, as well as the Arab provinces of the Near East; in 1922 it succumbed to a revolution that created a Turkish national republic.

The prewar Austro-Hungarian Empire, however anachronistic politically, had performed a valuable economic function by providing a large free trade area in the Danube basin. The new states that issued from the break-up of the empire were jealous of one another and fearful of great power domination. They therefore asserted their nationhood in the economic sphere by trying to become self-sufficient. Although complete self-sufficiency was manifestly impossible because of their small size and backward economies, their efforts to achieve it hindered the economic recovery of the entire region and added to its instability. The height of absurdity came with the disruption of transportation. Immediately after the war, with borders in dispute and continued border skirmishes, each country simply refused to allow the trains on its territory to leave. For a time trade came almost to complete standstill. Eventually agreements overcame these extremes of economic nationalism, but other types of restrictions remained.

Economic nationalism was not limited to the new states that emerged from the break-up of empires. During its civil war Russia simply disappeared from the international economy. When it reemerged under the Soviet regime its economic relations were conducted in a manner completely different from any previously experienced. The state became the sole buyer and seller in international trade. It bought and sold only what its political rulers regarded as strategically necessary or expedient.

In the West, countries that had formerly been highly dependent on international trade resorted to a variety of restrictions, including not only protective tariffs but also more drastic measures such as physical import quotas and import prohibitions. At the same time they sought to stimulate their own exports by granting export subsidies and other measures. Great Britain, formerly the champion of free international trade, had imposed tariffs during the war as a measure of war finance and to save shipping space. They remained (and were increased in both number and rate) after the war, at first on a "temporary" basis, but after 1932 as official protectionist policy. Britain also negotiated numerous bilateral trade treaties in which it abandoned the principle of the most favored nation that had done so much to extend trade in the nineteenth century.

The United States, which already had relatively high tariffs before the war, raised them to unprecedented levels thereafter. The Emergency Tariff Act of 1921 placed an absolute embargo on imports of German dyestuffs. (The dyestuff industry had not even existed in the United States before the war; it began with the confiscation of German patent rights during the war.) The Fordney-McCumber Tariff Act of 1922 contained the highest rates in American tariff history, but even those were surpassed by the Smoot-Hawley Tariff in 1930, which President Hoover signed into law in spite of the published protests of more than a thousand economists.

The adverse consequences of this neomercantilism, as such policies were called, did not stop with the immediate application of the laws in question. Each new measure of restriction provoked retaliation by other nations whose interests were affected. For example, after the passage of the Smoot-Hawley Tariff dozens of other nations immediately responded by raising their tariffs against American products. Although total world trade had more than doubled in the two decades before the war, it rarely achieved the prewar level in the two decades that followed. During the same period the foreign trade of European countries, which had also doubled in the two prewar decades, equaled the prewar figure in but a single year, 1929. In 1932 and 1933 it was lower than it had been in 1900. Such exaggerated economic nationalism produced the opposite of what its formulators intended—lower instead of higher levels of production and income.

The monetary and financial disorders caused by the war and aggravated by the peace treaties eventually led to a complete breakdown of the international economy. The problem of reparations was at the heart of these disorders, but the "reparations tangle" was, in reality, a complex problem involving inter-Allied war debts and the whole mechanism of international finance. The insistence of Allied statesmen, especially Americans, on treating each question in isolation, instead of recognizing relationships, was a major factor in the subsequent debacle.

Until 1917 Britain was the chief financier of the Allied war effort. By that year it had loaned about $4 billion to its allies. When the United States entered the war it took over the role of chief financier from Britain, whose financial resources were almost exhausted. Altogether, by the end of the war inter-Allied debts amounted to more than $20 billion, about half of which had been loaned by the U.S. government. (The latter included more than $2 billion advanced by the American Relief Agency between December 1918 and 1920.) Britain had advanced about $7.5 billion, roughly twice as much as it received from the United States, and France about $2.5 billion, roughly equal to the amount it had borrowed. Among the European allies the loans had been in name only; they expected to cancel them at the end of the war. They naturally regarded the American loans in the same light, all the more in that the United States had been a latecomer to the war, had contributed less in both manpower and materials, and had suffered negligible war damage. The United States, however, regarded the loans as commercial propositions. Although it agreed after the war to reduce the rate of interest and lengthen the period of repayment, it insisted on repayment of the principal in full.

At this point the reparations issue intruded. France and Britain demanded that Germany pay not only damages to civilians (reparations proper), but also the entire cost incurred by the Allied governments in prosecuting the war (an indemnity). President Wilson made no claims for the United States and tried to dissuade the others from pressing theirs; but his argument was not strong inasmuch as he insisted that the Allies should repay their war debts. The French wanted the United States to cancel the war debts but insisted on collecting reparations. Lloyd George, the British prime minister, suggested that both reparations and war debts be canceled, but the Americans stubbornly refused to recognize any relationship between the two. The American attitude was summed up in a remark subsequently made by President Coolidge: "They hired the money, didn't they?" The eventual compromise required Germany to pay as much as the Allies thought they could possibly extract, but in deference to Wilson the entire amount was called "reparations."

Meanwhile the Germans had begun to pay in cash and in kind (coal, chemicals, and other goods) as early as August 1919, even before the treaty was signed, and long before the total bill was known. These payments were to be credited toward the final amount. Finally, at the end of April 1921, only a few days before the deadline of May 1, the Reparations Commissions informed the Germans that the total would amount to 132 billion gold marks (about $33 billion), a sum greater than twice the German national income.

In fact, with the weakened European economies and the precarious state of the international economy, France, Britain, and the other Allies could repay the United States only if they received an equivalent amount in reparations. But Germany's capacity to pay reparations depended ultimately on its ability to export more than it imported to gain the foreign currency or gold in which the payments had to be made. The economic restrictions imposed on it by the Allies, however, made it impossible for Germany to obtain a surplus adequate for the annual payments. In the late summer of 1922 the value of the German mark began to decline disastrously as a result of the heavy pressure of reparations payments (and also as a result of the actions of speculators). By the end of the year the pressure was so great that Germany ceased payments altogether.

French and Belgian troops occupied the Ruhr in January 1923, took over the coal mines and railroads, and attempted to force the German mineowners and workers to deliver coal. The Germans replied with passive resistance. The government printed huge quantities of paper money for compensation payments to Ruhr workers and employers, setting in motion a wave of uncontrolled inflation. The German gold mark was valued at 4.2 to the dollar in 1914. At war's end the paper mark stood at 14 to the dollar; by July 1922 it had fallen to 493, and by January 1923 to 17,792. Thereafter the fall in the value of the mark proceeded exponentially until November 15, 1923, when the last official transaction recorded an exchange value for the dollar of 4.2 *trillion* (4,200,000,000,000)! The mark was literally worth less than the paper on which it was printed. At that point the German monetary authorities demonetized the mark and substituted a new monetary unit, the *rentenmark,* equal in value to 1 trillion of the old marks.

The adverse consequences of the inflation could not be confined to Germany. All of the successor states of the old Habsburg Monarchy, Bulgaria, Greece, and Poland suffered similar runaway inflations. The par value of the Austrian crown was five to the dollar; in August 1922 it was quoted at 83,600, at which time the League of Nations sponsored a stabilization program that succeeded by 1926, with the introduction of a new currency unit, the schilling. Even the French franc suffered; before the war the gold franc exchanged at 5 to the dollar, but in 1919 it had fallen by more than half, to 11 to the dollar. During the French occupation of the Ruhr it rose at first, then fell abruptly as it became obvious that the occupation was not achieving its purpose. After reaching a low of 40 to the dollar, the government finally stabilized the franc at 25.5 in 1926.

As Keynes had predicted, the international economy was confronted with a grave crisis. The French withdrew from the Ruhr at the end of 1923 without having accomplished their objective, the resumption of German reparations. A hastily convoked international commission under the chairmanship of Charles G. Dawes, an American investment banker, recommended a scaling down of annual reparations payments, reorganization of the German Reichsbank, and an international loan of 800 million marks (about $200 million) to Germany. The so-called Dawes Loan, most of which was raised in the United States, enabled Germany to resume reparations payments and return to the gold standard in 1924. It was followed by a further flow of American capital to Germany in the form of private loans to German municipalities and business corporations, who borrowed extensively in the United States and used the proceeds for technical modernization and "rationalization." In the process the German government obtained the foreign exchange it needed to pay reparations.

The disastrous inflation left deep scars on German society. The unequal incidence of inflation on individuals resulted in drastic redistributions of income and wealth. While a few clever speculators gained enormous fortunes, most citizens, especially the lower middle classes and those living on fixed incomes (pensioners, bondholders, many salaried employees), saw their modest savings wiped out in a matter of months or weeks, and suffered a severe decline in their standard of living (Fig. 14-1). This made them susceptible to the appeals of extremist politicians. Significantly, both Communists and Nationalists made large gains at the expense of the moderate democratic parties in the Reichstag elections of 1924.

Economic problems loomed large in postwar Britain. Even before the war Britain's unusually great dependence on international trade and overcommitment to lines of industry that were rapidly becoming obsolete had guaranteed that the British would face a difficult period of readjustment in the twentieth century. During the war they lost foreign markets, foreign investments, a large part of their mercantile marine, and other sources of overseas income. Yet they depended as much as ever on imports of food and raw materials, and they found themselves with even greater

FIGURE 14–1. Soup line. Soup is doled out to the poor in Berlin after World War I at a municipal lodging house. (The Bettman Archive.)

worldwide responsibilities as the strongest of the victors in Europe and as the administrator of new territories overseas. Export they must, yet factories and mines lay idle while unemployment mounted. In 1921 more than 1 million workers— about one-seventh of the labor force—had no work; in the 1920s the rate of unemployment rarely fell below 10 percent, and in the worst years of the depression it mounted to more than 25 percent.

The government's measures to deal with its economic problems were timid, unimaginative, and ineffective. Its only solution for unemployment was the dole, a system of relief payments that was entirely inadequate to support the families of the unemployed while it placed a heavy burden on an already overstrained budget. For the rest, government economic policy consisted mainly in paring expenditures to the bone, thus depriving the nation of urgently needed expansion and modernization of its schools, hospitals, highways, and other public works. The single forthright initiative taken by the government in the economic sphere resulted in disaster.

Britain had abandoned the gold standard in 1914 as a measure of war finance. Given London's prewar position as the undisputed center of the world's financial markets, strong pressures existed for a quick return to the gold standard to prevent further erosion, begun during the war, of its financial preeminence. The major unresolved questions were (1) how soon could it return, and (2) at what value for the pound sterling? The answer to the first question depended on the accumulation of gold reserves by the Bank of England; by general consensus, these were deemed adequate by the mid-1920s. The answer to the second was more controversial. Under the prewar system the pound was equal to \$4.86, but the United States had remained on the gold standard throughout the war. Britain had experienced a higher rate of inflation than the United States. To return to gold at the prewar parity would place British industry at a competitive disadvantage with respect to the United States and to other countries that maintained parity with the dollar, or adopted an even lower rate of exchange. On the other hand, the British had always manifested a strong desire to maintain traditions, especially in such important questions as finance. Moreover, since most British foreign investments were denominated in gold or sterling, to return to gold at a lower rate would penalize the owners of those investments. In 1925 the chancellor of the exchequer, Winston Churchill, who had earlier switched his allegiance from the Liberals to the Conservatives, resolved to return Britain to the gold standard at the prewar parity. To keep British industry competitive, this necessitated a fall in prices of approximately 10 percent, which in turn necessitated an equivalent fall in wages. The overall effect was a redistribution of income in favor of the rentiers and at the expense of workers.

The coal industry was one of the most severely affected by the loss of foreign markets and higher costs. Coalminers were among the most radical of British workers; they had already staged several major strikes in the early postwar years. When faced with a wage cut as a result of the return to the gold standard, the miners went out on strike on May 1, 1926, and persuaded many other trade unions to join them in what was intended to be a general strike. About 40 percent of British trade union members joined them, mainly those in public utilities and similar industries, but the strike lasted only ten days, ending in defeat for the unions. Middle-class volunteers manned essential services, and the trade union leaders gave up rather than risk civil war in the face of the government's strong opposition. Brief as it was, the general strike left a bitter legacy of class division and hatred, which made concerted national action against both domestic and international problems even more difficult.

In spite of Britain's problems, most of Europe prospered in the late 1920s. For five years, from 1924 to 1929, it seemed that normality had indeed returned. Reconstruction of physical damage had been largely achieved; the most urgent and immediate postwar problems had been solved; and under the newly created League of Nations a new era in international relations apparently had dawned. Most countries, especially the United States, Germany, and France, experienced a period of prosperity. Yet the basis of that prosperity was fragile, depending on the continued voluntary flow of funds from America to Germany.

The Great Contraction, 1929–33

Unlike Europe, the United States emerged from the war stronger than ever. In economic terms alone, it had converted from a net debtor to a net creditor, had won new markets from European producers both at home and abroad, and had established a highly favorable balance of trade. With its mass markets, growing population, and rapid technological advance it seemed to have found the key to perpetual prosperity. Although it experienced a sharp depression in 1920–21 along with Europe, the drop proved to be brief, and for almost a decade its growing economy experienced only minor fluctuations. Social critics who insisted on revealing the disgraceful conditions in urban and rural slums, or who pointed out that the new prosperity was shared most unequally between the urban middle classes, on the one hand, and factory workers and farmers, on the other, were dismissed by the former as cranks who did not share the American dream. For them the "new era" had arrived.

In the summer of 1928 American banks and investors began to cut down their purchases of German and other foreign bonds in order to invest their funds through the New York stock market, which accordingly began a spectacular rise. During the speculative boom of the "great bull market" many individuals with modest incomes were tempted to purchase stock on credit. By the late summer of 1929 Europe was already feeling the strain of the cessation of American investments abroad, and even the American economy had ceased to grow. The U.S. gross national product peaked in the first quarter of 1929, then gradually subsided; U.S. auto production declined from 622,000 vehicles in March to 416,000 in September. In Europe, Britain, Germany and Italy were already in the throes of a depression. But with stock prices at an all-time high, American investors and public officials paid scant heed to these disturbing signs.

On October 24, 1929—"Black Thursday" in American financial history—a wave of panic selling on the stock exchange caused stock prices to plummet and eliminated millions of dollars of fictitious paper values. Another wave of selling followed on October 29, "Black Tuesday." The index of stock prices, which peaked at 381 on September 3 (1926 = 100), fell to 198 on November 13. . . and kept on falling. Banks called in loans, forcing still more investors to throw their stocks on the market for whatever prices they would bring. Americans who had invested in Europe ceased to make new investments and sold existing assets there to repatriate the funds. Throughout 1930 the withdrawal of capital from Europe continued, placing an intolerable strain on the entire financial system. Financial markets stabilized, but commodity prices were low and falling, transmitting the pressure to producers like Argentina and Australia.

The stock market crash was *not* the cause of depression—that had already begun, in the United States as well as Europe—but it was a clear signal that the depression was underway. Monthly automobile production in the United States fell to 92,500 in December, and unemployment in Germany rose to 2 million. By the first quarter of 1931 total foreign trade had fallen to less than two-thirds of its value in the comparable period of 1929.

In May 1931 the Austrian Creditanstalt, of Vienna, one of the largest and most important banks in central Europe, suspended payments. Although the Austrian government froze bank assets and prohibited the withdrawal of funds, the panic spread to Hungary, Czechoslovakia, Rumania, Poland, and especially to Germany, where a large-scale withdrawal of funds took place in June, resulting in several bank failures. Under the terms of the Young Plan, which had replaced the Dawes Plan in 1929 as a method of settling the reparations problem, Germany was obliged to make a further reparations payment on July 1. In the United States President Hoover, forced by circumstances to recognize the interdependence of war debts and reparations, proposed on June 20 a one-year moratorium on all intergovernmental payments of war debts and reparations, but it was too late to stem the panic. France temporized, and the panic spread to Great Britain where, on September 21, the government authorized the Bank of England to suspend payments in gold.

Several countries hard hit by the decline in prices of their primary products, including Argentina, Australia, and Chile, had already abandoned the gold standard. Between September 1931 and April 1932 twenty-four other countries officially departed from the gold standard and several others, although nominally still on it, had actually suspended gold payments. Without an agreed-upon international standard, currency values fluctuated wildly in response to supply and demand, influenced by capital flight and the excesses of economic nationalism, as reflected in retaliatory tariff changes. Foreign trade fell drastically between 1929 and 1932, inducing similar, though less drastic, falls in manufacturing production, employment, and per capita income (see Fig. 14-2).

A principal characteristic of the economic policy decisions of 1930–31 had been their unilateral application: the decisions to suspend the gold standard and to impose tariffs and quotas had been undertaken by national governments without international consultation or agreement, and without considering the repercussions on or responses of the other affected parties. This accounted in large part for the anarchic nature of the ensuing muddle. Finally, in June 1932, representatives of the principal European powers gathered in Lausanne, Switzerland, to discuss the consequences of the end of the Hoover moratorium: should Germany resume reparations payments, and if so under what conditions? Should the European debtors resume war debt payments to the United States? Although the Europeans agreed on a virtual end of reparations, and with it an end to the war debts, the agreement was never ratified because the United States insisted the two issues were entirely separate. Thus, both reparations and war debts simply lapsed; it was left to Hitler in 1933 to declare an end to "interest slavery." Only tiny Finland repaid its small debt to the United States.

The last major effort to secure international cooperation to end the economic crisis was the World Monetary Conference of 1933. Officially proposed by the League of Nations in May 1932 and adopted as a resolution at the Lausanne Conference in July of that year, the draft agenda for the conference looked to agreements to restore the gold standard, reduce tariff and import quotas, and implement other forms of international cooperation. The role of the United States, then engaged in a presidential election, in such a conference was universally regarded as essential. Because of the election, and the unwillingness of the candidates, Hoover and Roosevelt, to commit themselves in advance, the conference was postponed to the spring of 1933, and then again until June to allow Roosevelt to organize his administration. Roosevelt took office in the very depths of the depression; one of his first official actions was to declare an eight-day "bank holiday" to allow the banking system time to reorganize, and most of the measures of the famous "hundred days" involved emergency action to prop up the domestic economy. Among others these included taking the United States off the gold standard, something that the First World War had been unable to do. When the conference finally convened

A. INDEX OF MANUFACTURING PRODUCTION
(1925 – 1929 = 100)

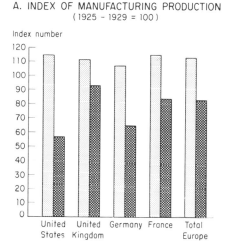

B. INDEX OF PER CAPITA INCOME
(1925 – 1929 = 100)

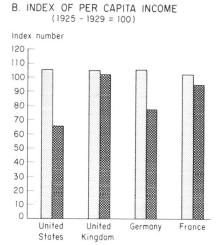

C. INDEX OF EMPLOYMENT
(1937 = 100)

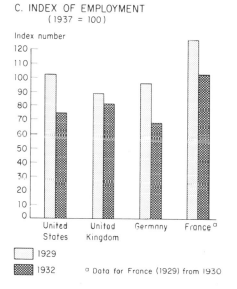

1929

1932   ᵃ Data for France (1929) from 1930

D. FOREIGN TRADE ᵃ

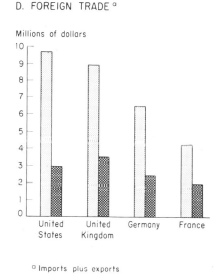

ᵃ Imports plus exports

FIGURE 14–2.   Economic collapse, 1929–1932.

in London in June, Roosevelt sent word that the American government's first responsibility was to restore domestic prosperity, and that he could not enter into any international commitments that would interfere with that task. Dispirited, the delegates to the conference listened to a few meaningless speeches and adjourned in July without taking any meaningful action. Once again, international cooperation had failed.

What caused the depression? After more than sixty years there is still no general consensus on this question. For some the causes were primarily monetary—a drastic decline in the quantity of money in the major industrial economies, the United

States in particular, which spread its influence to the rest of the world. For others the causes are to be found in the "real" sector: an autonomous fall in consumption and investment expenditure, which propagated itself throughout the economy, and the world, by means of the multiplier-accelerator mechanism. Still other explanations have been offered: the prior depression in agriculture, the extreme dependence of Third World countries on unstable markets for their primary products, a shortage or misallocation of the world's stock of gold, and so on. An eclectic view is that no single factor was responsible but an unfortunate concatenation of events and circumstances, both monetary and nonmonetary, occurred to produce the depression. One can further assert that these events and circumstances can be traced in part (perhaps in large part) to World War I and the peace settlement that followed. The breakdown of the gold standard, the disruption of trade, which was never fully restored, and the nationalistic economic policies of the 1920s all have a place in the explanation.

Whatever the precise cause (or causes) of the depression, there is more general agreement on the reasons for its severity and length. They relate to the relative positions and policies of Great Britain and the United States. Before the war Great Britain, as the world's leading commercial, financial, and (until late in the nineteenth century) industrial nation, had played a key role in stabilizing the world economy. Its free trade policy meant that commodities from all over the world could always find a market there. Its large foreign investments enabled countries with sizable deficits in their balances of trade to obtain the resources to balance their payments. Its adherence to the gold standard, together with London's preeminence as a money market, meant that nations with temporary balance of payments problems could obtain relief by discounting bills of exchange or other commercial paper. After the war Britain was no longer able to exercise such leadership—although that was not fully evident until 1931. The United States, clearly the world's dominant economy, was unwilling to accept the role of leader, as exemplified by its immigration policy, its trade (tariff) policy, its monetary policy, and its attitude toward international cooperation. Had the United States pursued more open policies both in the 1920s and especially in the crucial years of 1929 to 1933, the depression almost surely would have been both milder and briefer.

The long-run consequences of the depression also merit notice. Among them were a growth in the role of government in the economy, a gradual change in attitudes toward economic policy (the so-called Keynesian revolution), and efforts on the part of Latin American and some other Third World countries to develop import-substituting industries. The depression also contributed, through the suffering and unrest it caused, to the rise of extremist political movements of both the left and the right, notably in Germany, and thus contributed indirectly to the origins of World War II.

Rival Attempts at Reconstruction

When Franklin Roosevelt took office as the thirty-second president of the United States, on a cold, blustery day in March 1933, the nation lay in the grip of its worst crisis since the Civil War. With more than 15 million unemployed—almost half the industrial work force—industry had virtually shut down and the banking system was on the verge of complete collapse. Nor was the crisis solely economic. An "army" of about 15,000 unemployed veterans of World War I marched on Washington in 1932, only to be dispersed by the regular army under General Douglas MacArthur. In rural areas farmers sometimes took the law into their own hands to prevent the foreclosure of mortgages; and violence ruled in city streets.

In his campaign speeches Roosevelt had called for a "New Deal" for America. In the famous hundred days that followed his inauguration a willing Congress did his bidding, turning out new legislation at an unprecedented rate. In fact, for the four years of his first term the volume of legislation surpassed that of any previous administration. It dealt mainly with economic recovery and social reform in the areas of agriculture, banking, the monetary system, the securities markets, labor, social security, health, housing, transportation, communications, natural resources—in fact, every aspect of the American economy and society.

Perhaps the most characteristic enactment of the entire period was the National Industrial Recovery Act. It created a National Recovery Administration (NRA) to supervise the preparation of "codes of fair competition" for each industry by representatives of the industry itself. Although hailed at the time as a new departure in economic policy, it turned out to be very like the trade association movement that Herbert Hoover had promoted as secretary of commerce in the 1920s (without the element of coercion). It was even more like the wartime economic administration; a number of high government officials had, in fact, served in the wartime mobilization of the economy, including Roosevelt himself, as assistant secretary of the navy. The NRA also bore striking similarities to the Fascist system of industrial organization in Italy, although without its brutality and police-state methods. In essence, it was a system of private economic planning ("industrial self-government"), with government supervision to protect the public interest and guarantee the right of labor to organize and bargain collectively.

In 1935 the Supreme Court declared the NRA unconstitutional. In other areas in which the Court struck down his legislation Roosevelt achieved his goals by new laws, but with respect to industry he altered his stand and initiated a campaign of "trust busting" (also subsequently reversed with the approach of World War II). The industrial recovery had been disappointing, and in 1937 the economy suffered a new recession without having achieved full employment. The United States returned to war in 1941 with more than 6 million still unemployed. Although several of the New Deal reforms were valuable in themselves, the New Deal system as a whole was no more able to cure the depression than contemporary programs in Europe.

No Western nation suffered more from the war than France. Most of the fighting on the Western front had taken place in its richest area. More than half of France's prewar industrial production, including 60 percent of its steel and 70 percent of its coal, had been located in the war-devastated area, which was also among the most important agricultural regions. Most appalling was the loss of life: 1.5 million Frenchmen—half of the prewar male population of military age—had been killed, with half as many more permanently disabled. It is not surprising, therefore, that France demanded that Germany pay for the war.

Counting on German reparations to pay the cost, the French government undertook at once an extensive program of physical reconstruction in the war-damaged areas, which had the incidental effect of stimulating the economy to new production records. When German reparations failed to materialize in the expected amount, the ramshackle methods used to finance the reconstruction took their toll. The problem was compounded by the expensive and ineffective occupation of the Ruhr. The franc depreciated more in the first seven years of peace than during the war. Realizing at last that "the Boche" could not be made to pay, a coalition cabinet containing six former premiers stabilized the franc in 1926 at about one-fifth of its prewar value by undertaking drastic economies and stiff increases in taxation. This solution was more satisfactory than either of the extreme solutions adopted by Britain and Germany, but it alienated both the rentier class, which lost about four-fifths of its purchasing power in the inflation, and the working classes, which bore most of the burden of the increased taxation. Thus, as in Germany, the inflation contributed to the growth of extremism on both right and left.

The franc, when finally stabilized, was actually undervalued in relation to other major currencies. That stimulated exports, hindered imports, and led to an inflow of gold. Thus, the depression struck later in France than elsewhere—not until 1931—and was perhaps less severe, but it was longer lasting; the trough did not come until 1936, and the French economy was still floundering when war broke out in 1939. As it had in other countries, the depression spawned social protest and a new crop of extremist organizations. In 1936 three leftist political parties, the Communists, the Socialists, and the Radicals, formed a coalition, the Popular Front, and won the election of that year, forming a government under the premiership of the venerable Socialist politician, Leon Blum. The Popular Front government nationalized the Bank of France and the railways and enacted a number of reform measures affecting labor, such as a maximum forty-hour work week, compulsory arbitration of labor disputes, and paid vacations for industrial workers. On the larger question of economic recovery, however, the Popular Front was no more successful than previous French or other foreign governments had been, and it broke up in 1938 as foreign affairs increasingly dominated the political scene.

The smaller countries of western Europe, all heavily dependent on international trade, suffered accordingly during the depression, although not all to the same degree. In the 1920s, when Britain and France returned to the gold standard, many of the smaller countries, in eastern as well as western Europe, adopted the gold exchange standard. Their central banks, instead of maintaining reserves of gold with which to redeem their national currencies, maintained deposits with the central banks of the larger countries, which served the same purpose. After Britain's departure from gold in 1931 most of the countries that traded extensively with Britain also left the gold standard and aligned their currencies on the pound sterling. This constituted the "sterling bloc." Their ranks included most of the Commonwealth countries and Britain's colonies, several Middle Eastern countries, and, in Europe, Portugal and the Scandinavian countries. When the United States devalued the dollar in 1933 most of its major trading partners, mainly in Latin America and Canada, sought to align their currencies with the dollar. In Europe that left France at the center of the "gold bloc"—those nations trying to maintain convertibility into gold—which also included Switzerland, Belgium, and the Netherlands. They held out until 1936. (Germany, meanwhile, adopted a novel system of international trade and payments, discussed later.) When the French finally devalued the franc and cut its tie to gold, they did so following a limited resumption of international cooperation in monetary affairs. In the Tripartite Monetary Agreement of 1936 the governments of Britain, France, and the United States undertook to stabilize exchange rates among their respective currencies, to avoid competitive devaluations, and in other ways to contribute to a restoration of the international economy. It was a small step.

In central and eastern Europe, and also in Spain, political developments—the rise of Fascist dictatorships—overshadowed purely economic phenomena; but even they had their economic aspects. The earliest was Italy. Benito Mussolini came to office by legal means in 1922, but quickly consolidated his power with police-state methods. To bolster the ideological underpinnings of his regime Mussolini employed the philosopher Giovanni Gentile to provide a rationalization of Fascism, which was then publicized as Mussolini's own philosophy. Fascism glorified the use of force, upheld war as the noblest human activity, denounced liberalism, democracy, socialism, and individualism, treated material well-being with disdain, and regarded human inequalities as not only inevitable but desirable. Above all, it deified the state as the supreme embodiment of the human spirit.

As an attempted total reconstruction of society, Fascism needed a distinctive form of economic organization. Mussolini brought forth the corporate state, one of the most publicized and least successful innovations of his regime. In principle, the corporate state was the antithesis of both capitalism and socialism. Although it permitted the private ownership of property, the interests of both owners and workers were subordinated to the higher interests of society as a whole, as represented by the state. To accomplish this, all industries in the country were organized into twelve "corporations," corresponding to trade associations rather than business corporations. Workers, proprietors, and the state were represented, with party functionaries holding the key positions. All previously existing labor unions were suppressed. The functions of the corporations included regulating prices, wages, and working conditions and providing social insurance. In practice, insofar as the corporations functioned at all, they acted mainly as capitalistic trade associations whose aim was to increase the income of businessmen and party administrators at the expense of workers and consumers. Other aspects of Fascist economic policy were no more successful. In spite of large public works and armaments programs, Italy suffered severely during the depression; even the argument used by American apologists for Fascism, that "Mussolini made the trains run on time," was untrue.

More successful in combatting the depression than Italy—indeed, more successful than the Western democracies—Nazi Germany was the first major industrial nation to achieve complete recovery. (Among the smaller nations, Sweden had the lowest unemployment rate of any country throughout the 1930s.) From 6 million unemployed in 1933—one-fourth of the labor force—the German economy reached the point in 1939 of having more jobs than workers to fill them. This result was achieved primarily by a large-scale public works program that melded gradually into a rearmament program. In the process Germany developed the first modern highway system (the famed autobahns) and greatly strenthened and expanded its industries, which gave it a decided advantage over its enemies in the early years of World War II.

In place of the voluntary trade unions, suppressed in 1933, the Nazis established compulsory membership in the National Labor Front. They abolished collective bargaining between workers and employers, substituting boards of labor "trustees" with full power to determine wages, hours, and conditions of work. Industrialists were persuaded to cooperate with the new industrial regime by the promise of an end to labor problems if they did and the threat of confiscation and imprisonment if they did not. Unlike the totalitarian regime in Russia, the Nazis did not resort to wholesale nationalization of the economy (although confiscated Jewish enterprises were frequently turned over to party members); they relied on coercion and controls to achieve their objectives.

One of the principal economic objectives of the Nazis was to make the German economy self-sufficient in the event of war. They recalled the crippling effects of the Allied blockade in World War I and wished to be immune to such difficulties in the future. They directed their scientists to develop new *ersatz* or synthetic commodities, both consumer goods and military supplies, that could be manufactured from raw materials available in Germany. The policy of *Autarkie* (self-sufficiency) also determined the character of German trade relations with other nations. Already in 1931, before the advent of the Nazis, Germany had resorted to exchange controls to prevent the flight of capital; Dr. Hjalmar Schacht, Hitler's economic adviser, devised several new, intricate financial and monetary controls to give Germany the advantage in its dealings with other nations. Among these were trade agreements with Germany's neighbors in eastern Europe and the Balkans providing for the barter of German manufactured goods for foodstuffs and raw materials, thus avoiding the use of gold or scarce foreign currencies. Very few German goods were actually shipped, but the policy successfully tied eastern Europe into the German war economy.

Spain, having avoided involvement in World War I, escaped many of the problems and dilemmas that beset other European countries. Its industry actually benefited somewhat from wartime demand, but it was still a predominantly agrarian nation plagued by low-productivity agriculture. During the dictatorship of Miguel Primo de Rivera, from 1923 to 1930, the economy participated in the international prosperity of the period, but the ensuing depression was a factor in the demise of the monarchy and the establishment of the Second Republic in 1931. The international climate of those years was scarcely favorable to the reforms the republicans sought to bring about. In 1936 General Francisco Franco began a bloody, destructive civil war that ended in the overthrow of the republic in 1939 and the institution of an autarkic regime similar in some respects to those of Fascist Italy and Nazi Germany, but without the advanced technology of the latter.

The Russian Revolutions and the Soviet Union

Imperial Russia entered the First World War in the expectation of a quick victory over the Central Powers. That illusion was soon shattered, and as the war wore on the traditional Russian nemeses, inefficiency and corruption, took their toll. By the beginning of 1917 the economy was in shambles. In early March strikes and riots broke out in Petrograd (the renamed St. Petersburg), some soldiers joined the demonstrators and gave them arms, while railway workers prevented other troops from coming in to restore order. On March 12 the leaders of the strikers and soldiers were joined by representatives of the various socialist parties in a Soviet (council) of Workers' and Soldiers' Deputies. On the same day a committee of the Duma (parliament) decided to form a provisional government, and on March 15 obtained the abdication of the tsar. Thus ended the long reign of the Romanovs, in a brief, almost leaderless, and nearly bloodless revolution.

The Provisional Government was a motley collection of aristocrats, intellectuals, and parliamentarians; it contained but one (middle-class intellectual) socialist, Alexander Kerensky. Moreover, it had to share governance (in Petrograd, at least) with the Petrograd Soviet. (Other Soviets were also organized in Moscow and in several provincial cities.) The new regime immediately proclaimed freedom of speech, press, and religion, announced that it would undertake social reform and land redistribution, and promised to summon a constituent assembly to determine Russia's permanent form of government. It also attempted to continue the war against Germany; that proved to be its undoing.

V. I. Lenin, the leader of the Bolshevik faction of Russia's socialist parties, who had spent most of his adult life in exile, returned to Petrograd in April 1917 with the connivance of the German government, which expected him to contribute to social unrest and political chaos. Little did it imagine that he would become head of government! Lenin quickly established his dominance in the Petrograd Soviet and carried on a relentless campaign against the Provisional Government. The latter, riven by internal disputes and unable to establish its authority in either the army or the country at large, offered little resistance when a mob calling itself the Red Guards occupied the Winter Palace, the seat of government, on October 25, 1917 (November 7 in the Western calendar, which the Soviet Union adopted on January 1, 1918). The next day Lenin formed a new government, called the Council of Peoples' Commissars.

Nearly four years of bitter civil strife and war followed the October Revolution. In March 1918 the government ended the war with Germany in the Treaty of Brest-Litovsk (subsequently voided by the Treaty of Versailles), but still faced determined opposition from several so-called White armies, who were for a time aided by the Western Allies, and in 1920 it went to war with newly independent Poland. In their efforts to survive and stay in power the Bolsheviks, now calling themselves Communists, introduced a drastic policy called War Communism. It included nationalization of the urban economy, confiscation and distribution of land to the peasants, and a new legal system. Its outstanding characteristic, however, was its introduction of a single-party government, the "dictatorship of the proletariat," with Lenin as its voice.

In the elections to the long-awaited constitutional assembly the Social Revolutionaries (SRs), opponents of the Bolsheviks, won a large majority. It met briefly in January 1918, but Lenin sent troops to dissolve it after one session. The SRs then revived their traditional tactic of assassination, and succeeded in wounding Lenin in August 1918. The Communists thereupon adopted a deliberate reign of terror, murdering their political opponents while maintaining control of the central government, located in Moscow after March 1918.

The government granted Finland's demand for independence soon after the October Revolution. During the civil war and afterwards it faced demands from other regions for independence or at least autonomy. Although it acceded to these demands from the Baltic states of Estonia, Lithuania, and Latvia, it resisted those from the Ukraine, Transcaucasia, and elsewhere. The status of the non-Russian nationalities remained unclear for two years after their reconquest. Then in 1922 Lenin decided to create a federation, in name at least, against the advice of his specialist on nationality problems, the russified Georgian Joseph Stalin. On December 30, 1922, the Union of Socialist Soviet Republics (USSR) came into being. It consisted of the Russian Soviet Federated Socialist Republic (RSFSR), including most of European Russia plus Siberia, and the republics of the Ukraine, White Russia, and Transcaucasia. Subsequently other republics, in central Asia and elsewhere, were added to the facade; but the reality was that the whole was ruled by a small group of men in Moscow, who controlled the machinery of both the Communist party and the government.

By March 1921, when the Treaty of Riga brought peace with Poland, the Communists no longer faced active opposition to their rule either at home or abroad. But the economy was in shambles. The policy of War Communism, with its strong element of terrorism, had sufficed to defeat the enemy, but it clearly could not serve as a long term basis for the economy. Industrial production had fallen to less than a third of its 1913 level, and the government's agriculture policy produced no better results. The peasants, whose land seizures the Bolsheviks had legitimized, refused to deliver their produce at the artificially low prices set by the government. As early as August 1918 the government had sent troops and detachments of armed industrial workers into the countryside to confiscate the harvest, and black markets were rife. At the end of February 1921 a mutiny at the naval base of Kronstadt, caused by the abysmal conditions of the sailors, convinced Lenin that a new policy was necessary.

Faced with economic paralysis and the possibility of a major peasant revolt, Lenin radically reversed directions with the so-called New Economic Policy (NEP), a compromise with capitalist principles of economy that Lenin called ''a step backward in order to go forward.'' A special tax in kind of agricultural produce replaced compulsory requisitions, allowing peasants to sell their surpluses at free market prices. Small-scale industries (employing fewer than twenty workers) were returned to private ownership and allowed to produce for the market; foreign entrepreneurs leased some existing plants and obtained special concessions to introduce new industries. But the so-called commanding heights of the economy (large-scale industries, transportation and communication, banking and foreign trade) remained under state ownership and operation. The NEP also included a vigorous program of electrification, the establishment of technical schools for engineers and industrial managers, and the creation of a more systematic organization for the state-owned sectors of the economy. Despite some further difficulties with the peasants, output increased in both industry and agriculture, and by 1926 or 1927 the prewar levels of output had been substantially regained.

Meanwhile, important changes were occurring in the Communist party leadership. In May 1922 Lenin suffered the first of a series of paralytic strokes from which he never fully recovered before his death in January 1924. In spite of his power, Lenin refrained from explicitly designating his successor. In fact, in a unique "political will" he pointed out both the strengths and the faults of all his close associates and possible successors.

Two of the main contenders were Leon Trotsky and Joseph Stalin. Trotsky had served as war commissar and claimed credit for defeating the White armies during the civil war. A gifted orator, he had a large following both within and outside the party. But his late conversion to the Bolshevik cause (1917) and his penchant for making tactless remarks about his colleagues made him suspect to the Old Bolsheviks. Stalin, on the other hand, was a faithful adherent of Lenin and the Old Bolsheviks. Although he was not seriously considered as Lenin's successor immediately following the latter's death, Stalin used his position as general secretary of the Central Committee of the party (from 1922) to form coalitions within the party to dispose of his rivals, Trotsky first of all.

Fundamental differences on both domestic and foreign policy separated the contenders. Whereas Trotsky advocated world revolution, Stalin eventually sided with those who favored building a strong socialist state in the Soviet Union: "Socialism in one country." After Stalin succeeded in demoting, exiling, and eventually assassinating Trotsky, he turned on his former allies, accusing some of being "Left Deviationists," others of "Rightist Opportunism." By 1928 Stalin's control over both party and country was virtually complete.

Stalin's program of "Socialism in one country" implied a massive build-up of Russian industry to make the country both self-sufficient and powerful in the face of a largely hostile world. The means for achieving this was comprehensive economic planning, which from Stalin's view had the further advantage of increasing the state's control over the lives of its subjects and thus preventing attempts to overthrow the regime. In 1929, as soon as he was firmly in control of the party apparatus and the organs of the state, he launched the first of the five-year plans. This event is sometimes called "the second Bolshevik revolution."

All the resources of the Soviet government were directly or indirectly used in the effort. For purely technical matters the State Planning Commission (Gosplan) had overall responsibility for formulating plans, setting output goals, and sending directives to various subsidiary agencies. Without regard for costs, profits, or consumer preferences, the planning mechanism replaced the market. Instead of representing workers and protecting their interests, trade unions were used to preserve labor discipline, prevent strikes and sabotage, and encourage productivity. The ideal of "workers' control" of industry, held by trade union leaders before the final triumph of Stalin, had no place in the five-year plans.

Agriculture was one of the Soviet Union's most difficult and persistent problem areas. During the NEP the peasants had strengthened their traditional attachment to their own soil and livestock, but Stalin insisted that they be organized in state farms. The state owned all the land, livestock, and equipment and appointed a professional manager; the peasants who worked the land formed a pure agricultural proletariat. They bitterly resisted collectivization, in many instances burning their crops and slaughtering their livestock to prevent them from falling into the hands of the government. Faced by such determined resistance, even Stalin backed off for a time. As a compromise with the peasants the government sometimes allowed them to form cooperative farms, on which most of the land was tilled in common; each household, however, was allowed to keep small plots for its own use. The state supplied advice and machinery from state-owned Machine-Tractor Stations, which could also be used for inspection, propaganda, and control.

The objectives of the First Five-Year Plan were officially declared to have been achieved after only four and a quarter years. In fact, the plan was far from a complete success. Although output in some lines of industry had grown prodigiously, most industries had failed to make their quotas, which had been set unrealistically high. In agriculture about 60 percent of the peasants had been collectivized, but agricultural output had actually fallen, and the number of livestock declined to between half and two-thirds of the 1928 level (which was regained only in 1957). The costs of the Five-Year Plan were enormous, especially the human costs. In the collectivization of agriculture alone, millions died of starvation or were executed.

In 1933 the government inaugurated the Second Five-Year Plan, in which the emphasis was supposed to be on consumer goods; in fact, the government continued to devote an extraordinary proportion of its resources to capital goods and military equipment. In spite of great increases in industrial production, the country remained mostly agrarian, and agriculture was its weakest sector. A notable feature of the Second Five-Year Plan occurred in 1936–37—the Great Purge. Thousands of individuals, from lowly workers to high party and military leaders were placed on trial (or executed without trial) for alleged crimes ranging from sabotage to espionage and treason. Naturally, this had a significant effect on output.

The Third Five-Year Plan, launched in 1938, was interrupted by the German invasion of 1941, and the Soviet Union fell back on something like War Communism.

Economic Aspects of World War II

The Second World War was by far the most massive and destructive of all wars. In some respects it represented merely an extension and intensification of features that had manifested themselves in World War I, such as increasing reliance on science as the basis of military technology, the extraordinary degree of regimentation and planning of the economy and society, and the refined and sophisticated use of propaganda both at home and abroad. In other respects it differed markedly from all previous wars.

Truly a global war, it directly or indirectly involved the populations of every continent and almost every country in the world. Unlike its predecessor, which had been primarily a war of position, it was a war of movement—on land, in the air, and at sea. Aerial warfare, an incidental feature of World War I, became a critical element in the second. Naval operations, especially the use of carrier-based aircraft, became far more important. Science-based technology accounted for many of the special new weapons, both offensive and defensive, ranging from radar to rocket bombs, jet-propelled aircraft, and atomic bombs. The economic and especially the industrial capacities of the belligerents acquired new importance. Mere numbers counted for less than ever before, although size was still a factor in assessing the relative power of the opposing sides. In the final analysis the production line became as important as the firing line. The ultimate secret weapon of the victors was the enormous productive capacity of the American economy.

The pecuniary costs of the war have been estimated at more than 1 trillion dollars (contemporary purchasing power) for direct military expenditure, and that is a lower-bound estimate. It does not include the value of property damage, which has not been accurately estimated but was certainly much larger; nor does it include interest on war-induced national debt, pensions to wounded and other veterans, or—most appalling of all, and most difficult to evaluate in pecuniary terms—the value of lives lost or mangled, civilian as well as military.

Rough estimates place the number of war-related deaths at about 15 million in western Europe: 6 million military and more than 8 million civilians, including between 4.5 and 6 million Jews murdered by the Nazis in the Holocaust. Millions more were wounded, made homeless, and died of starvation or nutrition-related diseases. For Russia it is estimated that more than 15 million died, more than half civilian casualties. China suffered more than 2 million military deaths, and untold millions of civilians as a result of both enemy action and war-induced famine and disease. The Japanese lost more than 1.5 million military personnel and, again, millions of civilians; more than 100,000 died as a direct result of the atomic bombs dropped on Hiroshima and Nagasaki, and other Japanese cities were equally devastated by conventional bombs.

Property damage was far more extensive than in World War I, largely because of aerial bombardment. The U.S. Air Force prided itself on its strategic bombing, targeted on military and industrial installations rather than civilians; but the postwar Strategic Bombing Survey of Germany showed that only about 10 percent of industrial plants had been permanently destroyed, while more than 40 percent of civilian dwellings had been knocked out. About 9000 tons of bombs were dropped on Hamburg in July 1943, virtually leveling the city. The same happened to Dresden near the war's end, leaving unknown numbers of casualties. Many other cities on both sides—Coventry, England, and Rotterdam in the Netherlands, for example—suffered similar fates. Leningrad was virtually destroyed by artillery bombardment, but it never capitulated.

Transportation facilities, especially railways and ports and docks, proved tempting targets. Every bridge over the Loire River, separating northern from southern France, was destroyed, as were all but one on the Rhine—the famous Remagen bridgehead that enabled Allied soldiers to penetrate the heart of Germany.

All combatants resorted to economic warfare, a new phrase for an old policy. As in World War I and even the Napoleonic Wars, Britain (later assisted by the United States) imposed a blockade, to which Germany retaliated with unrestricted submarine warfare. In addition to its ersatz commodities, such as gasoline made from coal, Germany could command the resources of the occupied countries. In 1943 it extracted more than 36 percent of French national income, and in 1944 almost 30 percent of its industrial labor force consisted of non-Germans, virtual slave laborers.

At the end of the war in Europe the economic outlook was extremely bleak. Industrial and agricultural output in 1945 was half or less than it had been in 1938. In addition to the property damage and human casualties, millions of people had been uprooted and separated from their homes and families, and millions more faced the prospect of starvation. To make matters worse, the institutional framework of the economy had been severely damaged. Reconstruction would be no easy matter.

15

Rebuilding the World Economy

At the end of the war Europe lay prostrate, almost paralyzed. All belligerent countries except Britain and the Soviet Union had suffered military defeat and enemy occupation. Large areas of the Soviet Union had been effectively occupied by the Germans and fought over foot by foot, twice or even more often. Although Britain had not been occupied (except by Americans), it suffered severe damage from aerial bombardment of its densely populated cities and from acute shortages of food and other necessities. Only the few European neutrals escaped direct damage, but even they suffered from many war-induced shortages.

Before the war Europe had imported more than it exported, foodstuffs and raw materials in particular, and paid for the difference with the earnings of its foreign investments and shipping and financial services. After the war, with merchant marines destroyed, foreign investments liquidated, financial markets in disarray, and overseas markets for European manufactures captured by American, Canadian, and newly arisen firms in formerly underdeveloped countries, Europe faced a bleak prospect merely to supply its population with basic needs. Millions faced the threat of death from starvation, disease, and the lack of adequate clothing and shelter. Victors and vanquished were alike in their misery. The urgent need was for emergency relief and reconstruction.

Relief came through two main channels, most of it originating in America. As the Allied armies advanced across western Europe in the winter and spring of 1944–45 they distributed emergency rations and medical supplies to the stricken civilian population, enemy as well as liberated. Because the Allies had committed themselves to a policy of unconditional surrender, after the cessation of hostilities they had to assume the burden of policing defeated Germany, which included the continuation of emergency rations for the helpless civilian population.

The other channel of relief was the United Nations Relief and Rehabilitation Administration (UNRRA). In 1945–46 it spent more than 1 billion dollars and distributed more than 20 million tons of food, clothing, blankets, and medical supplies. The United States bore more than two-thirds of the cost, other United Nations members the remainder. Altogether, between July 1, 1945, and June 30, 1947, by means of grants to UNNRA and other direct emergency aid, the United States made available about 4 billion dollars to Europe and almost 3 billion dollars to the rest of the world. After 1947 the work of UNRRA was continued by the International Refugee Organization, the World Health Organization, and other specialized agencies of the United Nations, as well as by voluntary and official national agencies.

In contrast to Europe, the United States emerged from the war stronger than ever. To a lesser extent, Canada, the other Commonwealth nations, and several countries of Latin America did as well. Spared from direct war damage, their industries and agriculture benefited from high wartime demand, which permitted full use of capacity, technological modernization, and expansion. Many American economists and government officials feared a severe depression after the war, but after the removal of rationing and price controls, which had held prices at artificially low levels during the war, the pent-up consumer demand for war-scarce commodities created a postwar inflation that doubled prices by 1948. In spite of the hardships that the inflation brought to people living on fixed incomes, it kept the wheels of industry turning and enabled the United States to extend needed economic aid for the rebuilding of Europe and other war-devastated and poverty-stricken lands.

Planning for the Postwar Economy

One of the most urgent tasks facing the peoples of Europe after their survival requirements were satisfied was to restore normal law, order, and public administration. In Germany and its satellites Allied military governments assumed these functions pending peace settlements. Most of the countries that had been victims of Nazi aggression had formed governments in exile in London during the war. These governments returned to their homelands in the wake of Allied armies and soon resumed their normal functions.

Their return, however, did not imply a mere "return to normalcy," the chimera of the 1920s. Memories of the economic distress of the 1930s lingered through the ordeal of war, and no one wanted a repetition of either experience. On the Continent the leadership of the underground opposition to Nazi Germany played a large role in postwar politics, and the comradeship of those movements, in which Socialists and Communists had figured prominently, did much to overcome prewar class antagonisms and bring new men and women to positions of power. In Britain the participation of the Labour party in Churchill's wartime coalition government gave its leaders great prestige and influence, and enabled them to lead it to its first clearcut electoral victory soon after the end of the war in Europe. Finally, the very magnitude of the task of reconstruction indicated a much larger role for the state in economic and social life than had been characteristic of the prewar period.

In all countries the consequence of these various tendencies was widespread public demand for political, social, and economic reforms. The response to these demands in the economic sphere took the form of nationalization of key sectors of the economy, such as transportation, power production, and parts of the banking system; extension of social security and social services, including retirement pensions, family allowances, free or subsidized medical care, and improved educational opportunities; and assumption by governments of greater responsibilities for maintaining satisfactory levels of economic performance. Even the United States passed the Employment Act of 1946, which created the President's Council of

Economic Advisers and pledged the federal government to maintain a high level of employment.

At the international level planning for the postwar had begun during the war itself. Indeed, as early as August 1941, at their dramatic meeting on board a battleship in the North Atlantic (actually, Placentia Bay in Newfoundland), Franklin Roosevelt and Winston Churchill signed the Atlantic Charter, which pledged their countries (and subsequently other members of the United Nations) to undertake restoration of a multilateral world trading system in place of the bilateralism of the 1930s. Of course, this was only a statement of intentions and did not commit the parties to any concrete actions; but at least it was a statement of *good* intentions.

Subsequently, in 1944, at an international conference at the New Hampshire resort of Bretton Woods in which the American and British delegates played the leading roles, the bases were laid for two major international institutions. The International Monetary Fund (IMF) was to have the responsibility for managing the structure of exchange rates among the various world currencies, and also for financing short-term imbalances of payments among countries. The International Bank for Reconstruction and Development (IBRD), also known as the World Bank, was to grant long-term loans for reconstruction of the war-devastated economies and, eventually, for the development of the poorer nations of the world. These two institutions did not become operational until 1946 and, for reasons to be noted later were not fully effective for several years after that; but at least a beginning had been made toward rebuilding the world economy.

The conferees at Bretton Woods also envisaged the creation of an International Trade Organization (ITO) that would formulate rules for fair trade among nations. Further conferences were held to this end, but the best that could be obtained was a much more limited General Agreement on Tariffs and Trade (GATT), signed at Geneva in 1947. The signatories pledged themselves to extend most-favored-nation treatment to the others (i.e., not to discriminate in trade), to seek to reduce tariffs, not to resort to quantitative restrictions (quotas), and to remove those that existed, and to consult mutually before making major policy changes. These provisions were much less than had been hoped for from ITO, and they were not always observed in practice; but a number of international tariff reduction conferences were held under GATT's auspices, which did much to reduce trade barriers. Membership in GATT grew from twenty-three in 1947 to more than eighty two decades later.

The Marshall Plan and Economic "Miracles"

By the middle or end of 1947 most nations of western Europe except Germany had regained their prewar levels of industrial production. (Germany will be dealt with later.) But of course the prewar levels of production had been far from satisfactory. Moreover, the winter of 1946–47 was extremly severe, and was followed by a long drought over the greater part of Europe, making the agricultural harvest of 1947 the worst in the twentieth century. Clearly, much remained to be done.

In the monetary and financial chaos of the 1930s virtually all European and many other countries adopted exchange controls; that is, their currencies were not convertible into others except with a license issued by the monetary authorities. A counterpart of this was the bilateral balancing of commodity trade, a major cause of its greatly reduced volume. These controls, to which others were added, were of necessity continued during the war. After the war shortages of all kinds— foodstuffs, raw materials, replacement parts, and so on—seemed to dictate the continuance of the controls. The remedy for the shortages was to be found mainly overseas, especially in North and South America, but dollars were required for their purchase, and in Europe the greatest shortage of all was dollars.

American relief and rehabilitation grants, noted previously, helped ease the "dollar shortage" during the first two postwar years. In addition, the United States and Canada jointly loaned Great Britain 5 billion dollars in December 1945, which helped not only that country but, through its expenditures on the Continent, other countries as well. It was nevertheless becoming clear in the late spring of 1947 that the immediate postwar recovery was in serious danger of aborting. Moreover, the growing "cold war" between the United States and the USSR, and the role of Communist parties in the politics of several west European countries, notably France and Italy, gave American authorities cause for concern over the political stability of western Europe. On June 5, 1947, General George C. Marshall, who had been named U.S. secretary of state by President Truman, gave a commencement address at Harvard University in which he announced that if the nations of Europe would present a unified, coherent request for assistance the U.S. government would give a sympathetic response. This was the origin of the so-called Marshall Plan.

The French and British foreign ministers immediately conferred and invited their Soviet counterpart to meet with them in Paris to discuss a European response to Marshall's proposal. (Marshall had specifically included the Soviet Union and other countries of eastern Europe in his proposal. He may or may not have expected that the Soviet Union would refuse to cooperate. In any case, the Soviet foreign minister, although he came to Paris, soon left, charging that Marshall's proposal was an "imperialist plot.") With unaccustomed alacrity for diplomatic affairs, representatives of sixteen nations met in Paris on July 12, 1947, dubbing themselves the Committee of European Economic Cooperation (CEEC). These included all the democratic nations of western Europe (and Iceland), even neutral Sweden and Switzerland, as well as Austria (still under military occupation), undemocratic Portugal, and Greece and Turkey (to which the United States had already granted military aid to fight Communist subversion). Finland and Czechoslovakia indicated an interest in participating, but were called to heel by the Soviet Union; neither the Soviet Union nor any other East European country was represented. Franco's Spain was not invited and Germany, still subject to military occupation, had no government to be represented.

The American people and Congress still had to be persuaded that further economic assistance to Europe was in their interest. The Truman Administration launched a strong lobbying program to that end, and in the spring of 1948 Congress passed the Foreign Assistance Act, which created the European Recovery Program (ERP), to be administered by the Economic Cooperation Administration (ECA). At the same time there was less than total unanimity in Europe on the goals of the program. British officials had hoped to get more bilateral aid from the United States, rather than having it channeled through a European organization. (That is probably the main reason the Russians walked out of the initial planning meeting; they also hoped, vainly, for bilateral aid.) The French were uneasy about the future role of Germany in whatever organization might be set up. The smaller countries also had their particular concerns. Nevertheless, after the U.S. Congress acted the CEEC converted itself into the Organization for European Economic Cooperation (OEEC), which was responsible, jointly with the ECA, for allocating the American aid. Members of the OEEC also had to put up counterpart funds in their own currencies to be allocated with the consent of the ECA.

Altogether, including some interim aid sent to France, Italy, and Austria at the end of 1947 on an emergency basis, the ERP funneled about 13 billion dollars in economic assistance in the form of loans and grants from the United States to Europe by the beginning of 1952. This enabled the OEEC countries to obtain imports of scarce commodities from the dollar zone. Almost one-third (32.1 percent) consisted of food, feed, and fertilizer, mainly in the first year or so of the program. Thereafter the priority shifted to capital goods, raw materials, and fuel to enable European industries to rebuild and export.

Germany at first occupied an anomalous position in the European Recovery Program. After its defeat in May 1945 the heads of government of the United States, the United Kingdom, and the USSR met in July at Potsdam, near Berlin, to determine Germany's fate, but decided merely to prolong the military occupation. (France, although not represented at Potsdam, was allowed by Britain and the United States to occupy parts of Germany immediately adjacent to its territory.) The decision was not intended as a permanent division of the country, but simply for temporary convenience. As events unfolded, the disagreements between Russia and the Western Allies led the latter to give greater and greater measures of autonomy to the Germans in their zones of occupation. The Soviet authorities responded with similar nominal concessions in the eastern zone, although they maintained strict control through their puppets and the presence of Soviet troops. The ultimate result was the division of Germany into two separate states: the German Federal Republic (West Germany) and the German Democratic Republic (East Germany). Berlin, although deep inside the Soviet zone, was also divided into four sectors, later reduced to two: East Berlin, the capital of the GDR, and West Berlin, affiliated with the FRG. In the absence of a German government the Allied Control Council served as the nominal supreme authority, although in fact each occupying power administered its zone independently.

The Potsdam Conference had condoned the dismantling of German armaments and other heavy industries (already begun by the Russians), reparations to the victors and to the victims of Nazi aggression, strict limitations on German productive capacity, and a vigorous program of denazification, including the trial of Nazi leaders as war criminals. In fact, only the last aim was realized as originally intended. The Soviet authorities dismantled many factories in their zone and carried them to Russia as reparations. After a brief attempt by the Western powers to collect physical reparations and to break up large industrial combines in their zones, they realized that the German economy would have to be kept intact not only to support the German people but also to assist in the economy recovery of western Europe. They reversed their policy and, instead of limiting German production, took steps to facilitate it. One means of doing this was to provide for economic reunification, a process initiated with the creation of Bizonia, a union of the American and British zones of occupation at the end of 1946, to which the French zone was subsequently added. Just as the Zollverein served as the precursor of the German Empire, the economic unification of the western zones of occupation delineated the future German Federal Republic.

Meanwhile new difficulties arose. To ensure the survival of the population in their zones of occupation (which swelled rapidly with the influx of refugees from the east), the American military government between 1945 and 1948 financed about two-thirds of the essential imports, mainly food, of the western zones of occupation. To stimulate economic recovery in their zones, the Western powers carried out a reform of the German currency in June 1948, replacing the debased and despised Nazi Reichsmarks with Deutschemarks at a ratio of 1 new for 10 old marks. (The reform was facilitated by the fact that the populace had virtually deserted the old currency and returned to a barter system of exchange, with coffee, silk stockings, and especially cigarettes fulfilling the functions of both standards of value and media of exchange.) The response, immediate and overwhelming, was referred to as a *Wirtschaftswunder* (economic miracle). Goods previously hoarded or traded on the black market came into the open; stores were restocked, factories restarted, and western Germany began its remarkable economic revival.

The Soviet Union, which had not been consulted about the monetary reform, and which regarded it as a contravention of the Potsdam agreement (which it was), retaliated by closing off all road and rail links between the western zones of occupation and West Berlin. It hoped to force a withdrawal of Western forces from Berlin, or at least to secure concessions on disputed points; instead, the Western Allies responded promptly with a large-scale airlift of strategic supplies. In a tremendous operation lasting more than a year the U.S. Air Force and the RAF flew almost 300,000 flights into Berlin, transporting at its peak more than 8000 tons of supplies daily. The airlift supplied not only Western troops but also the 3 million inhabitants of West Berlin.

Meanwhile western Germany was being integrated into the European Recovery Program. At first, in 1948, aid for the western zones of occupation was received and allocated by the American military government. Subsequently the West German states were allowed to elect representatives to a constitutional convention, and in May 1949 the Federal Republic of Germany came into existence. Not to be outdone, the Soviet Union soon afterward set up the so-called German Democratic Republic. In September it lifted the Berlin blockade.

With West Germany now fully integrated into the OEEC and the Marshall Plan, the economic recovery of western Europe could be considered complete, but more and still better things were in store. The Marshall Plan came to an end in 1952; it had succeeded beyond the expectations of several of its participants and even those of some of its creators. It did not create a United States of Europe, as some had hoped, and many serious problems remained. But, beyond the fact that western

Europe had not only recovered and exceeded prewar levels of production, the OEEC and other newly created institutions remained and stimulated the economy to new heights.

One of the most important of those other new institutions was the European Payments Union (EPU). As recounted earlier, one of the major obstacles to increased trade in the immediate postwar years was the shortage of foreign exchange, especially dollars, and the consequent necessity for bilateral balancing of trade. Some attempts had been made to break out of this constraint, but they were awkward and not very effective. At length, in June 1950, the OEEC nations, with the assistance of a 500-million-dollar grant from the United States, inaugurated the EPU. This ingenious device allowed for free multilateral trade *within* the OEEC; precise accounts were kept of all intra-European trade, and at the end of each month balances were struck and canceled. Nations with deficits overall were debited on the central accounts, and if their deficits were large they had to pay a portion in gold or dollars; creditors, on the other hand, received credits on the central accounts; if their credits were very large they received a portion in gold or dollars, enabling them to import more from so-called hard-currency areas (mainly the dollar zone). This provided incentives for OEEC countries to increase their exports to one another and to lessen their dependence on the United States and other overseas suppliers.

The results were spectacular. In the two decades or so after the formation of the EPU world trade grew at an average annual rate of 8 percent, the highest in history apart from a few years after the trade treaties of the 1860s. Most of that growth, of course, took place in Europe, both within Europe and between Europe and overseas nations. The EPU was so successful that, in conjunction with the overall growth of trade, the OEEC countries were able to restore free convertibility of their currencies and full multilateral trade in 1958. In 1961 the OEEC itself metaphorphosed into the Organization for Economic Cooperation and Development (OECD), to which the United States and Canada (and later Japan and Australia) adhered: an organization of advanced industrial countries to coordinate aid to underdeveloped countries, to seek agreement on macroeconomic policies, and to discuss other problems of mutual concern.

The quarter-century or so after World War II witnessed the longest period of uninterrupted growth among the industrial countries of the world and at the highest rates in history (Fig. 15-1). For the industrial countries as a group (OEEC, the United States, Canada, and Japan) the average increase in gross domestic product per person employed from 1950 to 1973 amounted to about 4.5 percent per year. Rates for individual countries ranged from 2.2 percent for the United Kingdom to 7.3 percent for Japan. Growth was most rapid in those countries that had abundant supplies of labor, either from the reduction of the agrarian population (e.g., Japan, Italy, France) or from an influx of refugees (West Germany). Growth in the United States, Canada, and Great Britain, which had the highest per capita incomes at the end of the war, was slower than that of continental western Europe and Japan, but more rapid than in any prolonged period in their previous histories. At the same time countries with relatively low per-capita incomes within the industrial group— Italy, Austria, Spain, Greece, and Japan—grew more rapidly than the average.

The term "economic miracle," as noted earlier, was first applied to the remark-

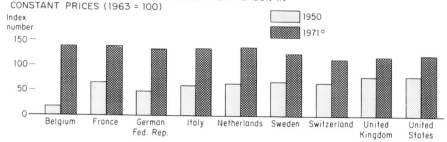

A. INDEX NUMBERS OF NATIONAL PRODUCT PER PERSON IN
   CONSTANT PRICES (1963 = 100)

ᵃ Data for France, Italy, and the Netherlands from 1970; Data for Switzerland from 1969.

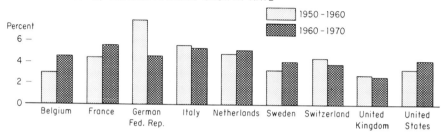

B. NATIONAL INCOME, ANNUAL AVERAGE GROWTH RATE

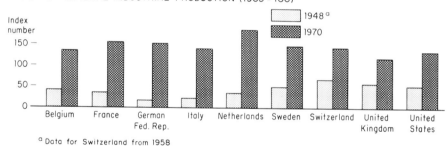

C. INDEX OF GENERAL INDUSTRIAL PRODUCTION (1963 = 100)

ᵃ Data for Switzerland from 1958

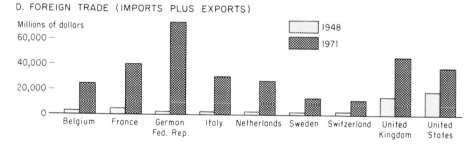

D. FOREIGN TRADE (IMPORTS PLUS EXPORTS)

FIGURE 15–1.   The economic recovery and growth of western Europe, 1948–71. (From
United Nations Statistical Yearbook, various years.)

378

able spurt in growth in West Germany after the currency reform of 1948. When the high rates of growth continued throughout the 1950s and 1960s it was used to refer to the entire era. It was then noted that several other nations, notably Italy and Japan, had growth rates as high as or higher than the German. Miracles abounded! Or did they? The high growth rates in most of the industrial countries were certainly remarkable, and unprecedented in history, but they were scarcely miraculous. There were solid reasons for them in every case.

American aid played a crucial role in sparking the recovery. Thereafter Europeans kept it going with high levels of savings and investments. At times the competition between consumption and investment spending caused severe inflationary pressures, but none so disastrous as the hyperinflations after World War I. Much of the investment went into equipment for new products and processes. During the depression years and the war a backlog of technological innovations had built up that only awaited capital and skilled labor to be employed. In effect, the European economies had stagnated for an entire generation. In addition to having forfeited their potential increment of growth, they operated with obsolete equipment and lagged behind the United States in technological progress. Thus, technological modernization both accompanied and was an important contributory factor to the so-called economic miracle.

Other major factors were the attitude and role of governments. They participated in economic life both directly and indirectly on a much larger scale than previously. They nationalized some basic industries, drew up economic plans, and provided a wide range of social services. Nevertheless, private enterprise was responsible for by far the largest part of economic activity. On average, between one-fourth and one-third of national income in western Europe originated in the government sector. Though this proportion was much greater than it had been before the war, it was less than half the contribution of the private sectors of the economy. The economic systems of postwar western Europe were equally far from the stereotyped old-style capitalism of the nineteenth century and from the doctrinaire socialized economies of eastern Europe. In the mixed or welfare state economies that became characteristic of the Western democracies, government assumed the tasks of providing overall stability, a climate favorable to growth, and minimal protection for the economically weak and underprivileged, but it left the main task of producing the goods and services desired by the population to private enterprise.

At the international level the relatively high degree of intergovernmental cooperation deserves major credit for the effectiveness of the economic performance. The cooperation was not always spontaneous—prodding by the United States was sometimes necessary to elicit it —and some promising projects failed for lack of it; but on the whole the contrast to the interwar years is conspicuous.

Finally, in the long term, much credit must go to Europe's wealth of human capital. Its high rates of literacy and specialized educational institutions, from kindergartens to *technische Hochschulen*, universities, and research establishments, provided the skilled personnel and brainpower to make the new technology work effectively. In the first flush of the success of the Marshall Plan many observers incorrectly deduced that physical or financial capital alone would suffice to bring about development, and several grandiose projects based on that false premise, such as the Alliance for Progress between the United States and the nations of Latin America, were undertaken, only to end in failure and disillusionment.

The Emergence of the Soviet Bloc

The Soviet Union suffered the greatest damage, in an absolute sense, of any nation engaged in the war. Estimates of the number killed, both military and civilian, range from a minimum of 10 million to more than 20 million. Twenty-five million were left homeless. Large areas of the most fertile agricultural land and some of the most heavily industrialized regions had been devastated. According to official estimates, 30 percent of the prewar wealth had been destroyed.

In spite of the sufferings of its people, the Soviet Union emerged as one of the two superpowers in the postwar world. Although it was poor on a per capita basis, its vast territory and population allowed it to play this role. To restore the devastated economy and boost output to new levels the government launched the Fourth Five-Year Plan in 1946. As previous plans had done, it emphasized heavy industry and armaments, with particular attention to atomic energy. The new plan also made extensive use of physical reparations and tribute from the former Axis countries and the USSR's new satellites.

Stalin, more powerful than ever, instituted a number of changes in the high offices of both government and the economy in the immediate postwar years. A constitutional revision in 1946 replaced the Council of People's Commissars with a Council of Ministers, in which Stalin assumed the position of chairman or prime minister. The ministries charged with the supervision and control of industry and agriculture experienced drastic purges of personnel on grounds of incompetence and dishonesty. Other high officials of government and party were dismissed on similar grounds, although there is reason to believe Stalin's real motive was distrust of their personal loyalty to him.

Stalin died in 1953. After two years of "collective leadership" and shifting alliances among the top leadership of the Communist party Nikita Khrushchev, who had succeeded Stalin as first secretary of the party, emerged as the paramount leader. At the Twentieth Party Congress in February 1956 Khrushchev gave a long speech in which he denounced Stalin as a ruthless, morbid, almost insane tyrant who had ordered the execution of countless innocent people, whose egotism had led him to make mistakes for which all citizens had suffered, who had caused the government to lose touch with the people, and who had established the "cult of personality" to glorify himself. Khrushchev carefully pointed out, however, that Stalin's despotism represented an aberration from an essentially correct policy and claimed that the new collective leadership had returned to proper Leninist principles. Supposedly secret, Khrushchev's speech leaked to the public, both within and outside the Soviet Union. It caused much confusion and ferment among the peoples of the Communist countries and did serious damage to the cause of Communism by confirming the evils of Stalin's regime. The government began an official program of "de-Stalinization" that included the removal of his body from the famed Lenin Tomb in Moscow's Red Square.

In spite of the change of leadership and a few superficial reforms, the basic nature of the Soviet economic system did not change. In 1955 the government announced the "fulfillment" of one five-year plan and the inauguration of another, even though high officials complained of widespread inefficiency and the failure of one-third of the industrial enterprises to meet their production targets. Soviet heavy industry continued to increase its output, but fell far short of the expressed intention to overtake the United States. The production of consumer goods, always given low priority in Soviet planning, continued to lag so that consumers were plagued by shortages and inferior goods.

Soviet agriculture remained in a condition of almost unrelieved crisis throughout the postwar period, despite massive efforts of the government to increase productivity. The collective farm system did not offer enough incentive for the peasants. Instead, they concentrated their energies on the small private plots of up to one-half hectare (1.2 acres) that they were allowed to cultivate, part of whose produce they could sell on the market. These plots formed little more than 3 percent of the USSR's cultivated land, but produced as much as one-fifth of the country's milk and one-third of its fresh meat, as well as much fruit and vegetables. In 1954 Khruschev started a "virgin lands" project to bring great stretches of arid land in Soviet Asia into cultivation. The next year he launched a drive to increase the production of corn, and in 1957 he announced a campaign to overtake the United States by 1961 in the production of milk, butter, and meat. None of these programs came anywhere near reaching their stated goals. Despite threats of punishment and wholesale dismissals of agricultural officials, Khrushchev and his planners could not overcome bad weather, bureaucratic mismanagement, fertilizer shortages, and above all the peasants' lack of enthusiasm. Food shortages continued to characterize Soviet life. Russia, historically an exporter of grain, was obliged in the 1960s to begin importing grain from Western countries (Australia, Canada, the United States), for which it paid in gold.

Although the Allies found it inexpedient at first and subsequently impossible to agree on the terms of a peace treaty with Germany, they did succeed in signing treaties with Germany's satellites and in agreeing on the treatment of the victims of Nazi aggression in eastern Europe. The general terms of the East European settlement had been foreshadowed in wartime conferences, notably the one at Yalta. They had envisaged a major role for the Soviet Union—made more concrete by the occupation of the area by Soviet troops—although Stalin had promised to allow free elections and "broadly representative governments," a promise that proved to be meaningless. After a year and a half of postwar negotiations treaties were signed in February 1947 with Rumania, Hungary, Bulgaria, and Finland.

The resurrection of Czechoslovakia and Albania was taken for granted. Since those countries had never been at war with the Allies—had, in fact, been among the first victims of Axis aggression—there was no problem in restoring their independence. The manner in which they were liberated, however, guaranteed that they would be within the Soviet sphere of influence.

Czechoslovakia was liberated by Soviet troops and Eduard Benes, the internationally respected prewar president, was returned as president of the provisional government. In a relatively free election in May 1946 the Communists polled a third of the votes and seated the largest delegation in the new constituent assembly. Klement Gottwald, the Communist leader, became prime minister, but the assembly unanimously reelected Benes as president. The country continued under a coalition government, with Benes hoping to make Czechoslovakia a bridge between the USSR and the West, until the Communists seized power in February 1948.

During the war Churchill and Stalin, without consulting Roosevelt, agreed to exercise equal spheres of influence in Yugoslavia after the war. In fact, Yugoslav partisans led by Marshal Tito liberated the country with very little help from Russia and negligible aid from Britain, thus allowing the country a measure of independence. Elections in November 1945 gave Tito's Communist-dominated National Liberation Front a substantial majority in the new constituent assembly, which promptly overthrew the monarchy and proclaimed a Federal People's Republic. On paper the new constitution closely resembled that of the Soviet Union, and Tito governed the country in a manner similar to that of Stalin. He refused to accept dictation from the Soviet Union, however, and in 1948 publicly broke with the latter and its other Communist satellites.

The determination of Poland's postwar boundaries and form of government constituted one of the thorniest problems of peacemaking. In the closing phases of the war there had been two Polish provisional governments, one in London and one in Russian-occupied Poland. At Russian insistence and with Western acquiescence, the two groups merged to form a Provisional Government of National Unity, with a promise of early "free and unfettered elections." The coalition lasted until 1947, when the Communists ousted their partners and assumed complete control. The territorial settlement agreed on provisionally at Potsdam in effect moved Poland 300 miles to the west. The language of the agreement provided only that Poland should have "temporary administration" of the area east of the Oder-Niesse line, or about one-fifth of the area of prewar Germany, but the Poles, with Soviet support, regarded the settlement as definitive compensation for their cessions to the Soviet Union, which constituted almost half of prewar Poland. They forthwith expelled the millions of Germans who resided in the area to make room for other millions of Poles streaming in from the Soviet-occupied zone. These enormous transfers of population, together with similar transfers in other parts of eastern Europe, returned the ethnic boundaries to something resembling "the *status quo ante* 1200 A.D.," in the words of Arnold J. Toynbee.

The peace treaties with Germany's East European satellites, Rumania, Bulgaria, and Hungary, contained territorial provisions that fell into a well-established historical pattern. Rumania regained Transylvania from Hungary, but had to return Bessarabia and northern Bukovina to the Soviet Union and southern Dobrudja to Bulgaria. Hungary lost the most, for it gained nothing and had to cede a small area to Czechoslovakia, which lost Ruthenia to the Soviet Union. All three defeated nations had to pay reparations: 300 million dollars for Hungary and Rumania and 70 million dollars for Bulgaria (all calculated at 1938 prices, so the real amount was greater), the bulk of which went to the USSR. Under the protection of Soviet troops the Moscow-trained Communists in their popular front governments had few difficulties in disposing of their Liberal, Socialist, and Agrarian collaborators and soon established people's republics in the Soviet pattern. Finland lost some territory to

Russia and had to pay 300 million dollars in reparations, but it escaped the fate of other popular front governments on Soviet borders and maintained a precarious neutrality.

The peace treaties had nothing to say about the disappearance of the Baltic countries of Latvia, Lithuania, and Estonia. Part of the tsarist empire before 1917, they had been incorporated in the Soviet Union in 1939–40, then overrun by Germany in 1941. Reoccupied by the Red Army in 1944–45, they were quietly annexed to the Soviet Union as so-called autonomous republics. That they were not mentioned in the peace negotiations implied recognition that they again formed parts of the new Soviet empire.

In January 1949, following the initial successes of the European Recovery Program, the Soviet Union created the Council for Mutual Economic Assistance (COMECON) in an attempt to mold the economies of its East European satellites into a more cohesive union. It included Albania, Bulgaria, Rumania, Hungary, Czechoslovakia, Poland, and East Germany. Ostensibly intended to coordinate the economic development of the Communist countries and promote a more efficient division of labor among them, it was actually used by the Soviet Union to make its satellites more economically dependent on it. Instead of developing a multilateral trading system, as in western Europe, most trade both with the Soviet Union and between the other countries remained bilateral.

When Stalin died in 1953 the Soviet bloc in Europe presented an appearance of monolithic unity. Each of the satellites was more or less a small-scale replica of the Soviet Union, and all danced to the same tune—called in Moscow. Nevertheless, divisive tendencies hid behind the facade of unity. When Yugoslavia had earlier broken away from the Soviet bloc, although remaining a Communist nation, many people in the other satellites would have liked to do the same. Soon after Stalin's death a wave of restiveness swept over the satellite states. Strikes and riots broke out in several countries, and became so serious that the Soviet authorities still in occupation had to use military force to suppress them.

In 1956 in Hungary Imre Nagy, a "national Communist," became prime minister, promising widespread reforms including free elections. He also announced that Hungary would withdraw from the Warsaw Pact, a Soviet-sponsored military alliance, and requested that the United Nations guarantee the perpetual neutrality of Hungary on the same basis as that of Austria. That was too much for the Soviet Union. On November 4 at 4 A.M., Soviet tanks and bombers began a synchronized attack on Hungary that inflicted destruction as horrible as that of World War II. For ten days Hungarian workers and students fought heroically against overwhelming odds, with weapons furnished by their own soldiers. Even after the Russians regained control and established a new puppet government many continued guerilla activity in the hills, while more than 150,000 escaped across the open border to Austria and eventually sought refuge in the West. The Hungarian revolt showed clearly that even a de-Stalinized Russia was not prepared to give up its Communist empire.

The movement for genuinely democratic socialism went furthest in Czechoslovakia. In January 1968 the Czech Communist party under Alexander Dubcek dismissed the old-guard Stalinist leaders and instituted a far-reaching program of reforms that included a greater reliance on free markets in place of government-dictated prices, the relaxation of press censorship, and a considerable measure of personal freedom. The rulers in the Kremlin at first tried to persuade the Czech leaders to return to orthodox Communist policies, but without success. At length, in August 1968, the Soviet army and air force invaded Czechoslovakia and established martial law. Once again, as in East Germany in 1953 and Hungary in 1956, events proved that Russia's Communist empire could be held together only by force.

Although not a member of the Soviet bloc, the People's Republic of China was briefly allied with the Soviet Union. China, a poor country to begin with, had suffered enormously in World War II. During the war the Chinese Communists cooperated with the Nationalist leader, Chiang Kai-shek, in his resistance to the Japanese, but maintained an independent army in northern China, supplied by local requisitions and equipped by the Soviet Union. They also won a large following among the war-weary peasants. After the war they turned on Chiang and in 1949 drove him and his followers off the mainland to Taiwan. On October 1, 1949, the Communists under Mao Tse-tung and Chou En-lai formally proclaimed the People's Republic of China (PRC) with the capital in Peking (Beijing).

Relying on the techniques of modern one-party dictatorships, the Communists quickly extended their control over the entire country, achieving a degree of centralized power unprecedented even in the long history of Chinese despotism. With political control firmly established, the new government undertook to modernize the economy and restructure the society. At first it tolerated both peasant proprietorship in agriculture and limited private ownership of commerce and industry, but in 1953 it began to encourage the collectivization of agriculture and engaged in widespread nationalization of industry. Although significant results were achieved, they were not enough to satisfy the party leadership, which in 1958 embarked on a "great leap forward"—an intensive effort to catch up with the advanced industrial economies. Within a short time this ambitious program proved a catastrophic failure. The population was unable to sustain the effort and sacrifices demanded by its rulers, and a manmade famine cost millions of lives. In 1961 the government revised its objectives and growth resumed at a less frenzied pace.

A major objective of the Chinese Communist leadership was to restructure society and reform thought processes, behavior, and culture. The vestiges of "feudal" and "bourgeois" class structure were eliminated by the simple expedients of expropriation and execution. Enforcing loyalty and obedience within the vast bureaucracy, which still clung to ancient mandarin traditions, and among the thin stratum of intellectuals, scientists, and technicians, most of whom had been educated in the West, proved more difficult. At length, in 1966, Mao launched a "great cultural revolution" marked by three years of terrorism and violence, during which many intellectuals were obliged to work as peasants and common laborers.

The Soviet Union had extended economic, technical, and military assistance to the PRC from the beginning, but the Chinese refused to accept Soviet dictation. In 1960 the USSR cut off all aid and withdrew all of its advisers and technical assistants. Within a few years, after a series of border clashes, the two superpowers of the Communist world were on the verge of open warfare. In spite of the withdrawal of Soviet technicians and assistance China scored its greatest technological triumph in 1964 with the explosion of an atomic bomb.

To compensate for the hostility of the Soviet Union the Chinese undertook a rapprochement with the West, climaxed in 1971 when the United States withdrew its objections to the admission of the PRC to the United Nations. After the death of Mao in 1976 Western contacts increased, and in the 1980s, under the leadership of Deng Xiaoping, the government allowed a limited reintroduction of free markets and free enterprise.

The Soviet Union had three other satellites or client states in Asia. In 1924 the Mongolian People's Republic became the first Communist state outside the USSR. (Formerly known as Outer Mongolia, it had won independence from China in 1921 with Soviet assistance.) Semiarid and sparsely populated, its economy was primarily pastoral, although after World War II it began to exploit its mineral resources (copper and molybdenum) with aid from the Soviet Union and other Communist countries. It became a member of COMECON in 1962. In 1978 the first secretary of the Communist party announced that it had been transformed from an agricultural industrial country to an industrial–agrarian one.

After the defeat of Japan American and Soviet armies occupied Korea jointly. Their zones were separated only by the thirty-eighth parallel of latitude. Efforts to unite the country under a single regime failed, and in 1948 the Soviet and American authorities organized separate regimes in their respective zones and shortly thereafter withdrew their armed forces. The Democratic People's Republic of Korea, or North Korea, had a Soviet-type economy that was substantially industrialized relative to much of the rest of East Asia. Although damaged during the Korean War, its industry was quickly rebuilt with Soviet and Chinese aid.

The Socialist Republic of Vietnam was the outgrowth of the Democratic Republic of Vietnam, established on September 2, 1945, by Ho Chi Minh, the leader of the resistance movement against the Japanese, who had occupied the country during World War II. After the war the French attempted to reinstate themselves, but were defeated by the Vietnamese in 1954. The country was then divided into a Communist North Vietnam and an anti-Communist South Vietnam. In the tragic civil war that followed in the 1960s and 1970s the south was defeated in spite of massive military support and economic aid from the United States. The economy was traditionally agrarian, with rice and rubber its principal exports, but the French also developed a significant industrial area in the north based on the processing of mineral resources. Industrialization has been pushed by the government, which owns and operates virtually all enterprises.

The only avowedly socialist state allied with the Soviet Union in the Western Hemishere was the Republic of Cuba. Fidel Castro, the revolutionary leader who overthrew the oppressive dictator Fulgencio Batista on January 1, 1959, did not at first overtly proclaim himself a Marxist; but the anti-Castro policy of the United States, culminating in support for the disastrous invasion of the Bay of Pigs in 1961, drove him into the arms of the Soviet Union, delighted to find a base to spread its doctrines in the Western Hemisphere. Cut off from its traditional markets, principally in the United States, but still dependent on its traditional export, sugar, Cuba received the greater part of its manufactured goods (including armaments) from the Soviet bloc. In 1972 it became a member of COMECON.

The Economics of Decolonization

World War II dealt a death blow to European imperialism. The Philippines, the Dutch East Indies, French Indochina, and British Burma and Malaysia fell under the temporary control of Japan. Elsewhere in Asia and Africa the defeat of France, Belgium, and Italy and the preoccupation of the British with the war effort left their colonial dependencies largely on their own. Some dependencies proclaimed their independence at once; others witnessed the rise of independence parties that agitated against continued colonial rule. The wartime slogans of the Western Allies, calling for liberty and democracy throughout the world, strengthened the appeal of the independence movements by highlighting the contrast between Western ideals and the realities of colonialism. They also undermined the willingness of Europeans to tax themselves in order to dominate others. In the immediate postwar years the imperial powers temporarily regained control in most of their former colonies, but their own war-induced weaknesses, the growing strength of native independence movements, and the ambivalent position of the United States led to gradual abandonment of imperial controls. In a few cases colonial areas fought successful wars of independence against their former masters. Increasingly the imperial powers relinquished dominion voluntarily, if reluctantly, rather than experience the costs and hazards of war.

When Great Britain granted independence to the Indian subcontinent in 1947, not one but two new nations emerged—then a third and a fourth. India and Pakistan, the first predominantly Hindu in religion, the latter Islamic, both became independent on August 15, 1947. The following year the island of Ceylon also obtained independence, renaming itself Sri Lanka in 1972. Pakistan as originally constituted was divided into two widely separated parts: Urdu-speaking West Pakistan, along the Indus River, and Bengali-speaking East Pakistan across India on the Ganges River. The West Pakistanis dominated the more numerous East Pakistanis politically until 1971, when the latter revolted and established the independent state of Bangladesh. All four countries have extremely dense populations, few and poor natural resources, and low levels of literacy. They are also subject to racial and religious disturbances and unstable, frequently dictatorial governments. Most of the labor force in all is engaged in low-productivity agriculture. Not surprisingly, all of these countries are extremely poor. India is the least unprosperous. In the 1960s and 1970s it took advantage of the "green revolution" in agriculture and is now virtually self-sufficient in food supply. It also has more industry than the others. None of the countries is a doctrinaire socialist state, but in all the government plays a substantial role in the economy.

Burma, renamed Myanmar, governed before the war as a part of British India, gained independence from the British in 1948, Indonesia from the Dutch in 1949, and Laos and Cambodia, along with North Vietnam, from the French in 1954. In 1963 the former crown colonies of Singapore, Sarawak, and North Borneo joined the Federation of Malaya as the Federation of Malaysia, a fully self-governing dominion within the Commonwealth of Nations; but in 1965 Singapore, populated overwhelmingly by ethnic Chinese, withdrew and became an independent republic. The Philippines, slated for independence from the United States before World War II, actually became independent as the Republic of the Philippines on July 4, 1946. All of these countries, except Singapore, have many characteristics in common, including climate and topography. All are predominantly rural and agrarian, with the labor force divided between subsistence peasant farms and plantation agriculture producing for export. Some also have strategic minerals valued in world markets— oil in Indonesia and tin in Malaysia. All have low rates of literacy and high rates of population growth. Although nominally republics, the forces of democracy are weak; most have endured prolonged periods of dictatorship. And most are desperately poor. Singapore, however, is highly urbanized, highly literate, and relatively wealthy. Situated at the confluence of major trade routes, it has developed a sophisticated economy like Hong Kong, with commerce as its mainstay as well as related banking and financial services and even some industry.

The political map of Africa at the end of World War II differed little from that of the interwar years. The imperial powers of the past still ruled almost all of the continent. Superficially, the momentous events of the two previous decades appeared to have had little effect. Underneath the surface, however, powerful currents of change had been set in motion, which in the next two decades completely altered the face of the continent.

The former Italian colony of Libya became the first African nation to gain independence. The United Nations made the decision in 1949, and the new state came into existence at the end of 1951 as a constitutional monarchy. With its sparse population, apparent lack of natural resources, and backward economy, the future of the new nation was far from promising, but Western subsidies helped it to survive until the discovery of oil strengthened its economic base. In 1969 a junta of young military officers overthrew the aging pro-Western king and set up a fanatically nationalist Arab republic, which in the 1980s aided and abetted international terrorists.

Great Britain formally terminated its protectorate over Egypt in 1922, but retained control of its military and foreign affairs. In 1952 a military junta overthrew the indolent and pleasure-loving British puppet and set up a military dictatorship in the guise of a republic. In 1956 the dictator forced out the last British troops, ostensibly guarding the Suez Canal, and shortly afterward nationalized the canal. Although insisting on full independence for themselves, the Egyptians wished to maintain control of the Sudan, which they had administered jointly with the British. In a plebiscite in 1955, however, the Sudanese voted heavily in favor of an independent republic, which they proclaimed on January 1, 1956. With a large area but poor resources, and a mostly illiterate population, the Sudan has been unable to make either democracy or the economy function, and has been ruled by a series of military regimes.

French North Africa engaged in a long, difficult struggle for independence. Tunisia and Morocco had retained their traditional governments under French direction. Algeria, on the other hand, where the French had been established for more than a hundred years and which had more than a million inhabitants of European ancestry (about one-tenth of the population), was treated as part of France for some purposes. Strong nationalist and pan-Arab movements developed in all three countries after the war. The French government responded with nominal concessions for Tunisia and Morocco, but attempted to integrate Algeria more firmly with France. Neither policy worked.

France eventually granted Tunisia and Morocco full independence, but strengthened its control in Algeria. The Algerians replied with an intensive guerrilla war beginning in 1954, in which they frequently engaged in acts of terrorism against both the European population and Algerian collaborators. Unable to locate and destroy the top leadership of the rebels, who frequently took refuge in other Arab countries, the French army responded with terrorism of its own. The government in Paris neither gave the army full support nor exercised firm control over it. In May 1958, faced with the threat of an army coup d'état, the government of the Fourth French Republic abdicated its powers to General de Gaulle, who assumed virtually dictatorial power. At first de Gaulle seemed intent on keeping Algeria French, but after further years of bloodshed and fruitless attempts to reach an understanding with Algerian leaders for autonomy within the "French Community," he agreed in 1962 to full independence.

All three North African countries were predominantly agrarian, with Mediterranean-type agriculture (cereals, olives, citrus fruit, etc.), but they also contain important mineral deposits. In particular, Algeria's oil and natural gas deposits, discovered shortly after independence, have given it the means both to develop industry and to play a role in world politics. Before their independence all three countries were commercially oriented to France, and that orientation continued, although a trade agreement with the European Community in 1976 broadened their external markets. Algeria exported much of its liquid natural gas to the United States.

In the early 1950s most observers expected that a generation or more would be required for the black peoples of Africa to obtain independence; yet within a decade more than a score of new nations had arisen from the former British, French, and Belgian empires. The strength of the native independence movements only partly accounted for this striking development. Equally important were domestic difficulties of the imperial powers, which made them unwilling to bear the high cost (economic, political, and moral) of continuing to rule alien peoples against their will. Once the process of emancipation had begun, it continued like a chain reaction, with each new day of independence hastening the next.

After World War II the British government realized it would have to do a better job of preparing its African wards for self-government if it were to avoid costly colonial wars and a total loss of the economic benefits of empire. It began by establishing more schools, creating universities, and opening the civil service to Africans. In 1951 the Gold Coast and Nigeria obtained some local autonomy. The British intended more, perhaps after several decades. In the Gold Coast, however, Kwame Nkrumah, a remarkable political leader, demanded immediate independence and showed determination to win it even from his prison cell. Rather than risk a full-fledged revolt, the British agreed to most of Nkrumah's demands, and in 1957 the state of Ghana (named for a medieval African empire) emerged as the first black nation in the British Commonwealth. Ghana also became a member of the United Nations. With this precedent before it, Nigeria achieved independence in 1960, and other former British possessions followed suit in later years.

Paradoxically, the first British colonies in Africa to achieve full independence were among the least advanced economically and politically. Because they were populated almost entirely by black Africans, there was no problem of white minorities. In East Africa and the Rhodesias, however, British settlers had acquired vast tracts of property and enjoyed substantial local self-government. Deprived of political rights and economic opportunity, the Africans constituted sullen, rebellious majorities who sometimes resorted to violence, as in the Mau Mau terrorism in Kenya in the 1950s.

By 1965 Britain had granted independence to all its African colonies except Southern Rhodesia (or Rhodesia as it was known after Northern Rhodesia became Zambia in 1964). The exception resulted from the refusal of Rhodesia's white population to accord equal status to their black fellow citizens, who greatly outnumbered the whites. In 1965 the white-dominated government of Prime Minister Ian Smith made a unilateral declaration of independence—the first in the British Empire since 1776. The United Nations attempted to apply economic sanctions, but Rhodesia, with some assistance from South Africa and Portugal, successfully resisted them for several years. At length, in 1979, after free elections, the black majority triumphed and renamed the country Zimbabwe.

When announcing the constitution of the Fifth Republic in 1958, de Gaulle offered the French colonies, except Algeria, the option of immediate independence or autonomy within the new French Community, with the right to secede at any time. Although de Gaulle realized that it would probably be impossible to hold the colonies against their will, this remarkably liberal offer stands in marked contrast to the stubborn unrealism and blunders of earlier French colonial policy. Of fifteen colonies in black Africa (including Madagascar) only Guinea, led by the Communist Sekou Touré, chose independence. The others organized their own governments but allowed France to retain control of defense and foreign policy in return for economic and technical assistance. In 1960 a further constitutional change granted them full independence while permitting them special economic privileges. In 1963, under the Yaounde Convention (see later), those privileges were extended to include all members of the European Community.

The sudden achievement of freedom by France's colonies stirred the formerly placid subjects of the Belgian Congo to riot, pillage, and demand similar treatment. The Belgians had made no provision for self-government in the Congo, much less for independence. The disturbances took them by surprise, but early in 1960 the Belgian government decreed that the Congo was to become independent on June 30. Elections were hastily arranged, a constitution was drawn up, and the largely illiterate Congolese, none of whom had ever voted before, were called on to choose a complete set of government officials. When the day arrived many of them expected their standard of living to be magically raised to the level enjoyed by their affluent former masters. In their disappointment they again resorted to pillage and wanton destruction. The Congolese army mutinied, rival political groups attacked one another, and the mineral-rich Katanga province tried to secede from the new nation. The central government called for assistance from the United Nations to restore order, but rebellion and wild outbursts of violence continued to occur sporadically. Not until 1965 did a military dictator, General Mobutu, succeed in restoring order. After consolidating his control he changed the name of his country to the Republic of Zaire.

By the mid-1960s the former European colonial powers, except Portugal, had granted independence to almost all their Asian and African dependencies. Portugal scornfully rejected all suggestions that it prepare its colonies for eventual liberation. In 1974, however, a coup in Portugal overthrew the dictatorial regime and the new government promptly negotiated independence for its African colonies, Angola and Mozambique, the following year.

Although colonialism was dying, if not dead, it left a rueful legacy. With few exceptions, confined largely to areas of European settlement, the new nations were desperately poor. In three-quarters of a century of colonialism the nations of Europe had extracted vast fortunes in minerals and other commodities, but shared little of their wealth with the Africans. Only belatedly had some colonial powers made any effort to educate their subjects or prepare them for responsible self-government. Nevertheless, most new nations made at least a pretense of following democratic forms, and some made a valiant effort to achieve true democracy. As in many other parts of the world, however, the social and economic bases for stable, viable democracies did not exist. Many former colonies succumbed to one-party governments, frequently influenced by Russian or Chinese Communists. Some fell victim to anarchy and civil war, during which thousands of innocent civilians, especially children, died of malnutrition and disease as well as from hostilities. Most governments of the new nations were plagued by inefficiency and corruption. Even when their intentions were benign, few had the resources, especially in human capital, to carry them out successfully.

The Origins of the European Community

The dream of a united Europe is as old as Europe itself. Charlemagne's Holy Roman Empire closely approximated the boundaries of the original Common Market. Napoleon's French Empire and its satellites in the Continental System encompassed almost all of continental Europe. The concert of Europe, which grew out of the Vienna Congress of 1815, represented an attempt to coordinate policy at the highest levels of government. The League of Nations was a concert of the European victors in World War I. Hitler very nearly succeeded in creating a *Festung Europa* under Nazi domination. All such efforts failed, however, because of the inability of the would-be unifiers to maintain a monopoly of coercive power and the unwillingness of the members to submit voluntarily to their authority. In earlier times the difficulties of communication contributed to the fractionation of Europe. Then the idea of nationalism became so deeply entrenched in European thought, especially after the French Revolution, that sovereignty, that is, supreme authority or dominion, and nationhood became almost synonymous. Prior to World War II modern nations zealously opposed all proposals or attempts to infringe on or in any way diminish their sovereignty.

It is important to bear in mind the distinction between international and supranational organizations. International organizations depend on the voluntary cooperation of their members and have no direct powers of coercion. Supranational organizations require their members to surrender at least a portion of their sovereignty and can compel compliance with their mandates. Both the League of Nations and the United Nations are examples of international organizations. Within Europe the OEEC and most other postwar organizations of nations have been international rather than supranational. Continued successful cooperation may lead eventually to a pooling of sovereignties, which is the hope of proponents of European unity. Proposals for some kind of supranational organization in Europe have become increasingly frequent since 1945 and have issued from ever-more influential sources.

The proposals spring from two separate but related motives—political and economic. The political motive is rooted in the belief that only through supranational organization can the threat of war between European powers be permanently eradicated. Some proponents of European political unity further believe that the compact nation-state of the past is now outmoded; if the nations of Europe are to resume their former role in world affairs they must be able to speak with one voice and have at their command resources and a work force comparable to those of the United States. The economic motive rests on the argument that larger markets will promote greater specialization and increased competition, thus higher productivity and standards of living. The two motives merge in agreeing that economic strength is the basis of political and military power and that a fully integrated European economy would render intra-European wars less likely if not impossible. Because of the deeply entrenched idea of national sovereignty, most of the practical proposals for a supranational organization have envisaged economic unification as a preliminary to political unification.

The Benelux Customs Union, which provided for the free movement of goods within Belgium, the Netherlands, and Luxemburg, and for a common external tariff, grew out of the realization that under modern conditions of production and distribution the economics of the separate states were too small to permit them to enjoy the full benefits of mass production. Belgium and Luxemburg had, in fact, joined in an economic union as early as 1921, and the governments-in-exile of Belgium and the Netherlands had agreed in principle on the customs union during the war. Formal ratification of the treaty came in 1947. Statesmen of these countries have been the warmest advocates of a European common market and have continued to work for a closer economic integration of their own countries independently of broader European developments.

The OEEC resulted largely from American initiative and provided only for cooperation, not full integration. In 1950 French Foreign Minister Robert Schuman proposed the integration of the French and West German coal and steel industries and invited other nations to participate. Schuman's motives were as much political as economic. Coal and steel lay at the heart of modern industry, the armaments industry in particular, and all signs pointed to a revival of German industry. The Schuman Plan was a device to keep German industry under surveillance and control. Anxious to be admitted to the new concert of Europe, West Germany responded with alacrity, as did the Benelux nations and Italy, afraid of being left behind if they did not participate. Great Britain, with nationalized coal and steel industries at the time and still mindful of its empire, replied more cautiously and in the end did not participate. The treaty creating the European Coal and Steel Community (ECSC) was signed in 1951 and took effect early the following year. It provided for the elimination of tariffs and quotas on intracommunity trade in iron ore, coal, coke, and steel, a common external tariff on imports from other nations, and controls on production and sales. To supervise its operations several bodies of a supranational character were established: a High Authority with executive powers, a Council of Ministers to safeguard the interests of the member states, a Common Assembly with advisory authority only, and a Court of Justice to settle disputes. The community was authorized to levy a tax on the output of enterprises within its jurisdiction to finance its operations.

Soon after the community commenced operations the same nations attempted another giant step forward on the road to integration with a treaty for a European Defense Community. Developments such as the Korean War, the formation of NATO (North Atlantic Treaty Organization) in 1949, and the rapid economic recovery of Germany had demonstrated the importance of including German contingents in a western European military force, but proposals to do this naturally aroused the suspicion and hostility of people who had recently been victims of German aggression. After prolonged debates the French National Assembly rejected the treaty outright in August 1954. This setback to the movement for unification demonstrated once again the difficulty in securing agreement on proposals for the limitation of national sovereignty. The proponents of unity then resorted to more cautious tactics, once again in the area of economics.

In 1957 the participants in the Schuman Plan signed two more treaties in Rome, creating the European Atomic Energy Community (EURATOM) for the development of peaceful uses of atomic energy and, most important, the European Economic Community (EEC), or Common Market. The Common Market treaty provided for the gradual elimination of import duties and quantitative restrictions on all trade between member nations and the substitution of a common external tariff over a transitional period of twelve to fifteen years. Members of the community pledged to implement common policies respecting transportation, agriculture, social insurance, and a number of other critical areas of economic policy and to permit the free movement of persons and capital within the boundaries of the community. One of the most important provisions of the treaty was that it could not be unilaterally renounced and that, after a certain stage in the transition period, further decisions would be made by a qualified majority vote rather than by unanimous action. Both the Common Market and EURATOM treaties created high commissions to oversee their operations and merged other supranational bodies (councils of ministers, common assemblies, and courts of justice) with those of the ECSC. The Common Market treaty took effect on January 1, 1958, and within a few years the community confounded the pessimists (of which there were many) by shortening instead of lengthening the transitional period. In 1965 the high commissions of the three communities merged, providing a more effective agency for eventual political unification. On July 1, 1968, all tariffs between member states were completely eliminated, several years earlier than the date originally foreseen.

In the preliminaries to the treaties of Rome invitations were extended to other nations to join the Common Market. Britain objected to the surrender of sovereignty implied in the treaties and attempted to persuade the OEEC nations to create a free trade area instead. After signature of the Common Market treaty Britain, the Scandinavian nations, Switzerland, Austria, and Portugal created the European Free Trade Association (EFTA), the so-called outer seven, in contrast to the inner six of the Common Market. The EFTA treaty provided only for the elimination of tariffs on industrial products among the signatory nations. It did not extend to agricultural products, it did not provide for a common external tariff, and any member could withdraw at any time. It was thus a much weaker union than the Common Market.

In 1961 Britain signified its willingness to enter the Common Market if certain conditions could be met. If effected, this move would have entailed the membership of most of the EFTA partners also. Lengthy negotiations over the terms of entry ensued, but in January 1963 President de Gaulle of France in effect vetoed Britain's membership, an action he repeated in 1967. After de Gaulle's resignation in 1969 the French government took a more moderate attitude on the question of British membership, which other Common Market countries favored. Following further negotiations Britain, Eire, Denmark, and Norway were accepted for membership in 1972, effective January 1, 1973. Although Norway applied for and was accepted for membership, the government put the question to a popular referendum, which lost; thus in 1973 the original six became nine. Subsequently, after prolonged negotiations, Greece joined in 1981 and Spain and Portugal in 1986.

The proponents of European unity had far more in mind than a mere common market or customs union. For them the Common Market was but a prelude to the United States of Europe, in much the same way that the Zollverein preceded the German Empire. After the signature of the treaties of Rome they began to speak of the "European Communities" and, after the merger of the high commissions in 1965, of the "European Community." The merged Common Assemblies became the European Parliament. At first its members were elected by and from the members of the parliaments of the member states, and sat in national groupings. It had only consultative powers until the community obtained an independent source of revenue from its customs duties. (Previously the communities had been financed by a portion of the value-added taxes from each country.) Thereafter the Parliament had limited control over the budget. In 1979, after long negotiations and several postponements, the members of the Parliament were elected directly by the people and took their seats in party blocs rather than according to nationality.

Nevertheless, European unity was far from an accomplished fact. In the 1950s and 1960s, when the community was establishing itself and taking its first steps toward integration, the world economy was robust and expansive, contributing to the sense of optimism of the new endeavor and facilitating its progress. Subsequently, during the period of expansion of membership, the world economy was much less conducive to growth. Many problems arose within the European Community to impede its progress. The Common Agricultural Policy (CAP) was a major headache. According to its terms the community was obligated to assist the agricultural sector by providing remunerative prices for its products. As the policy actually functioned it led to enormous overproduction of commodities such as butter, wine, and sugar, the surplus of which had to be disposed of in ways that did not lower prices, at the expense of consumers and taxpayers. Another persistent problem was the European Monetary System (EMS), which envisaged the replacement of the separate national currencies by a single money of account, the ECU (European Currency Unit; the *écu* was also the name of an ancient French coin). Although projected as early as 1970, the implementation of the EMS was repeatedly postponed because of international currency disorders and the disparities between the monetary and budgetary situations of the member states.

Another major problem the community had to deal with was its relations with countries of the Third World, especially those former colonial possessions of the member states. In 1963 it signed a convention in Yaounde, Cameroon, offering commercial, technical, and financial cooperation to eighteen countries of black Africa, mostly former French and Belgian colonies. In 1975 it signed a convention in Lomé, Togo, with forty-six countries of Africa, the Caribbean, and the Pacific, granting them free access to the community for virtually all of their products, as well as providing industrial and financial assistance. In 1979 the Lomé convention was renewed and extended to a total of fifty-eight ACP countries, and in 1984 to sixty-five. The community also made similar agreements with Israel (1975), Tunisia, Algeria, and Morocco (1976), and with Egypt, Syria, Jordan, and Lebanon (1977).

16

The World Economy at the End of the Twentieth Century

Europe's long postwar economic boom had its counterparts in other regions of the world economy, notably Japan. Indeed, the Japanese boom was both longer and stronger. From the late 1940s to the early 1970s the growth rate of Japanese GNP was in excess of 10 percent annually, unique in the history of economic growth. In the relatively depressed 1970s and 1980s it was somewhat lower, but still higher than that of most other areas of the world economy. Although frequently referred to as a "miracle," there were, as in Europe, solid reasons for this performance. In the first place there was the phenomenon of technological catch-up. From the late 1930s to the late 1940s the Japanese economy had been isolated from that of the rest of the world, and there were many technological innovations that Japan could borrow at minimal cost. That is scarcely a sufficient reason for Japan's high growth rate, however; if it were, many other countries would have done likewise. More important was Japan's high level of human capital, which enabled Japan to take advantage of superior technology. Moreover, after Japan made up for its technological lag it became a leader in introducing new technology, especially in electronics and robotics. For this, it could draw on not only its stocks of human capital but also on the high levels of saving and investment of the Japanese people. Another significant factor is the sophistication of Japanese management, which realized the high payoffs of industrial research and development. Finally (but not exhaustively), and more speculatively, one might cite the spirit or mentality of the Japanese people—collectivist (in a general sense), cooperative, and given to team play. This is manifest in both the attitudes of employers toward their employees (and vice versa) and in government policy, which provoked the epithet "Japan, Inc.," which has been applied by some critical Western observers.

Other Asian countries, notably South Korea and Taiwan, also had extremely high growth rates for both total output and foreign commerce. Several of the reasons for Japanese economic success could also apply to them. Singapore and Hong Kong, previously mentioned, occupied very special positions in the international economy, although the provision of the 1984 treaty between the United Kingdom and the People's Republic of China, according to which Hong Kong will revert to Chinese sovereignty in 1997, caused much anxiety among that territory's Westernized population. Overall, the Pacific Basin area, including Australia and New

Zealand, which had been a marginal participant in the world economy before the mid-twentieth century, became a major actor in the last quarter of that century.

Such was not the case for Latin America. In the late nineteenth century and the first half of the twentieth the countries of Latin America had been important participants in the international division of labor, based on their comparative advantage in primary products. As late as the mid-twentieth century some of them, notably the countries of the "southern cone" (Argentina, Uruguay, and Chile), enjoyed per capita incomes comparable with those of Western Europe. Thereafter, under the misguided assumption that they were somehow second-class citizens of the world because of their specialization in primary products, several Latin American nations embarked on programs of "import substitution industrialization," attempting to produce for themselves the manufactured goods they had previously imported. Almost without exception these programs failed, for several reasons: (1) domestic markets were too small, in both numbers and purchasing power, to justify the most economical methods of production; (2) there was a lack of international cooperation in the region (e.g., the Latin American Free Trade Association, LAFTA, never really got off the ground); and (3) unlike Japan, the region lacked the human capital to use effectively the new technology developed elsewhere, much less to develop its own. Although total output, both industrial and agricultural, rose substantially in the postwar period, the increase on a per capita basis was substantially below that of the rest of the world, except Africa, and the region's share of total world trade declined steadily. The individual nations' unfavorable trade balances, especially those of Argentina, Brazil, and Mexico, gave rise to alarmingly high levels of international indebtedness in the 1980s, which threatened the entire system of international payments.

Economic conditions in Africa as the twentieth century neared its end were even more deplorable than in Latin America. The new nations that emerged with the end of European colonialism lacked resources, natural and especially human, to cope with the complexities of a modern economy. Political circumstances likewise hindered efforts of economic development; ethnic enmities engendered frequent civil wars and coups d'état; most nations fell under the sway of one-party governments with varying degrees of dictatorial control. In the Republic of South Africa, long-dominated by its white minority of British and (mainly) Boer descent, the black majority finally obtained something approaching political equality in the early 1990s; but ethnic rivalries and resorts to violence among factions of the majority threatened to vitiate any economic advantages that might have resulted. Several African nations, already among the poorest in the world, actually endured negative rates of growth of income and wealth.

Another region of the world that has acquired greatly increased economic significance in the latter part of the twentieth century is southwest Asia, or the Middle East. The reason for its increased economic importance can be succinctly stated in a single word: oil.

Oil was discovered in Iran (then called Persia) in the first decade of the twentieth century, and subsequently in several Arab states bordering the Persian Gulf—Iraq, Saudi Arabia, Kuwait, and the smaller emirates—but as late as 1950 the region accounted for little more than 15 percent of total world production. (The United

States at that time was still far and away the largest producer, with more than 50 percent of the total.) In 1960 the countries of the Middle East, along with Libya and Venuzuela, formed the Organization of Petroleum Exporting Countries, OPEC, which several other countries later joined. By 1970 OPEC nations accounted for more than one-third of world energy production. In 1973, in the wake of the fourth Arab-Israeli war, the OPEC countries acted in cartel fashion to increase sharply the price of crude petroleum, an action they repeated later in the decade, with the result that the world price rose from $3 per barrel in 1973 to $30 in 1980, a ten-fold increase. Given the world economy's high degree of dependence on petroleum by that time, the effect on the economies of both highly industrialized and developing nations was devastating. The developing nations suddenly faced much larger deficits in their balance of payments, forcing them more deeply into debt, whereas the industrial nations encountered ''stagflation''—stagnation of output and employment combined with inflationary price rises. The situation persisted throughout the 1970s and early 1980s, producing the highest levels of unemployment since the 1930s.

Meanwhile political and religious changes in the Middle East altered the economic balance of power. In 1979 a fanatical religious revolt in Iran drove out the Shah and established an Islamic republic. The following year Iraq's dictator, Saddam Hussein, sought to take advantage of his neighbor's presumed weakness and invaded Iran with the aim of annexing its oil-producing regions; but the Iranians put up an unexpected resistance. For nine years the countries engaged in an inconclusive border war in which millions died to no avail. In 1990, after a truce of sorts with Iran, Saddam suddenly invaded his smaller and defenseless neighbor, Kuwait, proclaiming it the ''nineteenth province'' of Iraq.

International reaction was immediate. The United States came to the defense of Saudi Arabia, its largest external supplier of oil (along with Kuwait), and the United Nations' Security Council, in a rare show of unanimity, condemned Iraq's action and instituted an embargo on the aggressor. After six months of waiting to see if the embargo would force Saddam to back down, a combined force of Arab and non-Arab nations invaded Iraq and liberated Kuwait, but did not overthrow Saddam, who remained in power.

The Collapse of the Soviet Bloc

In the latter half of 1989 a series of events unfolded in Eastern Europe that was as momentous as it was unexpected: the (mostly peaceful) overthrow of Communist regimes in one country after another. Poland and Hungary led the way, but few outside observers expected the other nations to follow. Amazingly, they did so: Czechoslovakia, East Germany, Bulgaria, finally Rumania and, belatedly, Albania.

A mixture of political and economic motives underlay the revolt of the masses in the once Communist-dominated lands. As recounted in the previous chapter, the regimes in those countries had been imposed by the Soviet Union without the consent—indeed, against the will—of the people. Had those regimes been able to deliver on their promises of improved material conditions and a higher standard of living, the people probably would have accepted their deprivation of liberty; but the regimes did not. On the contrary, the material circumstances, including living and working conditions, of the masses steadily deteriorated, in contrast to the ease and abundance that they could observe on television in their Western neighbors, and to the lavish life-style of the new ruling class, the upper echelons of the Communist party, word of which gradually filtered out.

The masses had shown their displeasure on several occasions in the past: East Germany in 1953, Hungary in 1956, Czechoslovakia in 1968, and Poland more than once. On those occasions the Soviet Union had used armed force to quell the rebellions (or, in the case of Poland, that regime's own armed forces). Why it did not do so in 1989 is an interesting question that will be explored shortly; but it did not.

The roots of popular discontent were thus deep, and although the sudden success of the revolts took the world—the Communist leaders as well as Western observers—by surprise, some harbingers could be observed. In 1980 Polish workers led by Lech Walesa, an electrician in the shipyards of Gdansk, formed a labor federation, Solidarity, independent of the state and of the Communist Party. The regime tolerated it for a time, but in December 1981 the government proclaimed martial law—ostensibly to prevent Soviet intervention—and imprisoned Solidarity's leaders. Unrest continued; in April 1989, in an effort to defuse the unrest, the government again legalized Solidarity and arranged for partially free elections in June. (Partially free, because a majority of the seats in the Sejm, or parliament, were reserved for government candidates.) When Solidarity won all but one of the seats that it was allowed to contest, one of its leaders became prime minister under a Communist president. (Walesa could have had the job, but chose not to.) Subsequently, after the Communist president resigned, Walesa was elected president of free Poland in December 1990.

After the failure of the 1956 rebellion in Hungary the new government installed by the Soviet Union cravenly followed the Soviet line in foreign policy—for example, it provided some token forces in the 1968 invasion of Czechoslovakia—but in return was given some limited freedom in domestic affairs. In 1968 it instituted a "New Economic Mechanism" that was a compromise between strict central planning and a free market system. It also developed closer relations, political and economic, with Western Europe. It allowed the formation of opposition political parties which, in 1989, negotiated with the government for a peaceful transition. In a free election in May 1990 a coalition of three former opposition parties won a clear majority in the new National Assembly.

With the example of the relatively peaceful transitions in Poland and Hungary before them, in 1989 students and workers in Czechoslovakia stepped up their protests and mass demonstrations. At first the government responded with violent repression, in which hundreds of citizens were killed and injured; but eventually it agreed to negotiations. In December 1989 Alexander Dubcek, the leader of the "Prague spring" of 1968, was elected chairman of the new parliament and Vaclav Havel, a writer who had been imprisoned for his human rights activism, became president.

The German Democratic Republic (East Germany) celebrated its fortieth anniversary in October 1989, an event attended by Soviet President Mikhail Gorbachev.

Shortly thereafter, however, the Central Committee of the East German Socialist Unity (Communist) party deposed its veteran leader, Erich Honecker, who was subsequently charged with various crimes, including misappropriation of government funds. Meanwhile thousands of East Germans, unable to migrate directly to the West, had begun streaming through Czechoslovakia to Hungary, hoping thereby to reach Austria and the West. The new Hungarian government obliged them by opening the border to Austria.

One of the most dramatic—and symbolic—events of 1989 was the destruction of the Berlin wall. The wall had been erected around West Berlin by the East German government in 1961 to prevent the escape of its subjects to the West. For almost three decades it stood as a symbol of Communist tyranny and repression. Spontaneously, on the night of November 9–10, without hindrance from the East German authorities, demonstrators from both East and West Berlin began destruction of the wall, and thousands of East Berliners streamed into the West.

Events moved rapidly thereafter. The West German authorities, caught by surprise no less than those of the East, made emergency arrangements to care for the stream of refugees, but also sought to persuade the East Germans to stay where they were. In July 1990 they created an economic and monetary union with the German Democratic Republic, which ceased to exist as a separate state on October 3, being incorporated with the German Federal Republic.

In Bulgaria Todor Zhivkov, the long-time president and Communist party leader, resigned both posts the day after the Berlin wall came down. The new president, although from the same party, promised free elections. In Rumania, Nicolae Ceaucescu, president since 1974 and a well-entrenched dictator, vowed not to give in to popular protests. In December he ordered his security forces to massacre thousands of demonstrators; but soon thereafter the army sided with the revolutionaries and executed Ceaucescu and his wife, who was also his deputy, on December 25. Bulgaria and Rumania differed from the other former Soviet satellites, however, in that their democratic opposition forces were less well organized. The former Communist parties changed their names and their ostensible policies, but many of the same personnel continued to rule.

Throughout the turmoil of 1989 and 1990 the small Stalinist state of Albania, separated from the other satellites by maverick Yugoslavia, maintained its policy of popular repression undeterred; but in 1991 its government gave in to democratic pressure and permitted free elections. Yugoslavia itself, although not a Soviet satellite, likewise endured several years of turmoil associated with attempts at political and economic reform. As a federation of distinct ethnic groups, its situation was complicated by separatist movements as well as movements for economic reform. In 1992, after months of destructive civil wars, Yugoslavia broke up into its constituent republics. Czechoslovakia did likewise, though without conflict.

The desire for liberty wafted across the breadth of Asia. As noted in the previous chapter, the government of the People's Republic of China had allowed a limited introduction of free markets and free enterprise in the 1980s. This policy had been notably successful in the rural sector, and agricultural output increased substantially along with the incomes of farmers and farm workers. The policy also generated demands for greater political freedom and democracy. For seven weeks in April and

May 1989, students and others staged daily demonstrations in historic Tiananmen Square in the heart of Beijing. For a time it appeared that China, like Eastern Europe, might undergo a peaceful transition toward more democratic forms. In the end, however, the hard-liners won the upper hand, and on June 4 armored columns gunned down the demonstrators by the hundreds or thousands, and smashed their symbol, a plastic replica of the Statue of Liberty.

The events of 1989 in Eastern Europe, as momentous for world history as those of 1789 in France, abolished the "old regime" of Communist power, but what did they put in its place? The new democratic governments were tenuous and fragile; some of them, indeed, were still staffed by holdovers from the old system, and there was no guarantee that they would not succumb to another wave of authoritarianism. The basic reason for the collapse of the Communist regimes was popular dissatisfaction with their economic performance. Although general agreement existed that the old system of state ownership and management had to be replaced, no consensus existed on how to replace it. Private ownership of the means of production and a market system of organization was one obvious alternative, but where would the owners come from? Who would organize the markets? The new regimes agonized over these and similar questions throughout the early 1990s, with no clear resolution. Many East European leaders hoped for closer relations with Western Europe, including private investments in joint ventures and similar devices. Some even hoped for membership in the European Community, although that was unlikely for several years, until they could establish more stable and functioning economies and polities.

The nature of international economic relationships posed another problem that had to be faced. The Council for Mutual Economic Assistance (COMECON), founded by the Soviet Union in 1949, had never been especially successful, and after the events of 1989 it was even less so. It was officially dissolved in 1991.

Western observers—and no doubt many in the Soviet Union and Eastern Europe as well—wondered why the Soviet Union did not use armed force to quell the rebellions in the satellites, as it had done before and as the Chinese government did against its own people in June 1989. A full explanation has not yet emerged, but when it does it will probably feature the economic debility of the Soviet Union itself as well as contemporaneous political developments.

In 1964 conservatives in the Communist Party hierarchy deposed the ebullient Nikita Khrushchev, putting in his place Leonid Brezhnev, who ruled for almost two decades. Under Brezhnev the Soviet economy stagnated; inefficiency and corruption flourished. A "reform" instituted in 1965 encountered the opposition of the entrenched bureaucracy (which expanded by 60 percent between 1966 and 1977) and was quietly shelved a few years later. Both the economic growth rate and productivity declined. When Mikhail Gorbachev—the first Soviet leader born after the October Revolution—came to power in 1985 the economy was in a state of crisis. Gorbachev undoubtedly realized that the Soviet Union was no longer in a position to enforce its will on its reluctant former satellites. Its greatest need was to reform itself, hence Gorbachev's program of *perestroika* (restructuring) and *glasnost* (openness).

Although Gorbachev placed greater emphasis on *perestroika*—he even published a book with that title that was translated into several languages—it was *glasnost* that had the most immediate effect. *Glasnost* in the Soviet context meant greater freedom of expression (for the press, in particular), an ability to discuss and debate both official policies and their alternatives, and even (to some extent) an ability to act independently of the party and the state in political matters. Partly as a consequence, the Baltic republics of Latvia, Lithuania, and Estonia declared their independence, which was confirmed in 1991. Others moved in the same direction, and even the huge Russian Republic under the popularly elected presidency of Boris Yeltsin began to act independently of the Communist Party.

One of the justifications of *glasnost* was to enlist the initiative and enthusiasm of the population for the tasks of *perestroika*, or economic restructuring. Nowhere, however, did Gorbachev spell out what he meant by restructuring, beyond some vague generalizations about improved cost accounting, devolution of decision making to the level of the enterprise (as against the planning authorities and ministries), the necessity of enterprises to make profits (i.e., the abolition of subsidies), and similar matters. In his book he referred to the importance of ''mass initiative,'' an apparent oxymoron, and described *perestroika* as a ''combination of democratic centralism and self-management,'' a flat contradiction in terms.

Gorbachev apparently favored a return to something like Lenin's New Economic Policy, in which the state would retain control of the ''commanding heights'' of the economy but allow limited private enterprise in the remainder (see p. 366). He was caught in a bind, however, between the conservatives of the party hierarchy, who favored maintenance of the status quo, and radical reformers who wanted to abolish the system of central planning altogether and move to a pure market economy. While the debate on the ultimate nature of the reform raged, a few limited reforms were actually effected. For example, many economic activities that had formerly taken place on black or gray markets—private handicraft production, petty trading, the production of various personal services—were legitimated, provided that the producers also held full-time jobs in state enterprises. Cooperatives could be formed to produce consumer goods or services, subject to the same restrictions. Individuals or families could lease land for agricultural production, again subject to some restrictions. The Soviet Union also allowed some foreign capital to enter into joint ventures with state-owned enterprises.

In August 1991, on the eve of a new treaty between the Soviet Union and some of its constituent republics that would have given far more power to the latter, a small group of hard-liners in the Communist party attempted a coup d'état. The coup leaders, including Gorbachev's hand-picked vice president, the head of the KGB, and the minister of defense, placed Gorbachev, then vacationing in the Crimea, under house arrest, suspended press freedom, and declared martial law. The Russian populace, however, especially the citizens of Moscow and Leningrad, refused to be cowed. Under the leadership of Yeltsin, and with the support of some military units that came to his defense, they openly defied the coup leaders, who quickly lost courage and fled, only to be arrested.

After three days a triumphant Gorbachev returned to Moscow, but, as he himself said, the Moscow to which he returned was not the one he left. Power relationships had changed drastically. Gorbachev retained his constitutional post as president of the Soviet Union, but resigned as general secretary of the Communist

Party, which was subsequently dissolved. Yeltsin, as president of the Russian Republic, greatly increased his stature and power. Most of the constituent republics declared their independence of the central government. In December the popularly elected presidents of most of the remaining republics created a Commonwealth of Independent States. Gorbachev resigned as president on December 25, and the Soviet Union was no more. In economic affairs the role of private enterprise and markets increased significantly. The Commonwealth structure functioned mainly as a free trade area or common market, resembling in some respects the European Community.

Some economists studying the Soviet Union in the 1960s and 1970s predicted the phenomenon of "convergence," that the Soviet and Western economies would become more alike. They scarcely envisaged the dramatic events of 1991, but the results of those events did, indeed, make the economy of the former Soviet Union more like that of the West. Meanwhile events in North America created yet another large market. The United States and Canada signed a free trade treaty, and entered into negotiations with Mexico to extend the free trade area to virtually all of North America. It appeared that convergence was occurring on a large scale throughout much of the world.

The Exfoliation of the European Community

After more than thirty years of existence the European Community still had not realized the dreams and visions of the most ardent proponents of European unity, a United States of Europe. Indeed, in spite of the removal of internal tariff barriers it had not even succeeded in removing all restrictions on intra-European trade, nor in abolishing internal customs frontiers. The monetary union was far from complete, and budgetary crises were a perennial problem. The admission of the less developed Mediterranean countries, Greece, Spain, and Portugal, introduced a host of new problems, especially in the agricultural sphere.

The ultimate goal of political union evolved into a struggle between two broad groups of partisans. On one hand the European Commission, with its headquarters in Brussels and its hosts of "Eurocrats," was joined by the European Parliament in seeking ever-greater measures of unity and an enhanced role for the Parliament. They were opposed by the various national governments (and occasionally by the national parliaments) which, while paying lip service to the ultimate goal, nevertheless found reasons for foot-dragging on specific issues. The governments were represented in the Council of Ministers, also known as the European Council, which had final authority on all matters not covered by the treaties that established the Community.

In the early 1970s the Belgian prime minister, Leo Tindemans, at the invitation of his fellow heads of government, prepared a report that envisaged completion of the union by 1980. That proved much too ambitious a goal, given the fundamental differences between the member states on the necessary reforms and the constitutional structure of the eventual union. The report was never implemented.

After a few years in the doldrums the movement to "relaunch" Europe gained new force in the 1980s. In 1985 the European Council (heads of state or government) decided in principle to proceed to greater union, and in February 1986 signed the "Single European Act" (SEA), which took the form of amendments and additions to the existing treaties. Specifically, the SEA required that the Community adopt more than 300 measures to remove physical, technical, and fiscal barriers in order to complete the internal market. These measures were to be completed by December 31, 1992, at which time it would become a Community "without frontiers." The SEA also provided for a Community flag, a circle of twelve gold stars on a blue ground; a Community anthem, Beethoven's "Ode to Joy"; and Community passports.

Concurrently, Jacques Delors, a former French government official and a strong proponent of European unity, became president of the EC Commission in 1985, a post to which he was reappointed for a four-year term in 1989.

The movement for unity got a boost from another direction in 1986 when the governments of France and the United Kingdom agreed to allow a railway tunnel to be built under the English Channel. Such a tunnel, sometimes called a "chunnel," had been proposed as early as the 1870s and periodically thereafter, but had never been realized. A remarkable feature of the new proposal was that it was to be financed entirely by private capital, without government subsidies. It was scheduled for completion in 1993, shortly after the entry into force of the Single European Act.

Another favorable development, also scheduled for 1993, was the creation of a European Economic Area (EEA) by a merger of the European Community and the European Free Trade Association, then consisting of Austria, Finland, Iceland, Norway, Sweden, Switzerland, and Liechtenstein. The latter countries would not necessarily become members of the Community—although that was a possibility—but would join it in a free-trade area, the world's largest, with 380 million consumers and 46 percent of world trade.

Meanwhile new measures to increase the strength and cohesion of the European Community came into being. Although the EMS (European Monetary System) had been established in 1979, the coordination of monetary policies—not to mention the creation of a single monetary policy—remained one of the greatest hurdles in the way of complete economic unity. For one thing, the United Kingdom under Prime Minister Margaret Thatcher (1979–1990) never consented to join the EMS. Some progress occurred in the early 1990s, however. After Mrs. Thatcher resigned her successor, John Major, brought the UK in. In 1991 the Community decided to create its own central bank by 1994, to be followed by a single currency by 1999. Such a development, with concomitant harmonization of monetary and fiscal policies, and with the possible or probable adherence of other countries, including some from Eastern Europe, would make European unity a reality by the beginning of the twenty-first century.

Limits to Growth?

In 1972 a research team associated with the Massachusetts Institute of Technology published a book, *The Limits to Growth*, in which they predicted that "the limits to growth on this planet will be reached sometime within the next one hundred years.'' They invoked "five major trends of global concern": accelerating industrialization, rapid population growth, widespread malnutrition, depletion of nonrenewable resources, and a deteriorating environment. The book received a great deal of publicity for a scholarly production, and it was the subject of much public discussion. Although many critics believed that the authors had overdramatized their conclusions—especially the one that predicted a rapid cessation of growth—almost all agreed that they had truly identified trends of "global concern," notably population growth and environmental degradation.

Most critics also agreed that the trends were interrelated. Malnutrition, for example, is especially prevalent in those Third World countries that are experiencing rapid population growth (but it is not unknown even in affluent countries such as the United States). An incidental aspect of the overthrow of Communist regimes in Eastern Europe was the revelation of the extensive environmental damage occasioned by their programs of headlong industrialization. Environmental damage is not confined to Soviet-style economies, as the controversy over "acid rain" testifies; but in the more affluent industrial economies there is greater awareness of it, and greater pressure by both public opinion and government to force polluters to cease or to pay the cost of environmental clean-up.

The problems are serious, but they are not unprecedented, as many otherwise intelligent people ignorant of history tend to assume. Throughout human history the pressure of population on resources condemned the great majority of people to live at a bare subsistence level. Some two hundred years ago Thomas Malthus provided a theoretical explanation of why that would always be the case (see p. 14). Ironically, at the very time that Malthus was writing a process—the process of rapid technological change—was under way that vitiated his assumptions. As already indicated (above, p. 191), the next two hundred years witnessed a rise in living standards in many parts of the world that Malthus and his contemporaries could not have imagined.

In terms of total numbers and the rate of growth over the last half-century, the population problem *is* without historical precedent—but numbers and growth rates are not the only relevant variables. As pointed out in Chapter 1, population must be viewed in relation to the resources available to support it. The volume of resources—animal, vegetable, and mineral—to support the world's population is also at an all-time high.

During the last hundred years or so the affluent nations of the world have experienced a demographic transition from a regime of high birth and death rates to one much lower, with a consequent reduction in the rate of population growth (above, p. 324). The expectation is that as other, poorer nations increase their level of material well-being they, too, will reduce birth rates and thus rates of population growth. Some experts even believe that the world as a whole reached an inflection point—from a rising to a declining rate of growth for total population—in the 1970s.

The "race" between population and resources leads to two related problems, the rate at which resources are being used (and used up), and the inequality in the distribution of resources. There is no doubt that the world—especially the affluent nations of the world—is utilizing resources at historically unprecedented rates. That in itself is a measure of its "success" in mastering the environment and solving the economic problem, but it has also given rise to fears of a total exhaustion of resources. Such fears are not unreasonable, but they are not well-grounded historically. After all, it was the shortage of timber for charcoal that led to the use of coke for smelting iron ore. There are many other instances in which temporary or localized shortages of particular resources have given rise to substitutes, which often prove to be more efficient or economical. In the nineteenth century coal replaced timber as a source of inanimate energy. In the twentieth century petroleum has largely displaced coal. Electricity, generated by water power, coal, petroleum, and eventually nuclear power, proved to be the most versatile and almost ubiquitous form of energy. Pessimists point out that coal and petroleum exist in finite amounts, and their supply may eventually be exhausted; water power is subject to physical limits on its use; nuclear power poses serious environmental hazards. Yet optimists claim that solar energy—the original source of both coal and petroleum—has scarcely been tapped directly. The technology does not yet exist to exploit solar energy directly in more than trivial amounts. But as the conventional sources of power become scarcer—that is, as their prices rise—the inducement to engage in research directed to solar energy will increase. That is the way the economic mechanism functions. The possibilities are limitless.

The inequality in the distribution of resources—among individuals, social groups, and nations—is at the very heart of the problem of economic development, as mentioned in the first paragraph of this book. Its solution will not be easy. It will require study, research, and widespread institutional change. That is the challenge facing both developed and underdeveloped nations. The history recounted in this book shows that the challenge can be met.

Annotated Bibliography

This bibliography is intended as a supplement to the text, to enable readers to find further information on topics that, of necessity, are dealt with only summarily in the book itself. It is therefore highly selective. It includes only books and a very few articles that are likely to be easily accessible to students and interested readers. Works in languages other than English are generally excluded, as are rare and recondite works, although translations of important works are included. With a few exceptions, items are listed only once, even though their contents may be relevant to two or more chapters or parts. Advanced students and researchers should consult the bibliographies in such general works as the *Cambridge Economic History of Europe* and the *Fontana Economic History of Europe*, as well as the specialized bibliographies in the *International Bibliography of Historical Sciences*, the *International Bibliography of the Social Sciences—Economics, The New Palgrave: A Dictionary of Economics, The New Palgrave Dictionary of Money and Finance*, and the American Economic Association's *Index of Economic Articles in Journals and Collective Volumes*, all of which should be available in any moderately good college or university library.

Although journal articles are usually not featured in this bibliography, browsing through the specialized journals is recommended as a good way to become familiar with the scope and nature of the literature. The leading English-language journals are the *Economic History Review* (United Kingdom, since 1927), the *Journal of Economic History* (U.S., since 1941), the *Scandinavian Economic History Review* (since 1953; contents broader than its title suggests), *Explorations in Economic History* (U.S., since 1963), and the *Journal of European Economic History* (published since 1972 by the Banco di Roma). Other journals, both more general and more specialized, that publish articles of interest to economic historians are *Agricultural History, Business History* (U.K.), the *Business History Review* (U.S.), *Comparative Studies in Society and History, Journal of Interdisciplinary History, Social Science History*, and *Technology and Culture*. This by no means exhausts the list of relevant journals, even in English.

No other work covering the full scope of this volume exists. *The Cambridge Economic History of Europe*, planned in the 1930s by Sir John Clapham and Eileen Power and edited by various distinguished authorities, now consists of ten volumes spanning the period from the decline of Rome until the latter part of the twentieth century. The later volumes include chapters on the United States and Japan as well as the major European countries. The contributors are (or were) mostly well-known authorities on their subjects, but the quality of the contributions is uneven, and there are some lapses in editorial planning and supervision.

The Fontana Economic History of Europe, planned and edited by Carlo M. Cipolla, contains roughly the same chronological and geographical coverage as the Cambridge series, but in nine somewhat smaller volumes. Authored by numerous specialists, it is aimed more directly at a student readership. *Essays in Economic History* (3 vols., London, 1954–62), edited by E. M. Carus-Wilson for the Economic History Society (U.K.), is a collection of outstanding journal articles covering the period from the Middle Ages to the late nineteenth century, but heavily weighted toward Great Britain. In contrast, *Essays in French Economic History*, Rondo Cameron, ed. (Homewood, IL, for the American Economic Association, 1970), consists entirely of articles translated from French journals and dealing with all eras and aspects of French economic history. Other similar anthologies dealing with shorter time spans or smaller areas are mentioned below.

The British can boast of two excellent series of small paperbacks of special interest to students of economic history. The Economic History Society sponsors "Studies in Economic and Social History" (formerly "Studies in Economic History") originally edited by the late M. W. Flinn, later by T. C. Smout, more recently by L. A. Clarkson, and published by the Macmillan Press, Ltd. These consist of short essays, generally 50 to 100 pages, on significant topics, mostly but not exclusively British, in which a recognized expert summarizes and synthesizes the existing literature. The other series, "Debates in Economic History," is published by Methuen & Co., Ltd., with Peter Mathias as general editor. The volumes are slightly longer, generally about 200 pages, and contain a selection of articles on debatable topics, also generally but not exclusively British, preceded by an introduction by a recognized authority placing the issues in context. A number of titles from both series appear here, indicated, respectively, by "Studies" and "Debates."

Before reviewing the literature pertinent to individual chapters, it will be useful to consider some general works devoted to the four major determinants of economic change, as discussed in Chapter 1 and subsequently.

Population. The best introduction to the subject of population as a whole is Carlo M. Cipolla, *The Economic History of World Population* (7th ed., Sussex and New York, 1978; also available in Penguin paperback). Relatively brief, nontechnical, and well written, it is also authoritative. *Population and History,* by E. A. Wrigley (London, 1969), is a slightly more technical but clearly presented and also relatively brief introduction to historical demography. *Population in History,* edited by D. V. Glass and D. E. C. Eversley (London and Chicago, 1965), is a large collection of essays (27 in all) by leading experts demonstrating the interrelations of population movements and historical events in a wide variety of circumstances. W. S. Woytinsky and E. S. Woytinsky, *World Population and Production: Trends and Outlook* (New York, 1953), is a still valuable compendium of data that links population to another fundamental determinant, resources. In a much briefer compass, Ester Boserup, *Population and Technological Change: A Study of Long-Term Trends* (Chicago, 1981), looks at the relation between population and technology from the ancient world to the contemporary Third World. David Grigg does likewise in *Population Growth and Agrarian Change: an Historical Perspective* (Cambridge, 1980).

Resources. Erich W. Zimmermann, *World Resources and Industries* (rev. ed., New York, 1951), a model study when first published, is now out of date but still contains useful historical data. The publications of the World Energy Conference, especially *World Energy: Looking Ahead to 2020* (New York, 1978), provide more up-to-date information, as does the annual *Yearbook of World Energy Statistics,* published by the United Nations Statistical Office. In *The Agricultural Systems of the World: An Evolutionary Approach* (Cambridge, 1974), D. B. Grigg describes the chief characteristics of the major agricultural regions of the world and explains historically how they came into existence. The same author's *Dynamics of Agricultural Change: The Historical Experience* (New York, 1983) is a synoptic treatment of technological change in agriculture through the ages. Elias H. Tuma, *Twenty Six Centuries of Agrarian Reform* (Berkeley and Los Angeles, 1965), is a useful compendium.

Consideration of the historical aspect of resources inevitably leads one to the study of geography. A model study that reaches back to prehistory is C. T. Smith, *An Historical Geography of Western Europe before 1800* (New York, 1967). This is continued, less satisfactorily, by N. J. G. Pounds, *An Historical Geography of Europe, 1800–1914* (Cambridge, 1985). The same author's *An Historical Geography of Europe* (New York, 1990) should be avoided as unreliable. Consideration of geography inevitably leads one to the study of maps and atlases, indispensable for a proper understanding of the spatial aspects of economic development. There are many good historical atlases; the best is probably *The Times Atlas of World History*, edited by Geoffrey Barraclough (London, 1979).

The European Miracle: Environments, Economies, and Geopolitics in the History of Europe and Asia, by E. L. Jones (2nd ed., Cambridge, 1987), is an outstanding analysis of the interrelations of resources, technology, and institutions that has much in common with this book. The same author's *Growth Recurring: Economic Change in World History* (Oxford, 1988) is also highly recommended.

Technology. The most ambitious and comprehensive collection devoted to technology (though not without flaws) is *A History of Technology*, edited by Charles Singer et al. (7 vols., Oxford, 1954–78.) On a slightly more modest scale is *Technology in Western Civilization*, edited by Melvin Kranzberg and Carroll W. Pursell, Jr. (2 vols., Madison, WI, 1967), designed especially for undergraduate students. Among the better recent contributions on this subject are George Basalla, *The Evolution of Technology* (Cambridge, 1988); Louis A. Girifalco, *Dynamics of Technological Change* (New York, 1991); Joel Mokyr, *The Lever of Riches: Technological Creativity and Economic Progress* (New York, 1990); and Arnold Pacey, *Technology in World Civilization* (Cambridge, MA, 1990). Mokyr has also written a relatively brief book, *Twenty-five Centuries of Technological Change: An Historical Survey* (New York, 1990). Two books based on BBC television series, both lavishly illustrated, deal with "technology and its consequences": Jacob Bronowski, *The Ascent of Man* (London, 1975), and James Burke, *Connections* (Boston, 1978). A study of fundamental importance for the history of technology is A. P. Usher, *A History of Mechanical Inventions* (rev. ed., Cambridge, MA, 1954), which contains a theory of invention. Nathan Rosenberg, *Perspectives on Technology* (Cambridge, 1976), is a collection of essays by one of the leading specialists on the economics of technological change.

Institutions. It is not easy to decide what titles to put in the category of institutions, not because there are so few candidates but because there are so many—and so many different approaches. From an economist's point of view one could do no better than to begin with Sir John Hicks's small classic, *A Theory of Economic History* (Oxford, 1969). From a quite different perspective, that of the anthropologist, A. L. Kroeber, in *Anthropology* (rev. ed., New York, 1948) and also in *An Anthropologist Looks at History* (Berkeley, CA, 1963), relates the economy to the larger culture. Douglass C. North, *Structure and Change in Economic History* (New York and London, 1981), and *Institutions, Institutional Change and Economic Performance* (Cambridge, 1990), have much in common with the overall aim of this volume. Clarence Ayers's *Theory of Economic Progress* (Chapel Hill, NC, 1944; rpt. 1978), cited in Chapter 1, elucidates the "institutionalist" viewpoint on the relationship of the economy to society. Many other perspectives are possible; some are mentioned in the bibliographies for individual chapters.

Chapter 1. Introduction

For further discussion of the definition, measurement, and spatial distribution of economic development in the modern world, see, in addition to the items mentioned in the notes, the numerous writings of Simon Kuznets, especially *Modern Economic Growth: Rate, Structure, and Spread* (New Haven, 1966) and *Economic Growth of Nations: Total Output and Production Structure* (Cambridge, MA, 1971). There is a substantial literature on the methodology of economic history. Interesting comparisons can be drwn from the articles on "Economic History" in the original *Encyclopedia of the Social Sciences* (1931–35) and in the more recent *International Encyclopedia of the Social Sciences* (1969). The "achievements" of three distinct "schools" can be compared in a series of articles in the March 1978 issue of the *Journal of Economic History* (vol. 38, no. 1): Donald N. McCloskey, "The Achievements of the Cliometric School'' (pp. 13–28); Jon S. Cohen, ''The Achievements of Economic History: The Marxist School'' (pp. 29–57); and Robert Forster, ''Achievements of the *Annales* School'' (pp. 58–76). McCloskey has also contributed a brief introduction to *Econometric History* for non-economists (London, 1987; ''Studies''). Further contrasts are evident in the debate between Robert W. Fogel and G. R. Elton, in *Which Road to the Past? Two Views of History* (New Haven, CT, 1983).

Chapter 2. Economic Development in Ancient Times

The evoluttion of humankind from its presapient ancestors is described in popular form with abundant illustrations by the prominent paleoanthropologist Richard Leakey, in *The Making of Mankind* (New York, 1981). R. B. Lee and I. DeVore, eds., *Man the Hunter* (New York, 1968) is a seminal work, as is Barbara Bender, *Farming in Prehistory: From Hunter-gatherer to Food-producer* (London, 1975). Other works dealing with early agriculture are David Rindos, *The Origins of Agriculture: An Evolutionary Perspective* (New York, 1984) and Graeme Barker, *Prehistoric Farming in Europe* (Cambridge, 1985). *Stone Age Economics*, by Marshall Sahlins (Chicago, 1972), presents the views of a Marxist anthropologist. Two books of the famous archeologist V. Gordon Childe, *Man Makes Himself* (4th ed., London, 1965) and *What Happened in History* (rev. ed., Baltimore, 1964), remain among the most vivid accounts of the transition from prehistory to the first civilizations. Stuart Piggott, *Ancient Europe, From the Beginnings of Agriculture to Classical Antiquity* (Chicago, 1965), focuses more closely on Europe. Jacquetta Hawkes, *The First Great Civilizations: Life in Mesopotamia, the Indus Valley, and Egypt* (New York, 1973), places economic life in a broad cultural setting. Two multi-volume works by M. I. Rostovzeff, *The Social and Economic History of the Hellenistic World* (3 vols., Oxford, 1941) and *The Social and Economic History of the Roman Empire* (2nd rev. ed., 2 vols., Oxford, 1963), are deservedly regarded as classics. Moses I. Finley, *The Ancient Economy* (2nd ed., London, 1985), is a recent, briefer, and authoritative account. Finley has also contributed many other important works, notably *The World of Odysseus* (1954 and later editions) and *Economy and Society in Ancient Greece* (1981). K. D. White, *Greek and Roman Technology* (Ithaca, NY, 1984), is comprehensive, well illustrated, and has an excellent bibliography. Robin Osborne, *Classical Landscape with Figures: The Ancient Greek City and Its Countryside* (New York, 1987), is an intriguing treatment of the problem of food supply in ancient Greece. John Boardman, *The Greeks Overseas: Their Early Colonies and Trade* (New York, 1980), is a well-illustrated history of the Greek diaspora. *The Economy and Society of Pompeii*, by Willem Jongman (Amsterdam, 1988), is a methodologically sophisticated piece of revisionism, whereas Robert Sallares' *The Ecology of the Ancient Greek World* (Ithaca, NY, 1991) is a boldly revisionist work that contradicts many classical sources.

Chapter 3. Economic Development in Medieval Europe

Robert S. Lopez, *The Birth of Europe* (Philadelphia, 1973), is a well-illustrated, magisterial survey that places economic development in its social and cultural setting. A briefer, more sharply focused treatment by the same author is *The Commercial Revolution of the Middle Ages, 950–1350* (Englewood Cliffs, NJ, 1971). Harry A. Miskimin, *The Economy of Early Renaissance Europe, 1300–1460* (Cambridge, 1975), provides a convenient survey of the later Middle Ages. Gino Luzzatto, *An Economic History of Italy from the Fall of the Roman Empire to the Beginning of the 16th Century* (London, 1961), is a brief synoptic history of the most important region in Europe in that period by a past master of the economic historian's craft. Of similar importance are *Feudal Society* (Chicago, 1961) and *French Rural History:*

An Essay on Its Basic Characteristics (Berkeley and Los Angeles, 1966), by the famous French historian Marc Bloch, and *The Medieval Economy and Society: An Economic History of Britain in the Middle Ages* (London, 1972), by M. M. Postan. Leopold Genicot, *Rural Communities in the Medieval West* (Baltimore, 1990), is a general survey by a respected Belgian medievalist. Frederic C. Lane, *Venice, A Maritime Republic* (Baltimore, 1973), is a masterful survey of the history of that major medieval economy.

Population problems in that prestatistical era are wrestled with by J. C. Russell in *British Medieval Population* (Albuquerque, 1948) and *Medieval Regions and Their Cities* (Bloomington, IN, 1972). The progress of population was closely bound up with the fortunes of agriculture. A good introduction to that subject is B. H. Slicher van Bath, *The Agrarian History of Western Europe*, A.D. *500–1850* (London, 1963). Another enlightening account is *The Early Growth of the European Economy: Warriors and Peasants from the Seventh to the Twelfth Centuries* (London, 1974), by the noted French scholar Georges Duby.

On commerce, the best accounts are the chapters by M. M. Postan and Robert S. Lopez in volume II of the *Cambridge Economic History of Europe* (2nd ed., 1987). Lopez and I. W. Raymond edited *Medieval Trade in the Mediterranean World* (New York, 1955), a collection of original documents. Philippe Dollinger, *The German Hansa* (Stanford, CA, 1970), is the best study of that important institution.

The notion of a "static" Middle Ages in the field of technology was first challenged by Lynn White, Jr., whose book *Medieval Technology and Social Change* (Oxford, 1962) is still the best starting point for a study of medieval technology. Even more enthusiastic is Jean Gimpel's *The Medieval Machine: The Industrial Revolution of the Middle Ages* (New York, 1976), which should be read critically but sympathetically. Steven A. Epstein, *Wage Labor and Guilds in Medieval Europe* (Chapel Hill, NC, 1991), is wide-ranging and pertinent.

Monetary phenomena are dealt with in Peter Spufford, *Money and Its Use in Medieval Europe* (Cambridge, 1988). A. P. Usher, *The Early History of Deposit Banking in Mediterranean Europe* (Cambridge, MA, 1943), is still a fundamental work. *The Rise and Decline of the Medici Bank, 1397–1494* (New York, 1966), by Raymond de Roover, is a good story about an important enterprise. Joseph Shatzmiller, *Shylock Reconsidered: Jews, Moneylending, and Medieval Society* (Berkeley, 1990), is a fascinating case study based on a trial in Marseilles.

The realities of the medieval economy as they affected ordinary people come alive in Eileen Power's often-reprinted *Medieval People* (10th ed., New York, 1963), as they do in a much more recent work, David Herlihy's *Medieval Households* (Cambridge, MA, 1985). In a similar vein but more detailed is David Herlihy and Christiane Klapisch-Zuber, *Tuscans and Their Families: A Study of the Florentine Catasto of 1427* (New Haven and London, 1985), an abridged translation of a much longer work in French. Herlihy also contributed *Opera Muliebria: Women and Work in Medieval Europe* (Philadelphia, 1990). Heather Swanson, *Medieval Artisans: An Urban Class in Late Medieval England* (Cambridge, MA, 1989), dealing with workers and guilds in York (and Bristol and Norwich), also stresses the role of women workers. Robert C. Davis, *Shipbuilders of the Venetian Arsenal: Workers and Workplace in the Preindustrial City* (Baltimore, 1991), is a detailed study of a major medieval institution which also extended into the early modern period.

Chapter 4. Non-Western Economies on the Eve of Western Expansion

A History of the Arab Peoples, by Albert Hourani (Cambridge, MA, 1991), is fundamental for all aspects of its subject. André Wink, *Al-Hind: The Making of the Indo-Islamic World* (New York, 1990), details the eastward expansion of Islam from the seventh to the eleventh centuries A.D. Archibald Lewis, *Naval Power and Trade in the Mediterranean*, A.D. *500–*

1100 (Princeton, 1951), is a competent survey, emphasizing the interaction of Byzantine, Muslim, and western Christian navies and merchants. S. D. Goitein, *Studies in Islamic History and Institutions* (Leiden, 1966), contains three chapters on the Islamic middle classes and workers in the Middle Ages. In *A Mediterranean Society* (4 vols., Berkeley, 1967–83), the same author, exploiting the abundant documentation of the Cairo Geniza, has provided the most detailed account available of a medieval community, that of the Jews within Islam, which stretched from Muslim Spain to India; volume I deals with "The Economic Foundations (969–1250)"; volume II, "The Community"; volume III, "The Family"; and volume IV "Daily Life." Dorothy M. Vaughan, *Europe and the Turk: A Pattern of Alliances 1350–1700* (Liverpool, 1954), has a chapter on "The Early Ottoman Empire as a Naval Power and Economic Force." Kemal H. Karpat, ed., *The Ottoman State and Its Place in World History* (Leiden, 1974), assembles brief comments by distinguished authorities on various aspects of Ottoman history, including a chapter by Charles Issawi on economic structures before 1700.

In *Before European Hegemony: The World System A.D. 1250–1350* (New York, 1989) Janet L. Abu-Lughod paints an intriguing picture of financial and commercial networks encompassing Europe, the Eastern Mediterranean, the Persian Gulf, the Indian Ocean, Southeast Asia, and China. Much the same area is the subject of two impressive books by K. N. Chaudhuri: *Trade and Civilisation in the Indian Ocean: An Economic History from the Rise of Islam to 1750* (Cambridge, 1985), and *Asia before Europe: Economy and Civilisation of the Indian Ocean from the Rise of Islam to 1750* (Cambridge, 1990). Volume I of the *Cambridge Economic History of India,* edited by Tapan Raychaudhuri and Ifran Habib (Cambridge, 1982), covers the period A.D. 1200 to 1750, and has an excellent bibliography.

A History of East Asian Civilization, vol. I, *The Great Tradition* (Boston, 1960), by Edwin O. Reischauer and John K. Fairbank, is the best place to begin study of that important area of the world, both for its compressive treatment (see especially Chapter 6 on the economy) and its bibliography. The multi-volume *Science and Civilization in China,* by Joseph Needham and his collaborators (in progress; vol. I, Cambridge, 1954), is a treasure chest of information on a variety of subjects, including Chinese technology, though imperfectly indicated by the title. *The Mongols,* by David Morgan (Oxford, 1987), is a lively, brief, largely sympathetic account of a people that, in general, have not enjoyed a good press. The founder of the Mongol empire is the subject of a definitive biography that has recently been translated into English and presented in an accessible manner: Paul Ratchnevsky, *Genghis Khan: His Life and Legacy* (Oxford, 1991).

The literature on Africa before the sixteenth century has recently been enriched by the opening chapters of two general works by acknowledged masters of the subject: Roland Oliver, *The African Experience: Major Themes in African History from Earliest Times to the Present* (London and New York, 1991); and Ralph Austen, *African Economic History* (London, 1987). There is also a collection of articles edited by Z. A. Konczacki and J. M. Konczacki, *An Economic History of Tropical Africa,* vol. I, *The Precolonial Period* (London, 1977). Philip Curtin, *Cross-Cultural Trade in World History* (Cambridge, 1984), explores many incidents of cross-cultural commerce, from Africa and ancient Mesopotamia to the North American fur trade, including Southeast Asia before the Europeans and pre-Columbian America. On the latter, the first six chapters of volume I of *The Cambridge History of Latin America,* edited by Leslie Bethell (Cambridge, 1984), deal with "America on the Eve of the Conquest." Individual volumes on the two most famous pre-Columbian empires of the Western Hemisphere include *Aztecs,* by Inga Clendinnen (Cambridge, 1991) and *The Inca Empire: The Formation and Disintegration of a Pre-Capitalist State,* by Thomas C. Patterson (New York and Oxford, 1991).

Chapter 5. Europe's Second Logistic

Carlo M. Cipolla, *Before the Industrial Revolution: European Economy and Society, 1000–1700* (2nd ed., New York, 1980), is an excellent textbook that covers both the medieval and the early modern periods. Ralph Davis, *The Rise of the Atlantic Economies* (Ithaca, NY, 1973), is another good textbook that emphasizes the discovery and peopling of the Americas as well as the economies of Western Europe. Two other texts that can be recommended are Harry A. Miskimin, *The Economy of Later Renaissance Europe, 1460–1600* (Cambridge, 1977), and Jan De Vries, *The Economy of Europe in an Age of Crisis, 1600–1750* (Cambridge, 1976). *Peasants, Landlords and Merchant Capitalists: Europe and the World Economy, 1500–1800,* by Peter Kriedte (Cambridge, 1983), is a short Marxist text. *Economy and Society in Early Modern Europe: Essays from Annales,* edited by Peter Burke (London, 1972), and *Essays in European Economic History, 1500–1800,* edited by Peter Earle (Oxford, 1974), are both collections of notable journal articles.

A major work by virtue of both its bulk and its scope is Fernand Braudel, *Civilization and Capitlism: 15th–18th Centuries* (3 vols., New York, 1982–84); it contains a wealth of factual information, mostly correct, but the brilliance of its author's rather idiosyncratic interpretation has been exaggerated by the popular press. Braudel's earlier work, which established his reputation, is also available in English: *The Mediterranean and the Mediterranean World in the Age of Philip II* (2 vols., New York, 1972; first published in French in 1949). Two books better known for their controversial (and contradictory) interpretations of the economic history of the early modern period than for their command of the facts are Douglass C. North and Robert Paul Thomas, *The Rise of the Western World: A New Economic History* (Cambridge, 1973), and Immanuel Wallerstein, *The Modern World-System: Capitalist Agriculture and the Origins of the European World Economy in the Sixteenth Century* (New York, 1974). Wallerstein's is the first of a four-volume series, two more of which have been published: *The Modern World-System II: Mercantilism and the Consolidation of the European World-Economy, 1600–1750* (1980; more appropriate for Chapter 6) and *The Modern World-System III: The Second Era of Great Expansion of the Capitalist World-Economy, 1730–1840s* (1989; for Chapter 7).

Textbooks and general works in English relating to individual countries are most plentiful and satisfactory for England or Great Britain, of course. Three that can be recommended without hesitation are D. C. Coleman, *The Economy of England, 1450–1750* (Oxford, 1977), L. A. Clarkson, *The Pre-Industrial Economy in England, 1500–1750* (London, 1971), and Charles Wilson, *England's Apprenticeship, 1603–1763* (London, 1965). These should be balanced with T. C. Smout, *A History of the Scottish People, 1560–1830* (Edinburgh, 1969). The closest equivalents in English for France are Emmanuel LeRoy Ladurie, *The Peasants of Languedoc* (Urbana, IL, 1974), and Pierre Goubert, *The French Peasantry in the Seventeenth Century* (Cambridge, 1986); several of the essays in Cameron, ed., *Essays in French Economic History,* mentioned earlier, and the volumes on France cited for Chapter 6 can also serve. For central and eastern Europe there is Hermann Kellenbenz's *The Rise of the European Economy: An Economic History of Continental Europe from the Fifteenth to the Eighteenth Century* (New York, 1976), the title of which is slightly misleading, for the focus is on central and eastern Europe.

By far the best introduction to almost any aspect of Dutch economic history in the seventeenth century is Charles Wilson, *The Dutch Republic and the Civilization of the Seventeenth Century* (London, 1969). Charles R. Boxer, *The Dutch Seaborne Empire, 1600–1800* (New York, 1965), is also excellent. Violet Barbour, *Capitalism in Amsterdam in the Seventeenth Century* (Baltimore, 1950; reprinted, Ann Arbor, MI, 1963), contains a wealth of information in small compass. Jonathan Israel, *Dutch Primacy in World Trade, 1585–1740* (Oxford, 1989), is the best recent contribution. See also Johannes Menne Postma, *The Dutch in the Atlantic Slave Trade, 1600–1815* (New York, 1990). For other countries, see the works listed for Chapter 6.

The population history of early modern Europe is felicitously encapsulated in Michael W. Flinn, *The European Demographic System, 1500–1820* (Baltimore, 1981), which also features an excellent bibliography. E. A. Wrigley and R. S. Schofield, *The Population History of England, 1541–1871* (London, 1981), is a methodologically novel and detailed analysis. The interrelations of population and agriculture are dealt with by Ester Boserup, *The Conditions of Agricultural Growth: The Economics of Agrarian Change under Population Pressure* (London, 1965) (which is not limited to early modern Europe), and by B. H. Schlicher van Bath, *Agrarian History of Western Europe,* previously mentioned.

Other aspects of agriculture and rural life are considered by Ann Kussmaul, *A General View of the Rural Economy of England, 1538–1840* (New York, 1990); John Chartres and David Hey, eds., *English Rural Society, 1500–1800* (Cambridge, 1990); and by *European Peasants and Their Markets: Essays in Agrarian Economic History,* edited by William N. Parker and Eric L. Jones (Princeton, 1975). Jan de Vries, *The Dutch Rural Economy in the Golden Age, 1500–1700* (New Haven, CT, 1974), is a model of its kind. Agrarian life in Eastern Europe is presented in marvelously detailed fashion by Jerome Blum in *Lord and Peasant in Russia from the Ninth to the Nineteenth Century* (Princeton, 1961), especially in Chapters 8–14 for the sixteenth and seventeenth centuries. The same author's *The End of the Old Order in Rural Europe* (Princeton, 1978) deals with the transition to modern class society.

The literature on exploration and discovery is vast, and has been greatly expanded by the Columbian quincentenary. Typical is *Columbus and the Age of Discovery* (New York, 1991), a lavishly illustrated companion volume to a television series by Zvi Dor-Ner. A handy earlier survey is Charles E. Nowell, *The Great Discoveries and the First Colonial Empires* (Ithaca, NY, 1954). Greater but still manageable detail is offered by J. H. Parry, *The Age of Reconnaissance: Discovery, Exploration, and Settlement, 1450–1650* (Cleveland, 1963). Samuel Eliot Morison's *The Great Explorers: The European Discovery of America* (Oxford, 1978) is a hefty abridgement of his two-volume *The European Discovery of America,* and is especially good on the personalities of the explorers. Kirkpatrick Sale paints an unflattering portrait in *The Conquest of Paradise: Christopher Columbus and the Columbian Legacy* (New York, 1990). Two books by Alfred W. Crosby likewise take a dim view of the consequences of the discoveries: *The Columbian Exchange: Biological and Cultural Consequences of 1492* (Westport, CT, 1972) and *Ecological Imperialism: The Biological Expansion of Europe, 900–1900* (Cambridge, 1986). Experts at the Tenth International Congress of Economic History looked at the consequences *for Europe* of the discoveries in Hans Pohl, ed., *The European Discovery of the World and Its Economic Effects on Pre-industrial Society, 1500–1800* (Stuttgart, 1990). Portugal's role is detailed in Christopher Bell, *Portugal and the Quest for the Indies* (New York, 1974), and C. R. Boxer, *Four Centuries of Portuguese Expansion, 1415–1825: A Succinct Survey* (Berkeley and Los Angeles, 1969), which is precisely what its subtitle indicates. Carlo M. Cipolla, *Guns and Sails in the Early Phase of European Expansion* (London, 1965), is a perceptive and stimulating exercise in historical judgment.

The *locus classicus* of the so-called price revolution is Earl J. Hamilton, *American Treasure and the Price Revolution in Spain, 1501–1650* (Cambridge, MA, 1934; reprinted, New York, 1965), although the same author's "American Treasure and the Rise of Capitalism," *Economica,* 9 (November 1929): 338–57, is both briefer and more pointed.

A handy collection of criticisms of the Hamilton thesis, from different points of view, is Peter H. Ramsey, ed., *The Price Revolution in Sixteenth-Century England* (London, 1971).

Among works dealing with commerce and commercial organization in the early modern period, Herman Van der Wee, *The Growth of the Antwerp Market and the European Economy* (3 vols., The Hague, 1963), is the definitive work on the rise and decline of Antwerp. A good introduction to Dutch trade is D. W. Davies, *A Primer of Dutch Seventeenth Century Overseas Trade* (The Hague, 1961). Ralph Davis, *The Rise of the English Shipping Industry in the Seventeenth and Eighteenth Centuries* (London, 1962), relates problems of trade to those of transport. Philip D. Curtin, *The Atlantic Slave Trade: A Census* (London, 1969), is the definitive work on that unusual branch of commerce. Other books dealing with commerce are listed in the bibliography for Chapter 6.

Some older works on the early modern period that can still be recommended include George Unwin, *Industrial Organization in the Sixteenth and Seventeenth Centuries* (1904; reprinted, London, 1957); Richard Ehrenberg, *Capital and Finance in the Age of the Renaissance: A Study of the Fuggers and Their Connections* (1928; reprinted, New York, 1963 and 1985); and H. M. Robertson, *Aspects of the Rise of Economic Individualism: A Criticism of Max Weber and His School* (1933; reprinted, New York, 1959).

Chapter 6. Economic Nationalism and Imperialism

Eli F. Heckscher's classic *Mercantilism* (2nd English ed., 2 vols., London, 1965) is still the starting point for discussions of economic policy in the early modern period. A collection of some of the criticisms (and defenses) of Heckscher's conception is contained in D. C. Coleman, ed., *Revisions in Mercantilism* (London, 1969). Joseph A. Schumpeter's magisterial *History of Economic Analysis* (Oxford, 1954) should also be consulted, especially Part II, Chapter 7. Paul Kennedy, *The Rise and Fall of the Great Powers: Economic Change and Military Conflict from 1500 to 2000* (New York, 1987), contains a challenging thesis that encompasses a wide time span.

The nature and consequences of economic policies can most usefully be surveyed on a country-by-country basis. For Spain *An Economic History of Spain,* by Jaime Vicens Vives (Princeton, 1969), is the best place to begin (see especially Part IV, Chapters 23–31). David Ringrose, *Madrid and the Spanish Economy, 1560–1850* (Berkeley and Los Angeles, 1983), presents a challenging hypothesis on the reasons for the long economic stagnation of Spain. Richard Herr, *Rural Change and Royal Finances in Spain at the End of the Old Regime* (Berkeley, 1989) is a masterpiece. Julius Klein, *The Mesta: A Study in Spanish Economic History, 1273–1836* (Cambridge, MA, 1920), has not been replaced or surpassed. Much the same can be said of C. H. Haring's *Trade and Navigation between Spain and the Indies in the Time of the Hapsburgs* (Cambridge, MA, 1918). The same author has also contributed *The Spanish Empire in America* (New York, 1947). A recent, well-written account of the clash between Spaniards and the Amerindians is *Ambivalent Conquest: Maya and Spaniard in Yucatan, 1517–1570* (Cambridge, 1987), by Inga Clendinnen.

For Portugal the best account is C. R. Boxer, *The Portuguese Seaborne Empire, 1415–1825* (New York, 1969). See also C. R. Boxer, *The Dutch in Brazil, 1624–1654* (Oxford, 1957), and the relevant chapters of the *Cambridge History of Latin America.*

Kellenbenz, in *The Rise of the European Economy,* previously mentioned, places a heavy emphasis on the role of the state. The focus and emphasis are even more pronounced in Hans Rosenberg's *Bureaucracy, Aristocracy, and Autocracy: The Prussian Experience, 1660–1815* (Cambridge, MA, 1958), Both authors are masters of their subjects as is Eli F. Heckscher, of *Mercantilism* fame, whose *An Economic History of Sweden* (Cambridge, MA,

1954) is an abridged translation of the four-volume Swedish original. For Italy, in addition to Lane's *Venice: A Maritime Republic,* previously mentioned, and his *Venice and History: The Collected Papers of Frederic C. Lane* (Baltimore, 1966), there is Brian Pullan ed., *Crisis and Change in the Venetian Economy in the Sixteenth and Seventeenth Centuries* (London, 1968), and Domenico Sella, *Crisis and Continuity: The Economy of Spanish Lombardy in the Seventeenth Century* (Cambridge, MA, 1979).

The role of the state in the French economy is unusually well documented, thanks to a trio of works by Charles W. Cole: *French Mercantilist Doctrines before Colbert* (New York, 1931), *Colbert and a Century of French Mercantilism* (2 vols., New York, 1939), and *French Mercantilism, 1683–1700* (New York, 1943). The drawback is that Cole's conception of mercantilism was quite conventional. A good antidote is Martin Wolfe, *The Fiscal System of Renaissance France* (New Haven, 1972). Warren C. Scoville, *The Persecution of the Huguenots and French Economic Development, 1680–1720* (Berkeley and Los Angeles, 1960), also gives a somewhat different picture of the role of government in France, as does J. F. Bosher, *French Finances, 1770–1795: From Business to Bureaucracy* (Cambridge, 1970). Fernand Braudel's *The Identity of France* (2 vols., New York, 1990), the final, unfinished masterpiece of a great historian, deals with much more than mere economic policy, and covers a much longer period than other works mentioned here, but it also deserves to be mentioned.

Kristof Glamann, *Dutch-Asiatic Trade, 1620–1740* (Copenhagen and The Hague, 1958), is the best work in English on the Dutch East India Company. Aspects of the company's activities in Asia are dealt with by John E. Wills, Jr., *Pepper, Guns and Parleys: The Dutch East India Company and China, 1622–1681* (Cambridge, MA, 1974), and Om Prakash, *The Dutch East India Company and the Economy of Bengal, 1630–1720* (Princeton, 1985), as do the volumes by K. N. Chaudhuri mentioned in the bibliography for Chapter 4.

Charles Wilson's *England's Apprenticeship, 1603–1673,* previously mentioned, contains much material relevant to the formation and execution of economic policy. The same author's *Profit and Power: A Study of England and the Dutch Wars* (London, 1957) is more narrowly focused. His *Economic History and the Historian* (London, 1969), a collection of his essays, contains several dealing with the formation, execution, and consequences of economic policy. Specific episodes are dealt with by J. D. Gould, *The Great Debasement: Currency and Economy in Mid-Tudor England* (Oxford, 1970); Astrid Friis, *Alderman Cockayne's Project and the Cloth Trade* (London, 1927), good despite its age; and L. A. Harper, *The English Navigation Laws* (New York, 1939). Joan Thirsk, *Economic Policy and Projects: The Development of Consumer Society in Early Modern England* (Oxford, 1978) is brilliant. A worthy complement is Carole Shammas, *The Preindustrial Consumer in England and America* (New York, 1990).

Chapter 7. The Dawn of Modern Industry

The literature on the so-called industrial revolution in Great Britain is enormous and still rapidly growing. Much of it (mainly books) is listed in *British Economic and Social History: A Bibliographical Guide,* compiled by W. H. Chaloner and R. C. Richardson (Manchester, 1976). What follows is extremely selective, limited to a few standard general works and some others chosen for their felicitous style or seminal ideas.

A basic source is B. R. Mitchell (with the collaboration of Phyllis Deane), *Abstract of British Historical Statistics* (Cambridge, 1962); also, Mitchell and H. G. Jones, *Second Abstract of British Historical Statistics* (Cambridge, 1971). The *Atlas of Industrializing Britain, 1780–1914,* edited by John Langton and R. J. Morris (London, 1986), is extremely useful for visualizing the spatial aspects of industrialization. Similarly, *The Archaeology of*

the Industrial Revolution, edited by Brian Bracegirdle (London, 1973), abundantly illustrated, enables the reader to visualize the technology of early industrialization. M. W. Flinn,
British Population Growth, 1700–1850 (''Studies,'' London, 1970), summarizes and ana-
lyzes the essential information in brief compass.

The literature on proto-industrialization is summarized by Franklin Mendels in ''Proto-
industrialization: Theory and Reality'' in Eighth International Economic History Congress,
Budapest 1982, *''A'' Themes,* pp. 69–107. If that item is too difficult to locate, try idem,
''Proto-industrialization: The First Phase of the Industrialization Process,'' *Journal of Economic History,* 32 (March 1972): 241–61, in which the term was first given an explicit
definition (since modified). See also Peter Kriedte et al., *Industrialization before Industrialization* (Cambridge, 1981). For a skeptical view, D. C. Coleman, ''Proto-Industrialization:
A Concept Too Many,'' *Economic History Review,* 2nd ser., 36 (Aug. 1983): 435–48.

A recent text, informed by the insights and skills of the cliometricians, is *The Economic
History of Britain since 1700,* edited by Roderick Floud and Donald McCloskey (3 vols., 2nd
ed., Cambridge, 1993). Another standard text is Peter Mathias, *The First Industrial Nation:
An Economic History of Britain, 1700–1914* (2nd ed., London, 1983). See also the same
author's *The Transformation of England* (London, 1979), which focuses on the eighteenth
century, and E. A. Wrigley, *Continuity, Chance and Change: The Character of the Industrial Revolution in England* (Cambridge, 1988).

The best brief synopsis of the rise of modern industry in Britain is probably still T. S.
Ashton, *The Industrial Revolution, 1760–1830* (Oxford, 1948), in spite of its unfortunate
title. See also the same author's *An Economic History of England: the 18th Century* (London,
1955) and *Economic Fluctuations in England, 1700–1800* (Oxford, 1959). Covering the
period of about 1750 to 1850 is Phyllis Deane's *The First Industrial Revolution* (2nd ed.,
Cambridge, 1979), which is based in part on her pioneering work with W. A. Cole, *British
Economic Growth, 1688–1959* (2nd ed., Cambridge, 1967). N. F. R. Crafts, *British Economic Growth during the Industrial Revolution* (Oxford, 1985), criticizes the Deane and Cole
estimates; this book is also relevant for Chapter 9. *The Causes of the Industrial Revolution in
England,* edited by R. M. Hartwell (''Debates,'' London, 1967), is a collection of seminal
articles by outstanding authors which, however, does not answer the question implicit in the
title. Hartwell has also edited another collection of articles by various authors entitled,
simply, *The Industrial Revolution* (Oxford, 1970), and has published a collection of his own
articles under the title *The Industrial Revolution and Economic Growth* (London, 1971).

The First Industrialists, by François Crouzet (Cambridge, 1985), investigates the social
origins of the pioneers of modern industry and concludes that ''a large majority'' came from
the middle classes. That enormous, ambiguous group is the subject of Peter Earle, *The
Making of the English Middle Class: Business, Society, and Family Life in London, 1660–
1730* (Berkeley and Los Angeles, 1989). Who was beneath the middle classes? John F. C.
Harrison answers that question in *The Common People. A History from the Norman Conquest
to the Present* (London, 1984); not limited to the eighteenth century, it is nevertheless highly
recommended. Adrian Randall, *Before the Luddites: Custom, Community, and Machinery in
the English Woolen Industry, 1776–1809* (Cambridge and New York, 1991), is a study of
England's leading industry before extensive mechanization. Jane Randall, *Women in an
Industrialising Society: England, 1750–1880* (Oxford, 1990), traces the varied roles of
women during the process of industrialization. Sidney Pollard, *The Genesis of Modern
Management: A Study of the Industrial Revolution in Great Britain* (London, 1965), looks at
the problems of managing the first large-scale industrial enterprises. S. D. Chapman, *The
Cotton Industry in the Industrial Revolution* (''Studies,'' London, 1972), is a competent brief
summary of a large literature on the single most important industry of the period. *The
Arkwrights: Spinners of Fortune,* by R. S. Fitton (New York, 1989), is the definitive work on

Richard Arkwright and his family. Clark Nardinelli, in *Child Labor in the Industrial Revolution* (Bloomington, IN, 1990), disputes the usual assumption that widespread exploitation of children characterized the period. *Science, Technology, and Economic Growth in the Eighteenth Century,* edited by A. E. Musson ("Debates," London, 1972), is a collection of important articles by various authorities. References for the so-called standard of living question are listed under Chapter 9.

J. D. Chambers and G. E. Mingay, *The Agricultural Revolution, 1750–1880* (London, 1966) is the standard work on agriculture in the early stages of industrialization; but see the booklet by J. V. Beckett, *The Agricultural Revolution* (Oxford, 1990), for a recent questioning of the utility of the term "revolution." E. L. Jones, ed., *Agriculture and Economic Growth in England, 1650–1815* ("Debates," London, 1967), is a collection of important articles by various authorities, whereas E. L. Jones, *Agriculture and the Industrial Revolution* (Oxford, 1974), is the same scholar's own mature reflections on the subject. G. E. Mingay, *English Landed Society in the 18th Century* (London, 1963), is a substantial original work; *Enclosure and the Small Farmer in the Age of the Industrial Revolution* ("Studies," London, 1968), by the same author, is an extremely useful brief summary of a large literature on an important question. His *A Social History of the English Countryside* (London, 1990) covers a broader canvas.

General surveys of the role of transportation in the early stages of British industrialization are contained in the initial chapters of P. S. Bagwell, *The Transportation Revolution from 1770* (New York, 1974), T. C. Barker and C. I. Savage, *An Economic History of Transport in Britain* (3rd rev. ed., London, 1974), and H. J. Dyos and D. H. Aldcroft, *British Transport: An Economic Survey from the Seventeenth Century to the Twentieth* (London, 1969). More specialized topics are treated admirably by W. A. Albert, *The Turnpike Road System of England, 1663–1844* (Cambridge, 1972), A. R. B. Haldane, *New Ways through the Glens* (London, 1962, about roadbuilding in the Scottish Highlands); J. R. Ward, *The Finance of Canal Building in Eighteenth Century England* (Oxford, 1974), and T. S. Willan, *The English Coasting Trade, 1600–1750* (1938; reprinted, with new preface, London, 1967).

Financial problems of the seventeenth and eighteenth centuries are studied by R. D. Richards, *The Early History of Banking in England* (1929; reprinted, London, 1958). P. G. M. Dickson, *The Financial Revolution in England: A Study in the Development of Public Credit, 1688–1756* (London, 1967), invites comparison with the financial histories of France—for example J. F. Bosher, *French Finances,* previously cited—and other continental countries to understand why Great Britain was more successful both militarily and economically in the eighteenth century. L. S. Pressnell, *Country Banking in the Industrial Revolution* Oxford, 1956), also contributes to that understanding, whereas François Crouzet, ed., *Capital Formation in the Industrial Revolution* ("Debates," London, 1972), shows why capital formation was not a major obstacle to economic growth in the eighteenth century. Sir John Clapham, *The Bank of England: A History,* vol. I (Cambridge, 1944), is an authorized history by a great historian that covers the period from the founding of the bank in 1694 until the Napoleonic Wars.

Chapter 8. Economic Development in the Nineteenth Century: Basic Determinants

The bibliographies in the *Cambridge Economic History of Europe* need to be supplemented by more recent sources. The *Bibliography of European Economic and Social History,* compiled by Derek H. Aldcroft and Richard Rodger (Manchester, 1984), is helpful in that respect, covering continental Europe for the period from 1700 to 1939. It can be used in conjunction with the *Bibliography of British Economic and Social History,* cited in the previous chapter's bibliography. The bibliographies in the *Cambridge History of Latin Amer-*

ica are reasonably up to date; vols. III and IV cover the nineteenth and twentieth centuries. Bibliographies of American economic history are legion.

The need for quantitative data is well served by B. R. Mitchell, *International Historical Statistics, Europe: 1750–1988* (New York, 1992), a companion to his *Abstract of British Historical Statistics,* and U.S. Department of Commerce, *Historical Statistics of the United States* (various editions, Washington, D.C.). *The Dictionary of Statistics,* compiled by Michael G. Mulhall (4th ed., London, 1899; reprinted, Detroit, 1969), contains a melange of quantitative data on a variety of subjects from all over the world but, in view of the negligible information provided on either sources or methods, it should be used with great caution if at all. The U.S. Department of Commerce has published *Long Term Economic Growth, 1860–1965* (Washington, 1966); Part IV contains some international comparisons. A similar publication for the United Kingdom is C. H. Feinstein, *Statistical Tables of National Income, Expenditure and Output of the U.K., 1855–1965* (Cambridge, 1966).

The "Essays" series of the Economic History Society includes *Essays in European Economic History, 1789–1914,* edited by François Crouzet et al. (New York and London, 1969); *Essays in Quantitative Economic History,* edited by Roderick Floud (Oxford, 1974), most selections of which deal with the last two centuries, but with a British bias; *Essays in Social History,* edited by M. W. Flinn and T. C. Smout (Oxford, 1974), with a similar coverage and bias; and *Essays in British Business History,* edited by Barry Supple (Oxford, 1977). *Economic Development in the Long Run,* edited by A. J. Youngson (London, 1972), is also a collection of essays by eminent scholars on the fundamental determinants of economic change (except population) with, in addition, chapters on Africa, India, and Japan. It should not be confused with *Economics in the Long View,* edited by Charles P. Kindleberger and Guido di Tella (3 vols., London, 1982), which is a collection of *Essays in Honour of W. W. Rostow,* also containing many items of interest on the nineteenth and twentieth centuries.

Rostow is one of the most celebrated economic historians of the second half of the twentieth century, although in recent years his influence has been on the wane. Among his many books, a few are of special relevance for this and subsequent chapters: *The Stages of Economic Growth: A Non-Communist Manifesto* (Cambridge, 1960; 3rd ed., 1991), his most famous; *The Process of Economic Growth* (Oxford, 1952; 2nd ed, 1960), more theoretical than historical; and *The World Economy: History and Prospect* (Austin & London, 1978), his magnum opus. For critical views and a defense, see W. W. Rostow, ed., *The Economics of Take-off into Sustained Growth* (New York, 1963), the proceedings of a conference of the International Economic Association that Rostow was invited to edit.

Alexander Gerschenkron was another economic historian of the third quarter of the twentieth century whose views on economic development in the nineteenth century were at one time very influential. He expressed them mainly in essays: *Economic Backwardness in Historical Perspective: A Book of Essays* (Cambridge, MA, 1962) and *Continuity in History and Other Essays* (Cambridge, MA, 1968). *Patterns of European Industrialization: The Nineteenth Century,* edited by Richard Sylla and Gianni Toniolo (London and New York, 1991), contains essays by former students of Gerschenkron and also by some of his critics. A pair of advanced textbooks by two other prominent economic historians are quite useful for topics covered in this and subsequent chapters: Alan S. Milward and S. B. Saul, *The Economic Development of Continental Europe, 1780–1870* (London, 1973), and *The Development of the Economies of Continental Europe, 1850–1914* (Cambridge, MA, 1977). Sidney Pollard, *Typology of Industrialization Processes in the Nineteenth Century* (Chur, Switzerland, 1990), is a brief, straightforward account dealing with the principal European countries, the United States, and Japan. *Dynamic Forces in Capitalist Development: A Long-run Comparative View,* by Angus Maddison (Oxford, 1991) is a dynamic book dealing comprehensively with the world economy of the last two centuries.

The basic data on population are presented in W. S. and E. S. Woytinsky, *World Population and Production,* previously cited. *Population Growth and Economic Development since 1750,* by H. J. Habakkuk (Leicester, 1972), is a brief interpretive essay that relates the population history of the industrialized West to the problems of contemporary underdeveloped economies. *Population in Industrialization,* edited by Michael Drake ("Debates," London, 1969), is a collection of notable articles relating mainly to British experience. E. A. Wrigly, *Industrial Growth and Population Change: A Regional Study of the Coalfield Areas of North-West Europe in the Later Nineteenth Century* (Cambridge, 1961), on the other hand, is a seminal study of the demographic history of the Austrasian coalfield.

The role of resources in nineteenth century industrialization is highlighted by N. J. G. Pounds with W. N. Parker, *Coal and Steel in Western Europe* (London, 1957), and by Pounds in *The Ruhr: A Study in Historical and Economic Geography* (London, 1952). See also Pounds, *An Historical Geography of Europe, 1800–1914,* previously mentioned.

The Unbound Prometheus: Technological Change and Industrial Development in Western Europe from 1750 to the Present, by David Landes (Cambridge, 1969), is much more than a mere history of industrial technology, relating technological change to the overall economic, institutional, and political changes of the last two centuries and more. A. G. Kenwood and A. L. Lougheed, *Technological Diffusion and Industrialization before 1914* (London, 1982), is more narrowly focused. Ian Inkster, *Science and Technology in History: An Approach to Industrial Development* (New Brunswick, NJ, 1991), stresses the importance of international diffusion, as does David J. Jeremy, ed., *International Technology Transfer: Europe, Japan, and the U.S.A., 1700–1914* (Aldershot, England, 1991), with case studies by experts in each area. *The Economics of Technological Change,* edited by Nathan Rosenberg (Baltimore, 1971), is a collection of major articles covering all aspects of its subject. Insight into the new phenomenon of the professional inventor is provided by Andre Millard in *Edison and the Business of Innovation* (Baltimore, 1990).

Institutional Change and American Economic Growth, by Lance E. Davis and Douglass C. North (Cambridge, 1971), is a seminal account of the interrelations of institutions and economic change. *The State and Economic Growth,* edited by H. G. J. Aitken (New York, 1959), consists of the papers presented at a conference of the Social Science Research Council on that subject; most of them deal with the nineteenth century. Bishop C. Hunt, *The Development of the Business Corporation in England, 1800–1867* (Cambridge, MA, 1936), Charles E. Freedeman, *Joint-Stock Enterprise in France, 1807–1867: From Privileged Company to Modern Corporation* (Chapel Hill, 1979), and Alfred D. Chandler, Jr., *Strategy and Structure: Chapters in the History of the Industrial Enterprise* (Cambridge, MA, 1962), detail the development of modern forms of enterprise in three important countries. Chandler has also authored two other works of seminal importance on business organization: *The Visible Hand: The Managerial Revolution in American Business* (Cambridge, MA, 1977) and *Scale and Scope: Dynamics of Industrial Capitalism* (Cambridge, MA, 1990).

Three quite different books attempt to show the interrelations of resources, technology, and institutions: William N. Parker, *Europe, America, and the Wider World: Essays on the Economic History of Western Capitalism,* (2 vols., Cambridge, 1984, 1991); Nathan Rosenberg and L. E. Birdzell, Jr., *How the West Grew Rich: The Economic Transformation of the Industrial World* (New York, 1986); and Rondo Cameron, *France and the Economic Development of Europe, 1800–1914* (Princeton, 1961).

Education and Economic Development, edited by C. Arnold Anderson and Mary Jean Bowman (Chicago, 1965), was a pioneering treatment of its subject. *Literacy and Development in the West,* by Carlo M. Cipolla (Harmondsworth, England, 1969) is both succinct and comprehensive. A worthy recent contribution is Gabriel Tortella, ed., *Education and Economic Development since the Industrial Revolution* (Valencia, Spain, 1990).

Chapter 9. Patterns of Development: The Early Industrializers

Good textbook treatments of the economic history of Britain in the nineteenth century include Mathias, *The First Industrial Nation*, and Floud and McCloskey, eds., *An Economic History of Britain since 1700*, both previously cited; also S. G. Checkland, *The Rise of Industrial Society in England, 1815–1885* (London, 1964), William Ashworth, *An Economic History of England, 1870–1939* (London, 1960), J. D. Chambers, *The Workshop of the World: British Economic History, 1820–1880* (2nd ed., Oxford, 1968), and R. H. Campbell, *Scotland since 1707: The Rise of an Industrial Society* (London, 1964). The older tradition is represented by Sir John Clapham, *An Economic History of Modern Britain* (3 vols., Cambridge, 1926–38).

British Economic Growth, 1865–1973 by R. C. O. Mathews, C. H. Feinstein, and J. C. Odling-Smee (Oxford, 1982), is the near-definitive account of more than a century's economic history. *The Great Victorian Boom, 1850–1873* ("Studies," London, 1975) is a brief, lively treatment by Roy A. Church. Stung by some hostile criticism, the author organized a conference on *The Dynamics of Victorian Business: Problems and Perspectives to the 1870s* (London, 1980), the participants of which provided succinct surveys of the principal industries, which Church has edited. This appeared too late to be included in P. L. Payne's succinct survey, *British Entrepreneurship in the Nineteenth Century* ("Studies," London, 1974), but another conference provided another volume with a different approach—the cliometric—in *Essays on a Mature Economy: Britain after 1840*, edited by Donald McCloskey (London, 1971), in which most authors gave a favorable assessment of British entrepreneurship. McCloskey has also published a number of his own essays under the title *Enterprise and Trade in Victorian Britain* (London, 1981), in which he concluded that the British economy (and entrepreneurs) in the late nineteenth century did about as well as could have been expected. This conclusion has been sharply challenged by (among others) a British economic historian, M. W. Kirby, in *The Decline of British Economic Power since 1870* (London, 1981), and an American intellectual historian, Martin J. Wiener, in *English Culture and the Decline of the Industrial Spirit, 1850–1980* (Cambridge, 1981). The latter book was the subject of a special conference, the results of which are presented in *British Culture and Economic Decline*, edited by Bruce Collins and Keith Robbins (London, 1990). David Cannadine deals with a related subject in his delightfully written but heavy (800+ pages) *The Decline and Fall of the British Aristocracy* (New Haven and London, 1990). Alan Sked, *Britain's Decline: Problems and Perspectives* (Oxford, 1987) is rather cautious, whereas *Britain's Prime and Britain's Decline: The British Economy, 1870–1914*, by Sidney Pollard (New York, 1989) is more forthright. *New Perspectives on the Late Victorian Economy: Essays in Quantitative Economic History, 1860–1914*, edited by James Foreman-Peck (Cambridge, 1991), returns to the cliometric mode of analysis.

The "standard of living question" in British industrialization has been one of the most hotly debated topics since the 1830s. *The Standard of Living in Britain in the Industrial Revolution*, edited by Arthur J. Taylor ("Debates," London, 1975), presents views from both (or all) sides. Jeffrey G. Williamson, in *Did British Capitalism Breed Inequality?* (Boston, 1985), uses cliometric methods to argue that the standard of living of British workers rose, but that the distribution of income became more unequal until about the middle of the century. Other volumes on related topics are Arthur J. Taylor, *Laissez-faire and State Intervention in Nineteenth-century Britain* ("Studies," London, 1972), and A. W. Coats, ed., *The Classical Economists and Economic Policy* ("Debates," London, 1971).

Most readers of this book are already familiar with at least the outlines of American economic history—or soon will be. In view of the immense literature on the subject, it is impractical to list more than a few general works, and refer readers to them and their bibliographies. Among the better recent textbooks are Sidney Ratner, James H. Soltow, and

Richard E. Sylla, *The Evolution of the American Economy: Growth, Welfare, and Decision Making* (2nd ed., New York, 1993); Stanley Lebergott, *The Americans: An Economic Record* (New York, 1984); and Susan P. Lee and Peter Passell, *A New Economic View of American History* (New York, 1979). Somewhat older, but valuable because of the distinction of its authors (twelve in all!) is Lance E. Davis et al., *American Economic Growth: An Economist's History of the United States* (New York, 1972). Two books prepared with British readers in mind are A. W. Coats and R. M. Robertson, eds., *Essays in American Economic History* (London, 1969), and Peter Temin, *Causal Factors in American Economic Growth in the Nineteenth Century* ("Studies," London, 1975), with an excellent select bibliography. William Cronon, *Nature's Metropolis: Chicago and the Great West* (New York, 1991), is an outstanding recent example of urban history.

The relative scarcity of books in English on the economic history of Belgium means that more reliance must be placed on journal articles and chapters or passages in larger works. Joel Mokyr, *Industrialization in the Low Countries, 1795–1850* (New Haven, CT, 1976), is less useful than the title suggests, in that Mokyr was more interested in cliometric hypothesis-testing than in history-telling. The chapter on the Low Countries by Jan Dhondt and Marinette Bruwier in the *Fontana Economic History*, vol 4(1), is rather disappointing, as is the article by J. A. Van Houtte, "Economic Development of Belgium and the Netherlands from the Beginning of the Modern Era," *Journal of European Economic History, 1* (Spring, 1972): 100–120. A better overview can be obtained from the chapters on Belgium in Milward and Saul, *The Economic Development of Continental Europe, 1780–1870* and *The Development of the Economies of Continental Europe, 1850–1914* (Belgium is paired with Switzerland in the former and with the Netherlands in the latter). Jan Craeybeckx, "The Beginning of the Industrial Revolution in Belgium," in Cameron, ed., *Essays in French Economic History*, is informative on the French period, and Chapter XI in Cameron, *France and the Economic Development of Europe*, provides a general survey with special emphasis on the contributions of French entrepreneurs, engineers, and capital. The chapter on Belgium in Cameron et al., *Banking in the Early Stages of Industrialization*, gives more detail on the contribution of the Belgian banking system to industrialization. R. J. Lesthaeghe, *The Decline of Belgian Fertility, 1800–1970* (Princeton, 1978), is a rather specialized monograph.

François Caron, *An Economic History of Modern France* (New York, 1979), is a good book badly translated. Guy P. Palmade, *French Capitalism in the Nineteenth Century* (Newton Abbott, 1972), although perhaps not quite as good, had better luck with its translator (Graeme Holmes), who also provided a useful, lengthy introduction on "The Study of Entrepreneurship in Nineteenth-Century France." Tom Kemp, *Economic Forces in French History: An Essay on the Development of the French Economy, 1760–1914* (London, 1971), is old-fashioned and outdated. Patrick O'Brien and Caglar Keyder inaugurated a new era in the economic historiography of France with *Economic Growth in Britain and France, 1780–1914: Two Paths to the Twentieth Century* (London, 1978), by arguing that the French transition to industrial society was "more humane and perhaps no less efficient" than that of Britain. C. P. Kindleberger, on the other hand, in *Economic Growth in France and Britain, 1851–1950* (Cambridge, MA, 1964), accepted the conventional wisdom and tried to explain it. More recently François Crouzet has argued for British superiority in *Britain Ascendant: Comparative Studies in Franco-British Economic History* (Cambridge, 1990).

Agriculture remained a major sector of the French economy throughout the nineteenth century. Some studies of it include L. M. Goreux, *Agricultural Productivity and Economic Development in France, 1850–1950* (New York, 1977); W. H. Newell, *Population Change and Agricultural Development in Nineteenth Century France* (New York, 1977); and Roger Price, *The Modernization of Rural France: Communications Networks and Agricultural Market Structures in Nineteenth Century France* (New York, 1983). A rather special branch of agriculture is treated by Leo Loubère in *The Red and the White: The History of Wine in France and Italy in the Nineteenth Century* (Albany, 1978), and *The Wine Revolution in France: The Twentieth Century* (Princeton, 1990).

E. C. Carter et al., eds., *Enterprise and Entrepreneurs in Nineteenth and Twentieth Century France* (Baltimore, 1976), and J. M. Laux, *In First Gear: The French Automobile Industry to 1914* (Liverpool, 1976), demonstrate the poorly reported dynamic aspects of French entrepreneurship. Michael S. Smith, *Tariff Reform in France, 1860–1900* (Ithaca, NY, 1980), restores the balance to the often misrepresented protectionist stance of French entrepreneurs. Cameron, *France and the Economic Development of Europe*, presents examples of French entrepreneurship abroad.

Val R. Lorwin, *The French Labor Movement* (Cambridge, MA, 1954), is a sympathetic but balanced treatment of that subject. Other aspects of the labor movement are dealt with by M. P. Hanagan, *The Logic of Solidarity: Artisans and Industrial Workers in Three French Towns, 1871–1914* (Urbana, IL, 1980); E. C. Shorter and Charles Tilly, *Strikes in France, 1830–1968* (Cambridge, 1974); and Peter N. Stearns, *Paths to Authority: The Middle Class and the Industrial Labor Force in France, 1820–1848* (Urbana, IL, 1978).

Helmut Boehme, *An Introduction to the Social and Economic History of Germany: Political and Economic Change in the Nineteenth and Twentieth Centuries* (Oxford, 1978), is a brief overview by a German "revisionist" historian. Gustav Stolper et al., *The German Economy, 1870 to the Present* (New York, 1967), is an updated translation of a book by a well-known anti-Nazi German economist. W. O. Henderson, *The Rise of German Industrial Power, 1834–1914* (Berkeley and Los Angeles, 1975), is an old-fashioned, pedestrian history. The same author's *The Zollverein* (London, 1939; reprinted, 1959) is, unfortunately, the only reasonably complete account of that institution in English. Martin Kitchen, *The Political Economy of Germany, 1815–1914* (London, 1978), is a lively but not very sophisticated history. Knut Borchardt, *Perspectives on Modern German Economic History and Policy* (Cambridge, 1991) is a collection of essays by a leading German economic historian. W. J. Mommsen provides a comparative perspective in *Britain and Germany, 1800–1914: Two Development Paths toward Industrial Society* (London, 1986).

Frank B. Tipton, *Regional Variations in the Economic Development of Germany during the Nineteenth Century* (Middletown, CT, 1976), provides a refreshing contrast to the usual textbook treatments of uniform growth. Richard Tilly, *Financial Institutions and Industrialization in the Rhineland, 1815–1870* (Madison, WI, 1966), looks at the background of the "great banks." Fritz Stern, *Gold and Iron: Bismarck, Bleichroeder, and the Building of the German Empire* (New York, 1977), is a fascinating insider's view of Bismarck's personal, and the German Empire's semiofficial, banker. Paul Hohenberg, *Chemicals in Western Europe, 1850–1914* (Chicago, 1967), is a good account of the rise of the organic chemical industry in which Germany played such a crucial role. Thorstein Veblen, *Imperial Germany and the Industrial Revolution* (New York, 1919: reprinted, 1939); although hopelessly misinformed and outdated, is nevertheless interesting as an example of a contemporary view by an acute observer.

Chapter 10. Patterns of Development: Latecomers and No-Shows

It is ironic (or symptomatic?) that Switzerland, the wealthiest country in Europe, has the least satisfactory literature in English on its economic history. It receives only scattered mention in volume 6 of the *Cambridge Economic History of Europe,* and virtually none in subsequent volumes. Milward and Saul give it a half-chapter in the first of their two-volume text (to 1870), but do not even mention it in the second volume. The chapter on Switzerland in the *Fontana Economic History* is the least satisfactory chapter in that collection. Readers of

French or German will appreciate Jean-François Bergier, *Die Wirtschaftsgeschichte der Schweiz: Von den Anfangen bis zur Gegenwart* (Zurich and Cologne, 1983), also available in French as *Histoire économique de la Suisse* (Lausanne and Paris, 1984), but others will have to be satisfied with Bergier's brief synopsis, "Trade and Transport in Swiss Economic History," in Cameron, ed., *Essays in French Economic History*, which deals more with the early modern period than with the nineteenth century, and with gleanings from other works, such as the partial chapters on Switzerland in Cameron, *France and the Economic Development of Europe* (for banks and railways) and Hohenberg, *Chemicals in Western Europe* (for chemicals). A welcome exception is Eric Schiff, *Industrialization without National Patents: The Netherlands, 1869–1912; Switzerland, 1850–1907* (Princeton, 1971), which finds that, for small, open economies, patent systems were not terribly important. Aldcroft and Rodger, *Bibliography of European Economic and Social History*, lists about thirty other items, more or less relevant. In 1991, in celebration of the seven hundredth anniversary of the Swiss Confederation, Jean-François Bergier and *many* others published, in English, French, and German, a semi-popular *1291–1991: The Swiss Economy, A Trilogy* (St. Sulpice, Switzerland, 1991).

English-language readers on the economic history of the Netherlands are slightly better served than those on Switzerland. The limitations of the coverage of the *Fontana Economic History*, and of the individual works by Mokyr and Van Houtte, mentioned in connection with Belgium, apply here as well, but Milward and Saul, *The Development of the Economies of Continental Europe, 1850–1914*, have a succinct survey of the Dutch economy in the second half of the century. H. R. C. Wright, *Free Trade and Protection in the Netherlands, 1816–1830: A Study of the First Benelux* (Cambridge, 1955), is a good place to begin. R. T. Griffiths, *Industrial Retardation in the Netherlands, 1830–1850* (The Hague, 1979), is perhaps overly pessimistic, but this is offset to some extent by J. A. de Jonge, "Industrial Growth in the Netherlands, 1850–1914," in *Acta Histioriae Neerlandicae*, 5 (1971). Business history is useful for the Netherlands—for example, the first two volumes of Charles Wilson, *The History of Unilever* (London, 1954), or P. J. Bouman, *Phillips of Eindhoven* (London, 1958), or the first two volumes of Frederick C. Gerretson, *History of the Royal Dutch* (Leiden, 1953–58). The Netherlands Economic History Archive in 1989 began publication in English of a new journal, *Economic and Social History in the Netherlands*.

The literature in English on Scandinavia is far more plentiful and of high quality. In addition to the excellent surveys by Karl-Gustaf Hildebrand in volume VII of the *Cambridge Economic History*, Leonnart Jörberg in volume 4 of the *Fontana Economic History*, and Milward and Saul in *The Economic Development of Continental Europe*, there are a number of both monographs and general studies: Jörberg's *Growth and Fluctuations of Swedish Industry, 1869–1912* (Stockholm, 1961); Sima Lieberman, *The Industrialization of Norway, 1800–1920* (Oslo, 1970); and Svend Aage Hansen, *Early Industrialization in Denmark* (Copenhagen, 1970). Additional bibliographical suggestions can be found in all of these. Finnish industry is the subject of two books by Timo Myllyntaus: *Finnish Industry in Transition, 1885–1920: Responding to Technological Challenges* (Helsinki, 1989), and *The Gatecrashing Apprentice: Industrialising Finland as an Adopter of New Technology* (Helsinki, 1990).

Until fairly recently there was a dearth of good literature in any language on the economic development of the Austro-Hungarian, or Habsburg, Empire. That gap has now been filled, especially in English, with several high-quality contributions. The best is undoubtedly David F. Good, *The Economic Rise of the Habsburg Empire, 1750–1914* (Berkeley and Los Angeles, 1984). Others that merit comparison are John Komlos, *The Habsburg Monarchy as a Customs Union: Economic Development in Austria-Hungary in the Nineteenth Century*

(Princeton, 1983), and Thomas Huertas, *Economic Growth and Economic Policy in a Multinational Setting* (New York, 1977). Komlos has also edited a collection of essays by a number of mostly younger scholars of various nationalities, *Economic Development in the Habsburg Monarchy in the Nineteenth Century* (New York, 1983). A larger work by two eminent Hungarian economic historians, Ivan T. Berend and Gyorgy Ranki, includes the Habsburg Monarchy along with eastern Germany, Poland, and the former Balkan territories of the Ottoman Empire: *Economic Development of East-Central Europe in the 19th and 20th Centuries* (New York, 1974). Berend and Ranki also contributed *Hungary: A Century of Economic Development* (New York, 1974). Among older works that still merit citation is Jerome Blum, *Noble Landowners and Agriculture in Austria, 1815–1848* (Baltimore, 1948).

Berend and Ranki also authored *The European Periphery and Industrialization, 1780–1914* (Cambridge, 1982), in which they included Scandinavia with southern and eastern Europe in the "periphery." The book is outstanding in concept, but synoptic and brief in execution; details for any given area must be sought elsewhere. For the Iberian peninsula (realistically, Spain, as there is virtually nothing on Portugal) details will be found in the pertinent chapters of Vicens Vives, *An Economic History of Spain;* Gabriel Tortella, *Banking, Railroads, and Industry in Spain, 1829–1874* (New York, 1977); Joseph Harrison, *An Economic History of Modern Spain* (New York, 1978); and Sima Lieberman, *The Contemporary Spanish Economy: A Historical Perspective* (London, 1982), which carries the story to the 1970s. For Italy the basic data are given by Giorgio Fuà, *Notes on Italian Economic Growth, 1861–1964* (Milan, 1965). Gianni Toniolo, *An Economic History of Liberal Italy, 1850–1918* (New York and London, 1990) is a good synoptic treatment by a well-known Italian scholar. J. S. Cohen, *Finance and Industrialization in Italy, 1894–1914* (New York, 1977), is competent on its limited subject. These may be supplemented by the surveys in *Fontana* and in Milward and Saul, which is also quite good on southeastern Europe. For the latter, the relevent chapters of John R. Lampe and Marvin Jackson's *Balkan Economic History, 1550–1950* (Bloomington, IN, 1982) are by far the best available.

For Russia, a good place to begin is M. E. Falkus, *The Industrialization of Russia, 1700–1914* ("Studies," London, 1972). William L. Blackwell, *The Beginnings of Russian Industrialization, 1800–1860* (Princeton, 1968), is a solid, comprehensive account of Russian industrialization to the eve of the Emancipation. The story is continued by Theodore von Laue, *Sergei Witte and the Industrialization of Russia* (New York, 1963). Paul R. Gregory, *Russian National Income, 1885–1913* (Cambridge, 1982) is of fundamental importance. Olga Crisp, *Studies in the Russian Economy before 1914* (London, 1976), is a collection of her essays that deal with all aspects of the economy from the peasantry to public finance; the first, "The Pattern of Industrialization in Russia, 1700–1914," is especially noteworthy. *Russian Economic History: The Nineteenth Century*, by Arcadius Kahan, edited by Roger Weiss (Chicago, 1989), is also a collection of essays by a distinguished scholar. John P. McKay, *Pioneers for Profit: Foreign Entrepreneurship and Russian Industrialization, 1885–1913* (Chicago, 1970), is especially enlightening on the role of foreign entrepreneurs. *The Corporation under Russian Law, 1800–1917: A Study in Tsarist Economic Policy*, by Thomas C. Owen (Cambridge, 1991), gives some clues to why Russian entrepreneurs were not more dynamic. Theodore H. Friedgut, in *Inzovka and Revolution*, vol. I, *Life and Work in Russia's Donbass, 1869–1924* (Princeton, 1989), chronicles the growth of Russia's largest mining and metallurgical region. In *Road to Power: The Trans-Siberian Railroad and the Colonization of Asian Russia, 1850–1917* (Ithaca, 1991), Steven G. Marks concludes that the Trans-Siberian was built more for political reasons than economic reasons. Works on Russian agriculture, of major importance in the nineteenth century, include Blum, *Lord and Peasant;* W. S. Vucinich, ed., *The Peasant in Nineteenth Century Russia* (Stanford, 1968); and Esther

Kingston-Mann and Timothy Mixter, eds., *Peasant Economy, Culture, and Politics in European Russia, 1800–1921* (Princeton, 1991). Christine D. Worobec, *Peasant Russia: Family and Community in the Post-Emancipation Period* (Princeton, 1991) effectively replaces G. T. Robinson, *Rural Russia under the Old Regime* (2nd ed., New York, 1962).

The literature in English on Japanese economic history and development, formerly miniscule, is now abundant. A pioneering venture was William W. Lockwood, *The Economic Development of Japan: Growth and Structural Change, 1868–1938* (Princeton, 1954). Although still valuable, it has been superseded for quantitative data by Takafusa Nakamura, *Economic Growth in Prewar Japan,* translated by Robert A. Feldman (New Haven, CT, 1983). An excellent analysis of the pre-Meiji period is provided by Susan B. Hanley and Kozo Yamamura, *Economic and Demographic Change in Preindustrial Japan, 1600–1868* (Princeton, 1977). Allen C. Kelley and Jeffrey G. Williamson, *Lessons from Japanese Development: An Analytical Economic History* (Chicago, 1974), is an exercise in counterfactual cliometric history. Michio Morishima, a distinguished Japanese mathematical economist, temporarily abandoned mathematics for history and sociology and, in *Why Has Japan "Succeeded"? Western Technology and the Japanese Ethos* (Cambridge, 1982), found the answer in Japan's unique ideology.

Chapter 11. The Growth of the World Economy

Two relatively brief textbooks that deal with the material of this chapter in somewhat greater detail are William Ashworth, *A Short History of the International Economy since 1850* (4th ed., London, 1987) and A. G. Kenwood and A. L. Lougheed, *The Growth of the International Economy, 1820–1960* (London, 1971). Even greater detail will be found in James Foreman-Peck, *A History of the World Economy: International Economic Relations since 1850* (London, 1983), which also contains theoretical explanations.

A fascinating account of the repeal of the Corn Laws, and of many other aspects of nineteenth-century international trade, is given in Charles P. Kindleberger, *Economic Response: Comparative Studies in Trade, Finance, and Growth* (Cambridge, MA, 1978). Other aspects of Corn Law repeal are dealt with by Lucy Brown, *The Board of Trade and the Free Trade Movement, 1830–1842* (Oxford, 1958), and William D. Grampp, *The Manchester School of Economics* (Chicago, 1960). For the century as a whole, see A. H. Imlah, *Economic Elements in the Pax Britannica: Studies in British Foreign Trade in the Nineteenth Century* (Cambridge, MA, 1958).

The standard source on the Cobden-Chevalier treaty is still A. L. Dunham, *The Anglo-French Treaty of Commerce of 1860 and the Progress of the Industrial Revolution in France* (Ann Arbor, MI, 1930). For a more concise and intelligible treatment, see Marcel Rist, "A French Experiment with Free Trade: The Treaty of 1860," in Cameron, ed., *Essays in French Economic History.* The German experience is related in Ivo N. Lambi, *Free Trade and Protection in Germany, 1868–79* (Wiesbaden, 1963). Further French experience is ably documented by Michael S. Smith, *Tariff Reform in France, 1860–1900* (Ithaca, NY, 1980). For Britain, see S. B. Saul, *Studies in British Overseas Trade, 1870–1914* (Liverpool, 1960). General trends are highlighted by W. Arthur Lewis, *Growth and Fluctuations, 1870–1913* (London, 1978), and, more briefly, by S. B. Saul, *The Myth of the Great Depression, 1873–1895* ("Studies," London, 1969).

A good introduction to the gold standard can be found in P. T. Ellsworth, *The International Economy: Its Structure and Operation* (3rd ed., New York, 1964). Barry Eichengreen, ed., *The Gold Standard in Theory and History* (London, 1985), is a judicious selection of articles on all aspects of the subject. Greater detail can be found in Arthur I. Bloomfield,

Monetary Policy under the International Gold Standard, 1880–1914 (New York, 1959). Peter H. Lindert, *Key Currencies and Gold, 1900–1913* (Princeton, 1969), is a short case study.

Migration statistics are given in Woytinsky, *World Population and Production*. See also M. L. Hansen, *The Atlantic Migration, 1607–1860* (Cambridge, MA, 1940); Brinley Thomas, *Migration and Economic Growth: A Study of Great Britain and the Atlantic Economy* (Cambridge, 1954); and Charlotte Erickson, *American Industry and the European Immigrant, 1860–1885* (Cambridge, MA, 1957).

One of the pioneering works on foreign investment is Herbert Feis, *Europe, the World's Banker, 1870–1914* (New Haven, CT, 1930; reprinted, 1965); its statistics need revision, but it still makes interesting reading. A more recent comprehensive view, not only of investment but also of migration, trade, and the diffusion of technology, is William Woodruff, *Impact of Western Man: A Study of Europe's Role in the World Economy, 1750–1960* (New York, 1967). The British experience as a lender is summarized by P. L. Cottrell, *British Overseas Investment in the Nineteenth Century* ("Studies," London, 1975), and various incidents are detailed in A. R. Hall, ed., *The Export of Capital from Britain, 1870–1914* ("Debates," London, 1968). America's experience as a borrower has received definitive treatment in Mira Wilkins, *The History of Foreign Investment in the United States to 1914* (Cambridge, MA, 1989). For France, see Cameron, *France and the Economic Development of Europe*. The estimates of all of these references are criticized by D. C. M. Platt, *Foreign Finance in Continental Europe and the USA, 1915–1870* (London, 1984), and *Britain's Investments Overseas on the Eve of the First World War: The Use and Abuse of Numbers* (London, 1986).

Books and articles on imperialism are legion. The best by far, as a supplement to the brief discussion in this volume, is D. K. Fieldhouse, *Economics and Empire, 1830–1914* (Ithaca, NY, 1973). *Africa and the Victorians*, by John T. Gallagher and Roland I. Robinson (New York, 1961), is a stimulating but controversial reinterpretation. Henri Brunschwig, *French Colonialism, 1871–1914* (New York, 1966), shows the importance of nationalism as an explanation for French imperial expansion. Daniel Headrick, in *The Tools of Empire: Technology and European Imperialism in the Nineteenth Century* (Oxford, 1981), argues for technological determinism; see also the same author's *The Tentacles of Progress: Technology Transfer in the Age of Imperialism, 1850–1940* (Oxford, 1988). V. I. Lenin, *Imperialism, the Highest Stage of Capitalism* (1916; numerous editions), is the standard Marxist text. A good antidote is Lance E. Davis and Robert A. Huttenback, *Mammon and the Pursuit of Empire: The Economics of British Imperialism* (Cambridge, 1988; an abridged edition is available). David Landes, *Bankers and Pashas: International Finance and Economic Imperialism in Egypt,* (London, 1958), reads like a novel.

Carl A. Trocki, *Opium and Empire: Chinese Society in Colonial Singapore, 1800–1910* (Ithaca, 1990), documents the Chinese diaspora. Loren Brandt, *Commercialization and Agricultural Development: Central and Eastern China, 1870–1937* (Cambridge and New York, 1989), deals with the Chinese at home. Philip C. C. Huang, *The Peasant Family and Rural Development in the Yangtse Delta, 1350–1988* (Stanford, 1990), takes a long view of Chinese poverty. Colin Newbury, *The Diamond Ring: Business, Politics, and Precious Stones in South Africa, 1867–1947* (New York, 1989), exposes a shining scandal. Isaria N. Kimambo, *Penetration and Protest in Tanzania: The Impact of the World Economy on the Pare, 1860–1960* (Athens, OH, 1991), provides an African point of view on the impact of trade on Africa. Hilda Sábato, *Agrarian Capitalism and the World Market: Buenos Aires in the Pastoral Age, 1840–1890* (Albuquerque, 1991), tells the story of a promising beginning with a disappointing continuation. Marshall C. Eakin, *British Enterprise in Brazil: The St. John d'el Rey Mining Company and the Morro Velho Gold Mine, 1830–1960* (Durham, NC, 1990), is a readable business history of a European company in Latin America.

Chapter 12. Strategic Sectors

Several general works relating to agriculture are listed earlier in this bibliography; others pertinent to specific countries are mentioned under Chapters 9 and 10 and are not repeated here. E. L. Jones, *The Development of English Agriculture, 1815–1873* ("Studies," London, 1968), is a handy survey of an important period for English agriculture. It is complemented by P. J. Perry, ed., *British Agriculture, 1875–1914* ("Debates," London, 1973), a collection of pertinent articles. R. Trow-Smith, *Life from the Land: The Growth of Farming in Western Europe* (London, 1967), and M. Tracy, *Agriculture in Western Europe: Crisis and Adaptation since 1880* (London, 1964), deal more broadly with Western Europe as a whole. E. L. Jones and S. J. Woolf have edited *Agrarian Change and Economic Development; The Historical Problems* (London, 1969), with contributed chapters on Italy, England, Japan, Africa, and Mexico. J. W. Mellor, *The Economics of Agricultural Development* (Ithaca, NY, 1966), is more analytical than strictly historical, but has historical applications. M. W. Rossiter, *The Emergence of Agricultural Science: Justus Liebig and the Americas, 1840–1880* (New Haven, CT, 1975), deals with an often-neglected topic.

Charles P. Kindleberger, *A Financial History of Western Europe* (2nd ed., New York, 1993), is a compendium of information delightfully presented on a variety of subjects—money, banking, public and private finance—mainly but not exclusively in the nineteenth and twentieth centuries. *Keynesianism vs. Monetarism and Other Essays in Financial History* (London, 1985) is another collection of wit and wisdom by the same prolific author. Rondo Cameron et al., *Banking in the Early Stages of Industrialization* (Oxford, 1967), is a comparative study of England, Scotland, France, Belgium, Germany, Russia, and Japan. Rondo Cameron, ed., *Banking and Economic Development: Some Lessons of History* (Oxford, 1972), has chapters on Austria, Italy, Spain, Serbia, Japan, and the United States. *International Banking, 1870–1914*, edited by Rondo Cameron and V. I. Bovykin (New York, 1991), is a massive collection of case studies from all areas of the world. *Banks as Multinationals*, edited by Geoffrey Jones (London and New York, 1990), was the subject of a conference organized by the editor. For individual countries, see, in addition to the preceding: C. A. E. Goodhart, *The Business of Banking, 1891–1914* (Great Britain) (London, 1972); John Sykes, *The Amalgamation Movement in English Banking* (London, 1926), older but still useful; Richard Tilly, *Financial Institutions and Industrialization in the Rhineland, 1815–1870* (Madison, 1966); Udo E. G. Heyn, *Private Banking and Industrialization: The Case of Frankfurt am Main, 1825–1875* (New York, 1981); Richard Rudolph, *Banking and Industrialization in Austria-Hungary* (Cambridge, 1976); Olle Gasslander, *History of Stockholms Enskilda Bank to 1914* (Stockholm, 1962); and K.-G. Hildebrand, *Banking in a Growing Economy: Svenska Handelsbanken since 1871* (Stockholm, 1971), an abridged translation of a much larger work in Swedish. On the important subject of industrial finance, see P. L. Cottrell, *Industrial Finance, 1830–1914: The Finance and Organization of English Manufacturing Industry* (London, 1980); also *Financing Industrialization*, a collection of journal articles compiled and edited by Rondo Cameron (Aldershot, England, 1992).

The role of the government, or the state, is dealt with more or less adequately in almost all general and collective works. Volume VIII of the *Cambridge Economic History* is devoted entirely to "The Development of Economic and Social Policies" in the industrial economies. Aitken, *The State and Economic Growth*, Taylor, *Laissez-faire and State Intervention in Nineteenth-Century Britain*, and Coats, *The Classical Economists and Economic Policy*, have already been mentioned. Manfred D. Jankowski, *Public Policy in Industrial Growth: The Case of the Ruhr Mining Region, 1776–1865* (New York, 1977), deals with the transition from the *Direktionsprinzip* to the *Inspektionsprinzip*.

Chapter 13. Overview of the World Economy in the Twentieth Century

The basic data on population in the first half of the twentieth century are given in Woytinsky, *World Population and Production.* These are updated annually in the United Nations *Demographic Yearbook.* A closer look at Europe's interwar population is provided by Dudley Kirk, *Europe's Population in the Interwar Years* (Geneva, 1946). Recent trends in the industrial nations are summarized in National Bureau of Economic Research, *Demographic and Economic Change in Developed Countries* Princeton, 1976). The United Nations Department of Economic and Social Affairs, *The Population Debate: Dimensions and Perspectives* (2 vols., New York, 1975), presents the papers of the 1974 World Population Conference, which debated the (then) present plight and prospects of Third World nations. Paul R. Ehrlich, Anne H. Ehrlich, and John P. Holdren, *Ecoscience: Population, Resources, Environment,* is a good textbook on present and future problems.

E. M. Kulischer, *Europe on the Move: War and Population Changes, 1917–1949* (New York, 1948), was an early look at the upheavals brought on by the wars. The same subject is viewed in broader perspective in *Human Migration: Patterns and Policies,* edited by W. H. McNeill and R. S. Adams (Bloomington, IN, 1978).

The references for resources listed at the beginning of this bibliography and under Chapter 8 are, for the most part, relevant here as well. J. Fredric Dewhurst et al., *Europe's Needs and Resources: Trends and Prospects in Eighteen Countries* (New York, 1961), is an encyclopedic accumulation of data and analysis whose contents are even broader than the title indicates. Dewhurst had earlier led a team in a similar study of *America's Needs and Resources* (New York, 1947). One of the publishing phenomena of the 1970s was the appearance of D. H. Meadows et al., *The Limits to Growth* (New York, 1972), under the sponsorship of the prestigious Club of Rome, which prophesied the exhaustion of critical world resources in the twenty-first century. Partly to study such concerns, the United Nations and its affiliates have undertaken numerous studies of the interactions of population, resources, technology, and environment; typical of the genre is *The Future of the World Economy: A United Nations Study,* by Wassily Leontief et al. (Oxford, 1977). Angus Maddison, *The World Economy in the Twentieth Century* (Paris, 1989), highlights the increasing disparity between rich and poor nations.

Volumes 6 and 7 of *A History of Technology,* edited by Trevor I. Williams (Oxford, 1978), cover the first half of the twentieth century. *Technology and Social Change in America,* edited by Edwin T. Layton, Jr. (New York, 1973), is a small collection of essays by eminent historians of technology. John G. Clark, *The Political Economy of World Energy: A Twentieth Century Perspective* (Chapel Hill, NC, 1990) surveys all forms of energy for the entire century. R. R. Nelson, M. J. Peck, and E. D. Kalacheck stress the interaction of technology and institutions in *Technology, Economic Growth, and Public Policy* (Washington, 1967). Harry G. Johnson noted the international dimensions of technology in *Technology and Economic Interdependence* (London, 1975). The American National Science Foundation thought it important to stress the *Interactions of Science and Technology in the Innovation Process* (Washington, 1976). Dennis Gabor, a distinguished physicist-engineer-inventor, in *Innovations: Scientific, Technological, and Social* (Oxford, 1970), predicted some one hundred important technological and biological innovations, a number of which have already been realized. Daniel Bell, a sociologist, has also predicted a number of changes in society as a result of the interaction of technology and institutions in *The Coming of Post-Industrial Society. A Venture In Social Forecasting* (New York, 1973; reprinted, 1976). The Institute of History of Natural Sciences and Technology and the Institute of Philosophy of the USSR Academy of Sciences, together with the Institute of Philosophy and Sociology of the

Czechoslovak Academy of Sciences, published *Man, Science, Technology: A Marxist Analysis of the Scientific and Technological Revolution* (Moscow and Prague, 1973).

The major institutional changes of the twentieth century are related, on the one hand, to the rapid development of science and technology and, on the other, to the massive type of warfare they have made possible. The interrelations of all these forces were nicely captured by the French sociologist Raymond Aron in *The Century of Total War* (New York, 1954) and, more recently, by the British historian Arthur Marwick in *War and Social Change in the Twentieth Century: A Comparative Study of Britain, France, Germany, Russia, and the United States* (London, 1974). The impact of war (and other changes) on economic policy making in one country are studied by Richard F. Kuisel, *Capitalism and the State in Modern France: Renovation and Economic Management in the Twentieth Century* (Cambridge, 1981). Twentieth-century changes in business organization and management are the subject of *Managerial Hierarchies: Comparative Perspectives on the Rise of the Modern Industrial Enterprise,* edited by Alfred D. Chandler, Jr., and Herman Daems (Cambridge, MA, 1980).

Chapter 14. International Economic Disintegration

Direct and Indirect Costs of the Great War, by E. L. Bogart (Oxford, 1919), represented an early attempt by a famous American economist to measure the costs of World War I. A famous British economist, A. L. Bowley, took somewhat more time to assess *Some Economic Consequences of the Great War* (London, 1930). Historians' views are represented by J. M. Cooper, *Causes and Consequences of World War I* (London, 1975), and Gerd Hardach, *The First World War, 1914–1918* (London, 1977). Charles Gilbert details *American Financing of World War I* (Greenwood, CT, 1970).

J. M. Keynes, *The Economic Consequences of the Peace* (London, 1919), written in a great hurry by a talented writer in high dudgeon, is nevertheless an historic document in itself and still makes good reading. Etienne Mantoux, *The Carthaginian Peace—or the Economic Consequences of Mr. Keynes* (New York, 1946), was written with equal or greater moral fervor by a young Frenchman who died in World War II. A sort of reconciliation of the two points of view can be found in Chapter 16 of Kindleberger, *Financial History of Western Europe.* An important study of the 1920s in Europe is Charles S. Maier, *Recasting Bourgeois Europe: Stabilization in France, Germany, and Italy in the Decade after World War I* (Princeton, 1975). A work of similar importance, although mistitled, is Steven A. Schuker, *The End of French Predominance in Europe: The Financial Crisis of 1924 and the Adoption of the Dawes Plan* (Chapel Hill, NC, 1976). Of the many studies of Germany's hyperinflation, the most recent and authoritative is Carl-Ludwig Holtfrerich, *The German Inflation, 1914–1923: Causes and Effects in International Perspective* (Berlin, 1986). The consequences of the inflation on a specific (and important) industry are detailed by Gerald D. Feldman, *Iron and Steel in the German Inflation, 1916–1923* (Princeton, 1977). Other economic aspects of the peace settlement are covered in Derek H. Aldcroft's general survey of the 1920s, *From Versailles to Wall Street, 1919–1929* (London, 1977). Anne Orde deals with *British Policy and European Reconstruction after the First World War* (Cambridge and New York, 1990).

The best, or at least the most readable, account of the 1930s is Charles P. Kindleberger, *The World in Depression, 1929–1939* (London, 1973). Another entertaining book on a dismal experience is John Kenneth Galbraith, *The Great Crash, 1929* (Boston, 1955; reprinted, 1962). One of the most influential interpretations of the causes of the depression, emphasizing the role of the U.S. Federal Reserve system, is Milton Friedman and Anna J. Schwartz, *The Great Contraction* (Princeton, 1966), a reprint of one chapter of their monumental *A Monetary History of the United States, 1867–1960* (Princeton, 1963). A contrary view is found in Peter Temin, *Did Monetary Forces Cause the Great Depression?* (New York, 1976), although Temin apparently changed his mind in *Lessons from the Great Depression* (Cambridge, MA, 1989). The depression is set in a larger context by Ingvar Svennilson, *Growth and Stagnation in the European Economy* (Geneva, 1954). The interwar British economy is the subject of both B. W. E. Alford, *Depression and Recovery? British Economic Growth 1918–1939* ("Studies," London, 1972), and Forrest Capie, *Depression and Protectionism: Britain Between the Wars* (London, 1983). *British Unemployment, 1919–1939: A Study in Public Policy*, by W. R. Garside (Cambridge and New York, 1990), is a comprehensive study of its limited but important subject. Retrospective views that treat the depression experience in a great many countries around the world are contained in Herman Van der Wee, ed., *The Great Depression Revisited: Essays on the Economics of the Thirties* (The Hague, 1972), and Ivan T. Berend and Knut Borchardt, eds., *The Impact of the Depression of the 1930s and its Relevance for the Contemporary World* (Budapest, 1986).

A broad survey of the 1930s in America is Broadus Mitchell, *Depression Decade: From New Era through New Deal, 1929–1941* (New York, 1947; reprinted, 1969). The French experience is illuminated by Stanley Hoffman, *Decline or Renewal? France since the 1930s* (New York, 1974). Charles F. Delzell, ed., *Mediterranean Fascism, 1919–1945* (London, 1971), contains contributions on the Fascist regimes of Italy, Spain, and Portugal. An important work on the advent of Nazism in Germany is Henry A. Turner, Jr., *German Big Business and the Rise of Hitler* (Oxford, 1985). The rearmament program is dealt with by Burton H. Klein, *Germany's Economic Preparations for War* (Cambridge, MA, 1959). *Hitler's Social Revolution: Class and Status in Nazi Germany, 1933–1939*, by David Schoenbaum (London, 1966), is of special interest.

Alec Nove, *An Economic History of the U.S.S.R.* (London, 1969; reprinted, 1986), is the best introduction to its subject. Nove has also published *The Soviet Economy* (3rd ed., London, 1969) and *Was Stalin Really Necessary?* (London, 1964). Of E. H. Carr's multi-volume *History of Soviet Russia*, those of greatest interest to economic historians are *Socialism in One Country* (London, 1958) and *Foundations of a Planned Economy, 1926–29*, with R. W. Davies (London, 1969–78). An abridgement of the fourteen volumes on the 1920s is also available: E. H. Carr, *The Russian Revolution: From Lenin to Stalin* (London, 1979). Also of interest: Alexander Erlich, *The Soviet Industrialization Debate, 1924–28* (Cambridge, MA, 1960); E. C. Brown, *Soviet Trade Union and Labour Relations* (Oxford, 1960); and Moshe Lewin, *Russian Peasants and Soviet Power* (London, 1968).

War, Economy, and Society, 1939–45, by Alan S. Milward (Berkeley and Los Angeles, 1977), is the most comprehensive and competent economic history of World War II. Other significant works by the same author: *The German Economy at War* (London, 1965); *The New Order and the French Economy* (Oxford, 1970); and *The Fascist Economy in Norway* (Oxford, 1972). America's role in the war is encapsulated in D. M. Nelson, *The Arsenal of Democracy* (New York, 1946). For Japan, see J. R. Cohen, *Japan's Economy in War and Reconstruction* (Minneapolis, 1949), and F. C. Jones, *Japan's New Order in East Asia: Its Rise and Fall, 1937–45* (Oxford, 1954). Alec Cairncross, *The Price of War: British Policy on German Reparations, 1941–1949* (London, 1986) is an insightful view on an important but neglected topic.

Chapter 15. Rebuilding the World Economy

The most recent, comprehensive, and authoritative history of the world economy since World War II is Herman Van der Wee, *Prosperity and Upheaval: The World Economy, 1945–1980* (Berkeley and Los Angeles, 1986). In *The Reconstruction of Western Europe, 1945–51* (London, 1984), Alan S. Milward provides a richly detailed if somewhat controversial description of the origins and exfoliation of the Marshall Plan, the European Payments Union, and the Schuman Plan. *Reconstruction in Post-war Germany: British Occupation Policy and the Western Zones, 1945–1955* (Oxford, 1989), edited by Ian D. Turner, is a better-than-average conference volume. Michael J. Hogan, *The Marshall Plan: America, Britain, and the Reconstruction of Western Europe, 1947–1952* (Cambridge, 1987), is reasonably up-to-date. *The Netherlands and the Economic Integration of Europe, 1945–1957* (Amsterdam, 1990), edited by R. T. Griffiths, is a series of essays based on archival sources showing the important role of the Netherlands in keeping the movement for European unity on track. William James Adams, *Restructuring the French Economy: Government and the Rise of Market Competition since World War II* (Washington, 1989), is a competent if somewhat controversial account of France's amazing economic turn-around. Vera Lutz, *Italy, A Study in Economic Development* (Oxford, 1962), convincingly accounts for Italy's equally amazing rebound. *Governments, Industries and Markets: Aspects of Government-Industry Relations in the UK, Japan, West Germany and the USA since 1945* (Aldershot, England, 1990), edited by Martin Chick, provides useful summaries of industrial policies in four major countries. *The Golden Age of Capitalism: Reinterpreting the Postwar Experience* (Oxford, 1990), edited by Stephen A. Marglin and Juliet B. Schor, is a comparative macroeconomic history of the six largest advanced industrial economies from a Marxist point of view.

The economic problems of Third World countries are the subject of Angus Maddison, *Economic Progress and Policy in the Developing Countries* (London, 1970); Gerald Helleiner, *A World Divided: The Less Developed Countries in the International Economy* (Cambridge, 1976); W. G. Barnes, *Europe and the Developing World* (London, 1967); J. N. Bhagwati, ed., *The New International Economic Order: The North-South Debate* (Cambridge, MA, 1977); and N. Islam, ed., *Agricultural Policy in Developing Countries* (London, 1974).

The origins, operations, and problems of the European Community are dealt with in Jeffrey Harrop, *The Political Economy of Integration in the European Community* (Aldershot, England, 1989); Clifford Hackett, *Cautious Revolution: The European Community Arrives* (New York, 1990); Dennis Swan, *The Economics of the Common Market* (Hammondsworth, 1975); and Stephen Frank Overturf, *The Economic Principles of European Integration* (New York, 1986).

Other international economic institutions also merit consideration. J. K. Horsefield, *The International Monetary Fund, 1945–1965* (Washington, 1969), is the official history of that institution, as is Edward S. Mason and Robert E. Asher, *The World Bank since Bretton Woods* (Washington, 1973). Jacob J. Kaplan and Gunther Schleiminger, *The European Payments Union: Financial Diplomacy in the 1950s* (Oxford, 1989), is also an official history; although based on primary sources, it is authored by civil servants rather than scholars, and suffers accordingly. Brian Tew, *The Evolution of the International Monetary System, 1945–1977* (New York, 1977), is more analytical, and F. L. Block, *The Origins of International Economic Disorder: A Study of United States International Monetary Policy from World War II to the Present* (Berkeley and Los Angeles, 1977), is frankly critical. K. Kock, *International Trade Policy and the GATT, 1947–1967* (Stockholm, 1969), is also rather critical.

The measurement of economic growth is the subject of Edward S. Denison, *Why Growth Rates Differ: Postwar Experience in Nine Western Countries* (Washington, 1967). Denison has also applied his growth-accounting technique to Japan and the United States in (with W. K. Chung) *How Japan's Economy Grew So Fast: The Sources of Postwar Expansion* (Washington, 1976), and *Accounting for Slower Growth: The United States in the 1970s* (Washington, 1979). Angus Maddison is also interested in measuring economic growth, but follows a quite different technique and takes a much longer time span in *Economic Growth in the West: Comparative Experience in Europe and North America* (New York, 1964) and *Phases of Capitalist Development* (Oxford, 1982).

Changes in the nature of capitalism have attracted the attention of a variety of scholars. One of the early ones was A. A. Berle, Jr., with *The 20th Century Capitalist Revolution* (New York, 1954). Another was J. K. Galbraith with *The Affluent Society* (London, 1958) and *The New Industrial State* (Boston, 1967). Andrew Shonfield weighed in with *Modern Capitalism: The Changing Balance of Public and Private Power* (Oxford, 1965), and John Cornwall with *Modern Capitalism: Its Growth and Transformation* (London, 1977).

Chapter 16. The World Economy at the End of the Twentieth Century

As this bibliography is being compiled no comprehensive study of the collapse of Communism in either Eastern Europe or the Soviet Union has yet appeared. Most of Chapter 16 was written with the help of daily newspapers, magazines, and even television news programs! In the late summer and fall of 1991 the manuscript was revised almost daily. Roman Laba presents a somewhat idealized view of *The Roots of Solidarity: A Political Sociology of Poland's Working-class Democratization* (Princeton, 1991), whereas the author of *Breaking the Barrier: The Rise of Solidarity in Poland* (New York, 1991), Lawrence Goodwyn, does not know Polish. *The Road to a Free Economy: Shifting from a Socialist System: The Example of Hungary* (New York, 1990) is not history at all, but a passionate argument for a total break with the past by a distinguished Hungarian economist, János Kornai, written when the transition was just beginning.

The Japanese economy is appraised from a variety of viewpoints in Hugh Patrick and Henry Rosovsky, eds., *Asia's New Giant: How the Japanese Economy Works* (Washington, 1976), and again in Hugh Patrick with the assistance of Larry Meissner, eds., *Japan's High Technology Industries: Lessons and Limitations of Industrial Policy* (Seattle and London, 1987). In *Australia in the International Economy in the Twentieth Century* (Melbourne, 1990) Barry Dyster and David Merideth provide a background for understanding the importance of a part of the Pacific Basin. *The International Debt Crisis in Historical Perspective* (Cambridge, MA, 1989), edited by Barry Eichengreen and Peter H. Lindert, was inspired by the international debt crisis of the 1980s, especially in Latin America.

Jill Crystal, *Oil and Politics in the Gulf: Rulers and Merchants in Kuwait and Qatar* (Cambridge, 1990), provides a historical background (not intended by the author) for the Gulf War of 1991.

William Molle, *The Economics of European Integration (Theory, Practice, Policy)* (Dartmouth, England, 1990), is a full and informative account of the European Community on the eve of implementation of the Single European Act.

Index

Geographical, political, and technical terms that are mentioned frequently and in passing—e.g., economic, England, Europe, Mediterranean, Rome, technology—are not, as a rule, separately indexed. Consult the Contents for major headings.